## Applied Algebra and Number Theory

Essays in Honor of Harald Niederreiter on the occasion of his 70th birthday

Harald Niederreiter's pioneering research in the field of applied algebra and number theory has led to important and substantial breakthroughs in many areas. This collection of survey articles has been authored by close colleagues and leading experts to mark the occasion of his 70th birthday.

The book provides a modern overview of different research areas, covering uniform distribution and quasi-Monte Carlo methods as well as finite fields and their applications, in particular cryptography and pseudorandom number generation. Many results are published here for the first time. The book serves as a useful starting point for graduate students new to these areas, or as a refresher for researchers wanting to follow recent trends.

GERHARD LARCHER is Full Professor for Financial Mathematics and Head of the Institute for Financial Mathematics at the Johannes Kepler University Linz.

FRIEDRICH PILLICHSHAMMER is Associate Professor in the Institute for Financial Mathematics at the Johannes Kepler University Linz.

ARNE WINTERHOF is Senior Fellow at the Johann Radon Institute for Computational and Applied Mathematics (RICAM) at the Austrian Academy of Sciences, Linz.

CHAOPING XING is Full Professor in the Department of Physical and Mathematical Sciences at Nanyang Technological University, Singapore.

Harald Niederreiter in Marseille, 2013

# Applied Algebra and Number Theory

Essays in Honor of Harald Niederreiter on the occasion
of his 70th birthday

*Edited by*

GERHARD LARCHER
*Johannes Kepler University Linz*

FRIEDRICH PILLICHSHAMMER
*Johannes Kepler University Linz*

ARNE WINTERHOF
*Austrian Academy of Sciences, Linz*

CHAOPING XING
*Nanyang Technological University, Singapore*

CAMBRIDGE
UNIVERSITY PRESS

# CAMBRIDGE
## UNIVERSITY PRESS

University Printing House, Cambridge CB2 8BS, United Kingdom

One Liberty Plaza, 20th Floor, New York, NY 10006, USA

477 Williamstown Road, Port Melbourne, VIC 3207, Australia

314-321, 3rd Floor, Plot 3, Splendor Forum, Jasola District Centre, New Delhi - 110025, India

103 Penang Road, #05-06/07, Visioncrest Commercial, Singapore 238467

Cambridge University Press is part of the University of Cambridge.

It furthers the University's mission by disseminating knowledge in the pursuit of education, learning and research at the highest international levels of excellence.

www.cambridge.org
Information on this title: www.cambridge.org/9781107074002

© Cambridge University Press 2014

First published 2014

*A catalogue record for this publication is available from the British Library*

*Library of Congress Cataloging in Publication data*
Applied Algebra and Number Theory : Essays in Honor of Harald Niederreiter on the occasion of his 70th birthday / edited by Gerhard Larcher, Johannes Kepler Universität Linz, Friedrich Pillichshammer, Johannes Kepler Universität Linz, Arne Winterhof, Austrian Academy of Sciences, Linz, Chaoping Xing, Nanyang Technological University, Singapore.
pages cm
Includes bibliographical references.
ISBN 978-1-107-07400-2 (hardback)
1. Number theory. I. Niederreiter, Harald, 1944- honoree. II. Larcher, Gerhard, editor.
QA241.A67 2014
512.7–dc23
2014013624

ISBN 978-1-107-07400-2 Hardback

# Contents

The color plates are situated between pages 294 and 295.

# Preface

Harald Niederreiter's pioneering research in the field of applied algebra and number theory has led to important and substantial breakthroughs in many areas, including finite fields and areas of their application such as coding theory and cryptography as well as uniform distribution and quasi-Monte Carlo methods. He is the author of more than 350 research papers and 10 books.

This book contains essays from close colleagues and leading experts in those fields in which he has worked. The essays contain short overviews of different research areas as well as some very new research results.

The articles focus on uniform distribution and quasi-Monte Carlo methods as well as finite fields and their applications, in particular cryptography and pseudorandom number generation.

The first chapter gives an overview of Harald's career and describes some scientific spotlights.

Linz and Singapore, January 2014
Gerhard Larcher, Friedrich Pillichshammer,
Arne Winterhof and Chaoping Xing

# 1

# Some highlights of Harald Niederreiter's work

*Gerhard Larcher and Friedrich Pillichshammer*
Johannes Kepler University Linz

*Arne Winterhof*
Austrian Acadamy of Sciences, Linz

*Chaoping Xing*
Nanyang Techological University, Singapore

*Dedicated to our teacher, colleague and friend, Harald Niederreiter, on the occasion of his 70th birthday.*

## Abstract

In this paper we give a short biography of Harald Niederreiter and we spotlight some cornerstones from his wide-ranging work. We focus on his results on uniform distribution, algebraic curves, polynomials and quasi-Monte Carlo methods. In the flavor of Harald's work we also mention some applications including numerical integration, coding theory and cryptography.

## 1.1 A short biography

Harald Niederreiter was born in Vienna in 1944 on June 7 and spent his childhood in Salzburg. In 1963 he returned to Vienna to study at the Department of Mathematics of the University of Vienna, where he finished his PhD thesis entitled "Discrepancy in compact Abelian groups" *sub auspiciis praesidentis rei publicae*[1] under the supervision of Edmund Hlawka in 1969. From 1969 to 1978 he worked as scientist and professor in the USA at four different institutes: Southern Illinois University, University of Illinois at Urbana-Champaign, Institute for Advanced Study, Princeton, and University of California at Los Angeles. From 1978 to 1981 he was Chair of Pure Mathematics at the University of the West Indies in Kingston (Jamaica). He

---

[1] The term "Promotion sub auspiciis praesidentis rei publicae" is the highest possible honor for course achievement at school and university in Austria.

returned to Austria and served as director of two institutes of the Austrian Academy of Sciences in Vienna, of the Institute for Information Processing until 1999 and then of the Institute of Discrete Mathematics. From 2001 to 2009 he was professor at the National University of Singapore. Since 2009 he has been located at the Johann Radon Institute for Computational and Applied Mathematics in Linz. From 2010 to 2011 he was professor at the King Fahd University of Petroleum and Minerals in Dhahran (Saudi Arabia).

Harald Niederreiter's research areas include numerical analysis, pseudorandom number generation, quasi-Monte Carlo methods, cryptology, finite fields, applied algebra, algorithms, number theory and coding theory. He has published more than 350 research papers and several books, including the following.

- (with L. Kuipers) *Uniform Distribution of Sequences*. Wiley-Interscience, 1974; reprint, Dover Publications, 2006.
- (with R. Lidl) *Finite Fields*. Encyclopaedia of Mathematics and its Applications, volume 20. Addison-Wesley, 1983; second edition, Cambridge University Press, 1997.
- (with R. Lidl) *Introduction to Finite Fields and their Applications*. Cambridge University Press, 1986; revised edition, 1994.
- *Random Number Generation and Quasi-Monte Carlo Methods*. CBMS-NSF Regional Conference Series in Applied Mathematics, volume 63. Society for Industrial and Applied Mathematics (SIAM), 1992.
- (with C. P. Xing) *Rational Points on Curves over Finite Fields: Theory and Applications*. London Mathematical Society Lecture Note Series, volume 285. Cambridge University Press, 2001.
- (with C. P. Xing) *Algebraic Geometry in Coding Theory and Cryptography*. Princeton University Press, 2009.

Furthermore he is editor or co-editor of the following proceedings.

- (with P. J.-S. Shiue) *Monte Carlo and Quasi-Monte Carlo Methods in Scientific Computing*. Springer-Verlag, 1995.
- (with S. D. Cohen) *Finite Fields and Applications*. London Mathematical Society Lecture Note Series, volume 233. Cambridge University Press, 1996.
- (with P. Hellekalek, G. Larcher and P. Zinterhof) *Monte Carlo and Quasi-Monte Carlo Methods 1996*. Springer-Verlag, 1998.
- (with C. Ding and T. Helleseth) *Sequences and their Applications*. Springer-Verlag, 1999.

- (with J. Spanier) *Monte Carlo and Quasi-Monte Carlo Methods 1998.* Springer-Verlag, 2000.
- (with D. Jungnickel) *Finite Fields and Applications.* Springer-Verlag, 2001.
- (with K.-T. Fang and F. J. Hickernell) *Monte Carlo and Quasi-Monte Carlo Methods 2000.* Springer-Verlag, 2002.
- *Coding Theory and Cryptology.* World Scientific, 2002.
- *Monte Carlo and Quasi-Monte Carlo Methods 2002.* Springer-Verlag, 2004.
- (with K. Feng und C. P. Xing) *Coding, Cryptography and Combinatorics.* Birkhäuser-Verlag, 2004.
- (with D. Talay) *Monte Carlo and Quasi-Monte Carlo Methods 2004.* Springer-Verlag, 2006.
- (with A. Keller and S. Heinrich) *Monte Carlo and Quasi-Monte Carlo Methods 2006.* Springer-Verlag, 2008.
- (with Y. Li, S. Ling, H. Wang, C. P. Xing and S. Zhang) *Coding and Cryptology.* World Scientific, 2008.
- (with A. Ostafe, D. Panario and A. Winterhof) *Algebraic Curves and Finite Fields: Cryptography and Other Applications.* de Gruyter, 2014.
- (with P. Kritzer, F. Pillichshammer and A. Winterhof) *Uniform Distribution and Quasi-Monte Carlo Methods: Discrepancy, Integration and Applications.* de Gruyter, 2014.

Some important methods are named after him, such as the Niederreiter public-key cryptosystem, the Niederreiter factoring algorithm for polynomials over finite fields, and the Niederreiter and Niederreiter–Xing low-discrepancy sequences.

Some of his honors and awards are

- full member of the Austrian Academy of Sciences
- full member and former member of the presidium of the German Academy of Natural Sciences Leopoldina
- Cardinal Innitzer Prize for Natural Sciences in Austria
- invited speaker at ICM 1998 (Berlin) and ICIAM 2003 (Sydney)
- Singapore National Science Award 2003
- honorary member of the Austrian Mathematical Society 2012
- Fellow of the American Mathematical Society 2013.

Niederreiter was also the initiator and, from 1994 to 2006, the co-chair of the first seven biennial *Monte Carlo and quasi-Monte Carlo meetings* which took place in

- Las Vegas, NV, USA (1994)
- Salzburg, Austria (1996)

- Claremont, CA, USA (1998)
- Hong Kong (2000)
- Singapore (2002)
- Juan-Les-Pins, France (2004)
- Ulm, Germany (2006)
- Montreal, Canada (2008)
- Warsaw, Poland (2010)
- Sydney, Australia (2012)
- Leuven, Belgium (2014).

In 2006 Harald Niederreiter announced his wish to step down from the organizational role, and a Steering Committee was formed to ensure and oversee the continuation of the conference series.

## 1.2  Uniform distribution theory and number theory

When we scroll over the more than 350 scientific articles by Niederreiter which have appeared in renowned journals such as *Mathematika*, *Duke Mathematical Journal*, *Bulletin of the American Mathematical Society* and *Compositio Mathematica*, we find that most of these papers have connections to topics from number theory or use techniques from number theory, and many of the articles deal with problems and solve open questions, or initiate a new field of research in the theory of uniform distribution of sequences. The later sections in this overview of Harald's work on coding theory, algebraic curves and function fields, pseudorandom numbers, finite fields, and quasi-Monte Carlo methods in a certain sense will also deal with number-theoretical aspects.

Let us give just one example: the analysis and the precise estimation of exponential sums $\sum_{k=0}^{N-1} e^{2\pi i f(k)}$ or, in particular, of character sums plays an essential role in many different branches of mathematics and especially in number theory. In particular, it plays a basic role in many questions concerning uniform distribution of sequences, discrepancy theory, quasi-Monte Carlo methods, pseudorandom number analysis, the theory of finite fields, and many more. In a variety of papers on exponential sums and their applications, Niederreiter has proven to be a leading expert in the analysis of exponential sums and has essentially developed a variety of important techniques.

In this section we want to pick out some of the most impressive of Niederreiter's work on topics in number theory and in uniform distribution theory that will not be described explicitly in subsequent sections.

In the first years after finishing his PhD thesis "Discrepancy in compact Abelian groups" under the supervision of Edmund Hlawka, Niederreiter was concerned with basic questions from the theory of uniform distribution, from discrepancy theory and from metrical uniform distribution theory. We want to highlight three papers of this first phase.

In the paper "An application of the Hilbert–Montgomery–Vaughan inequality to the metric theory of uniform distribution mod 1" [12] which appeared in 1976 in the *Journal of the London Mathematical Society*, Niederreiter used tools from the theory of bounded quadratic and bilinear forms, especially an inequality of Montgomery and Vaughan based on large sieve methods, to establish an analog of Koksma's metric theorem for uniform distribution modulo one with respect to a general class of summation methods.

One of the most powerful tools for estimating the discrepancy of sequences is the Koksma–Erdős–Turán inequality which bounds the discrepancy of a sequence by a weighted sum of the values of its Weyl sums. The joint paper with Walter Philipp, which appeared in the *Duke Mathematical Journal* in 1973, "Berry–Esseen bounds and a theorem of Erdős and Turán on uniform distribution mod 1" [29], gave a much more general result about distances of functions that contains the one-dimensional Koksma–Erdős–Turán inequality as a special case. The given theorem is an analog of the standard Berry–Esseen lemma for $\mathbb{R}^s$.

One of the highlights in this period, and of the work of Niederreiter in metric Diophantine approximation theory certainly, was the solution of a conjecture of Donald Knuth, together with Robert F. Tichy, in the paper "Solution of a problem of Knuth on complete uniform distribution of sequences" [37] which appeared in *Mathematika* in 1985. It was shown that for any sequence $(a_n)_{n\geq 1}$ of distinct positive integers, the sequence $(x^{a_n})_{n\geq 1}$ is completely uniformly distributed modulo one for almost all real numbers $x$ with $|x| > 1$. In the paper "Metric theorems on uniform distribution and approximation theory" [38], again in cooperation with Tichy, this result was even generalized to the following form: the sequence $(cx^{b_n})_{n\geq 1}$ is completely uniformly distributed modulo one for all $c \neq 0$ for almost all real numbers $x$ with $|x| > 1$ whenever $(b_n)_{n\geq 1}$ is any sequence of reals with $\inf b_n > -\infty$ and $\inf_{m\neq n} |b_n - b_m| > 0$.

In the analysis of the distribution properties of sequences and of point sets, especially of Kronecker sequences

$$(((\{n\alpha_1\}, \ldots, \{n\alpha_s\}))_{n\geq 0}$$

and of lattice point sets

$$\left(\left(\left\{n\frac{a_1}{N}\right\}, \ldots, \left\{n\frac{a_s}{N}\right\}\right)\right)_{n=0,\ldots,N-1}$$

in the $s$-dimensional unit cube, one is often led to questions from the theory of Diophantine approximations, of the geometry of numbers or to questions concerning continued fraction expansions. A famous still open problem in the theory of continued fractions is the following conjecture of Zaremba.

*There is a constant $c$ such that for every integer $N \geq 2$ there exists an integer $a$ with $1 \leq a \leq N$ and with $\gcd(a, N) = 1$ such that all continued fraction coefficients of $\frac{a}{N}$ are bounded by $c$. Indeed it is conjectured that $c = 5$ satisfies this property.*

In the paper "Dyadic fractions with small partial quotients" [14], Niederreiter proved that this result is true even with $c = 3$ if $N$ is a power of 2. He also proved the conjecture of Zaremba for $N$ equal to powers of 3 and equal to powers of 5. Only quite recently it was shown by Bourgain and Kontorovich that Zaremba's conjecture holds for almost all choices of $N$.

From Niederreiter's result it can be deduced, for example, that for all $N = 2^m$ there exists an integer $a$ such that the lattice point set

$$\left( \left( \left\{ n \frac{1}{2^m} \right\}, \left\{ n \frac{a}{2^m} \right\} \right) \right)_{n=0,\ldots,2^m-1}$$

has discrepancy $D_N \leq c' \frac{\log N}{N}$, i.e., has best possible order of discrepancy.

The investigation of certain types of digital $(t, m, s)$-nets and of digital $(\mathbf{T}, s)$-sequences (see also Section 1.5) in analogy leads to questions concerning non-Archimedean Diophantine approximation and to questions concerning continued fraction expansions of formal Laurent series. Such questions were analyzed, for example, in the papers [7, 8, 16, 21].

In an impressive series of papers together with Igor Shparlinski, powerful methods for the estimation of exponential sums with nonlinear recurring sequences were developed by Niederreiter, see also Section 1.4 below. In the paper "On the distribution of power residues and primitive elements in some nonlinear recurring sequences" [36] which appeared in the *Bulletin of the London Mathematical Society* in 2003, it was shown that these methods can also be applied to estimation of the sums of multiplicative characters. As a consequence, results were obtained in this paper on the distribution of power residues and of primitive elements in such sequences.

So consider a sequence of elements $u_0, u_1, \ldots, u_{N-1}$ of the finite field $\mathbb{F}_q$ obtained by the recurrence relation

$$u_{n+1} = au_n^{-1} + b,$$

where we set $u_{n+1} = b$ if $u_n = 0$. For a divisor $s$ of $q - 1$ let $R_s(N)$ be the number of $s$-power residues (i.e., the number of $w \in \mathbb{F}_q$ such that there are $z \in \mathbb{F}_q$ with $z^s = w$) among $u_0, u_1, \ldots, u_{N-1}$. Then

$$\left| R_s(N) - \frac{N}{s} \right| < (2.2)N^{1/2}q^{1/4}$$

for $1 \leq N \leq t$, where $t$ is the least period of the recurring sequence. The case of general nonlinear recurrence sequences was studied later [40].

In the present, Harald Niederreiter is still a creative and productive researcher in the field of number theory and uniform distribution of sequences. We want to confirm this fact by giving two recent examples of his impressive work in these fields.

In the joint paper "On the Gowers norm of pseudorandom binary sequences" [32] with Joël Rivat, the modern concepts of Christian Mauduit and András Sárközy concerning new measures for pseudorandomness and of William T. Gowers in combinatorial and additive number theory were brought together, and the Gowers norm for periodic binary sequences was studied. A certain relation between the Gowers norm of a binary function $f$ defined on the integers modulo $N$ and a certain correlation measure for the sequence $(f(n))_{n \geq 1}$ introduced in [11] was shown.

A quite new and challenging trend in the theory of uniform distribution of sequences is the investigation of the distribution of hybrid sequences. A hybrid sequence is defined as follows: take an $s$-dimensional sequence $(\mathbf{x}_n)_{n \geq 0}$ of a certain type and a $t$-dimensional sequence $(\mathbf{y}_n)_{n \geq 0}$ of another type and combine them as an $(s+t)$-dimensional *hybrid sequence*, i.e., with some abuse of notation,

$$(\mathbf{z}_n)_{n \geq 0} := ((\mathbf{x}_n, \mathbf{y}_n))_{n \geq 0}.$$

Well-known examples of such sequences are Halton–Kronecker sequences (generated by combining Halton sequences with Kronecker sequences) and Halton–Niederreiter sequences (a combination of digital $(t, s)$-sequences or of digital $(\mathbf{T}, s)$-sequences in different bases). Investigation of these sequences again leads to challenging problems in number theory. For example, with the papers [22, 23, 24, 25, 26], Niederreiter influenced the direction of research in this topic.

## 1.3 Algebraic curves, function fields and applications

The study of algebraic curves over finite fields can be traced back to Carl Friedrich Gauss who studied equations over finite fields. However, the real beginning of this topic was the proof of the Riemann hypothesis for algebraic curves over finite fields by André Weil in the 1940s. This topic has attracted the attention of researchers again since the 1980s due to the discovery of algebraic geometry codes by Valerii D. Goppa. This application of algebraic curves over finite fields, and especially of those with many rational points, created a much stronger interest in the area and attracted new groups of researchers such as coding theorists and algorithmically inclined mathematicians. Nowadays, algebraic curves over finite fields is a flourishing subject which produces exciting research and is immensely relevant for applications.

Harald Niederreiter started this topic from applications first. In the late 1980s, he found an elegant construction of $(t, m, s)$-nets and $(t, s)$-sequences (see Section 1.5). Then he realized that the construction can be generalized to global function fields [43, 44]. From this point, Harald Niederreiter investigated extensively algebraic curves over finite fields with many rational points and their applications. Algebraic curves over finite fields can be described in an equivalent algebraic language, i.e., global function fields over finite fields. For many of the applications, people are interested in algebraic curves over finite fields with many rational points or, equivalently, global function fields over finite fields with many rational places. Since the global function field language was usually used by Harald Niederreiter, we adopt this language from now onwards in this section.

Let $\mathbb{F}_q$ denote the finite field of $q$ elements. An extension $F$ of $\mathbb{F}_q$ is called an algebraic function field of one variable over $\mathbb{F}_q$ if there exists an element $x$ of $F$ that is transcendental over $\mathbb{F}_q$ such that $F$ is a finite extension over the rational function field $\mathbb{F}_q(x)$. We usually denote by $F/\mathbb{F}_q$ a global function field with the full constant field $\mathbb{F}_q$, i.e., all elements in $F \setminus \mathbb{F}_q$ are transcendental over $\mathbb{F}_q$. A place $P$ of $F$ is called *rational* if its residue field $F_P$ is isomorphic to the ground field $\mathbb{F}_q$. For many applications in coding theory, cryptography and low-discrepancy sequences, people are interested in those function fields with many rational places. On the other hand, the number of rational places of a function field over $\mathbb{F}_q$ is constrained by an important invariant of $F$, called the genus. If we use $g(F)$ and $N(F)$ to denote the genus and the number of rational places of $F/\mathbb{F}_q$, the well-known Hasse–Weil bound says that

$$|N(F) - q - 1| \leq 2g(F)\sqrt{q}. \qquad (1.1)$$

The above bound implies that the number of rational places cannot be too big if we fix the genus of a function field. Now the problem becomes to find the maximal number of rational places that a global function field over $\mathbb{F}_q$ of genus $g$ could have. We usually denote by $N_q(g)$ this quantity, i.e., $N_q(g) =$

$\max\{N(F): F/\mathbb{F}_q \text{ has genus } g\}$. Apparently, it follows from the Hasse–Weil bound that

$$|N_q(g) - q - 1| \leq 2g\sqrt{q} \qquad (1.2)$$

for any prime power $q$ and nonnegative integer $g$. For given $q$ and $g$, determining the exact value of $N_q(g)$ is a major problem in the study of global function fields. In general it is very difficult to determine the exact value of $N_q(g)$. Instead, it is sufficient to find reasonable lower bounds for most applications. Lower bounds on $N_q(g) \geq N$ are found either by explicit construction or by showing the existence of global function fields of genus $g$ with at least $N$ rational places. Investigation of this problem involves several subjects such as algebraic number theory and algebraic geometry and even coding theory. The method that Harald Niederreiter employed is class field theory in algebraic number theory. He found many record function fields through class field theory, i.e., global function fields with best-known number of rational places. Some of these record function fields are listed below (see [44, 45, 46, 47, 48, 49, 50, 53, 59]).

| $(q,g)$ | $(2,23)$ | $(2,25)$ | $(2,29)$ | $(2,31)$ | $(2,34)$ | $(2,36)$ | $(2,49)$ | $(3,6)$ | $(3,7)$ |
|---|---|---|---|---|---|---|---|---|---|
| $N_q(g)$ | 22 | 24* | 25 | 27 | 27 | 30 | 36 | 14* | 16* |

The entries with an asterisk are the exact values of $N_q(g)$, while the entries without an asterisk are lower bounds on $N_q(g)$.

For a fixed prime power $q$, to measure how $N_q(g)$ behaves while $g$ tends to infinity, we define the following asymptotic quantity

$$A(q) := \limsup_{g \to \infty} \frac{N_q(g)}{g}. \qquad (1.3)$$

It is immediate from the Hasse–Weil bound that $A(q) \leq 2\sqrt{q}$. Sergei G. Vlăduţ and Vladimir G. Drinfeld refined this bound to $A(q) \leq \sqrt{q} - 1$. Yasutaka Ihara first showed that $A(q) \geq \sqrt{q} - 1$ if $q$ is a square. Thus, the problem of determining $A(q)$ is completely solved for squares $q$. It still remains to determine $A(q)$ for nonsquare $q$. Like the case of $N_q(g)$, finding the exact value of $A(q)$ for nonsquare $q$ is very difficult. Although people have tried very hard, so far $A(q)$ has not been determined for any single nonsquare $q$. In particular, if $q$ is a prime, it is a great challenge to determine or find a reasonable lower bound on $A(q)$.

What Harald Niederreiter did for this problem was to find a new bound on $A(2)$ and an improvement on $A(q^m)$ for odd $m$. More precisely, he proved the following result [51, 52].

**Theorem 1.1** *One has* $A(2) \geq \frac{81}{317} = 0.2555\ldots$.

**Theorem 1.2** *One has the following bounds.*

(*i*) *If q is an odd prime power and m $\geq$ 3 is an integer, then*

$$A(q^m) \geq \frac{2q+2}{\lceil 2(2q+3)^{1/2} \rceil + 1}.$$

(*ii*) *If q $\geq$ 8 is a power of 2 and m $\geq$ 3 is an odd integer, then*

$$A(q^m) \geq \frac{q+1}{\lceil 2(2q+2)^{1/2} \rceil + 2}.$$

Harald Niederreiter has also been working on applications of algebraic curves over finite fields. These applications include low-discrepancy sequences, coding theory and cryptography, etc. For details on the application of algebraic curves over finite fields to low-discrepancy sequences, we refer to Section 1.5.

For applications to coding theory, Harald Niederreiter's contribution was the discovery of several new codes via the theory of algebraic curves over finite fields. Some of the new codes discovered by Harald Niederreiter are listed below (see [3]). In the table, $[n, k, d]_q$ is a $q$-ary code of length $n$, dimension $k$ and minimum distance $d$.

| | | | | | |
|---|---|---|---|---|---|
| $[108, 25, 44]_4$ | $[108, 26, 43]_4$ | $[113, 27, 45]_4$ | $[130, 29, 53]_4$ | $[27, 11, 13]_8$ | $[30, 7, 19]_8$ |
| $[30, 8, 18]_8$ | $[30, 9, 17]_8$ | $[36, 7, 23]_8$ | $[36, 8, 22]_8$ | $[36, 9, 21]_8$ | $[36, 10, 20]_8$ |

Harald Niederreiter has also done some significant work on asymptotic results of coding theory and cryptography via algebraic curves over finite fields.

## 1.4 Polynomials over finite fields and applications

Now we describe some of Harald Niederreiter's results on polynomials over finite fields and applications. We start with complete mappings and check digit systems.

Let $\mathbb{F}_q$ be the finite field of $q > 2$ elements and $f(X) \in \mathbb{F}_q[X]$ a permutation polynomial over $\mathbb{F}_q$. We call $f(X)$ a *complete mapping* if $f(X) + X$ is also a permutation polynomial. Existence results on complete mappings and their application to check digit systems were discussed in [33, 56].

It is easy to see that $f(X) = aX$ is a complete mapping whenever $a \notin \{-1, 0\}$.

Complete mappings are pertinent to the construction of orthogonal Latin squares, see [10], which can be used to design some agricultural experiments. However, here we will describe another application of complete mappings, namely, check digit systems.

A *check digit system* (defined with one permutation polynomial over $\mathbb{F}_q$) consists of a permutation polynomial $f(X) \in \mathbb{F}_q[X]$ and a control symbol $c \in \mathbb{F}_q$ such that each word $a_1, \ldots, a_{s-1} \in \mathbb{F}_q^{s-1}$ of length $s-1$ is extended by a check digit $a_s \in \mathbb{F}_q$ such that

$$\sum_{i=0}^{s-1} f^{(i)}(a_{i+1}) = c,$$

where $f^{(i)}$ is defined recursively by $f^{(0)}(X) = X$ and $f^{(i)}(X) = f(f^{(i-1)}(X))$ for $i = 1, 2, \ldots$.

An example of a check digit system is the international standard book number (ISBN-10) which consists of a string of 10 digits $x_1$–$x_2x_3x_4$–$x_5x_6$ $x_7x_8x_9$–$x_{10}$. The first digit $x_1$ characterizes the language group, $x_2x_3x_4$ is the number of the publisher, $x_5x_6x_7x_8x_9$ is the actual book number, and $x_{10}$ is a check digit. A correct ISBN satisfies

$$x_1 + 2x_2 + 3x_3 + 4x_4 + 5x_5 + 6x_6 + 7x_7 + 8x_8 + 9x_9 + 10x_{10} = 0 \in \mathbb{F}_{11}.$$

With the variable transformation $a_i = x_{2^{i-1} \bmod 11}$ we get a check digit system defined with one permutation polynomial $f(X) = 2X$. Note that $f(X) = 2X$ and $-f(X) = 9X$ are both complete mappings of $\mathbb{F}_{11}$.

For example, the ISBN-10 of the monograph on finite fields by Lidl and Niederreiter [9] is 0–521–39231–4.

Since $f(X)$ is a permutation polynomial, such a system detects all single errors $a \mapsto b$. Moreover it detects all

- neighbor transpositions $ab \mapsto ba$ if $-f(X)$ is a complete mapping, and
- twin errors $aa \mapsto bb$ if $f(X)$ is a complete mapping.

Niederreiter and Karl H. Robinson [33] found several nontrivial classes of complete mappings and proved in particular a generalization of the following result.

**Theorem 1.3** *Let $q$ be odd. Then $f_b(X) = X^{(q+1)/2} + bX$ is a complete mapping of $\mathbb{F}_q$ if and only if $b^2 - 1$ and $b^2 + 2b$ are both squares of nonzero* elements of $\mathbb{F}_q$. The number of $b$ such that $f_b(X)$ is a complete mapping is $\frac{q}{4} + O(q^{1/2})$.

For a survey on (generalizations of) complete mappings and some applications we refer to [58].

Harald Niederreiter also invented a *deterministic algorithm* based on linear algebra for *factoring a univariate polynomial* $f(X)$ over $\mathbb{F}_q$ which is efficient for small characteristic, see [20] for the initial article and [4] for a survey on factorization. The key step is to find a polynomial $h(X)$ which satisfies the differential equation

$$f^q (h/f)^{(q-1)} + h^q = 0,$$

where $g^{(k)}$ denotes the $k$th Hasse–Teichmüller derivative. Then $\gcd(f, h)$ is a nontrivial factor of $f$.

Harald Niederreiter contributed to cryptography not only via the above mentioned public-key cryptosystem named after him, but also in many other ways. For example he proved several results on the interpolation of the discrete logarithm [17, 39], showing that there is no low degree polynomial $f(X) \in \mathbb{F}_q[X]$ which coincides with the discrete logarithm on many values, that is for prime $q$, $f(g^x) = x$ for many $x$, where $g$ is a primitive element of $\mathbb{F}_q$. Hence, the discrete logarithm problem is not attackable via simple interpolation which is necessary for the security of discrete logarithm based cryptosystems such as the Diffie–Hellman key exchange.

Finally, he introduced and studied *nonlinear pseudorandom number generators*, i.e., sequences over $\mathbb{F}_q$ of the form

$$u_{n+1} = f(u_n), \quad n = 0, 1, \ldots$$

for some initial value $u_0 \in \mathbb{F}_q$ and a polynomial $f(X) \in \mathbb{F}_q[X]$ of degree at least 2. These sequences are attractive alternatives to linear pseudorandom number generators which are not suitable for all applications. For example, linear generators are highly predictable and are not suitable in cryptography.

As mentioned before, in joint work with Igor Shparlinski [34, 35], Niederreiter found a way to prove nontrivial estimates on certain character sums which in the simplest case are of the form

$$\sum_{n=0}^{N-1} \chi(f(u_n))$$

where $\chi$ is any nontrivial additive character of $\mathbb{F}_q$. For those character sums, the standard method for estimating incomplete character sums by reducing them to complete ones and then applying the Weil bound does not work. The method and result of [34] was later slightly improved in [41]. In particular, if $f(X) = aX^{q-2} + b$, i.e. $f(c) = ac^{-1} + b$ if $c \neq 0$, this method yields strong bounds on the exponential sums and leads to very good discrepancy bounds for corresponding sequences in the unit interval. For a survey on nonlinear recurrence sequences see [57].

## 1.5 Quasi-Monte Carlo methods

The quasi-Monte Carlo method has its roots in the theory of uniform distribution modulo 1 (see Section 1.2) and is nowadays a powerful tool in computational mathematics, in particular for the numerical integration of very high dimensional functions, with many applications to practical problems from biology, computer graphics, mathematical finance, statistics, etc. Here the integral of a function $f : [0, 1]^s \to \mathbb{R}$ is approximated by a quasi-Monte Carlo (QMC) rule which computes the arithmetic mean of function values over a finite set of sample nodes, i.e.,

$$\int_{[0,1]^s} f(\boldsymbol{x})\, d\boldsymbol{x} \approx \frac{1}{N} \sum_{n=0}^{N-1} f(\boldsymbol{x}_n)$$

with fixed $\boldsymbol{x}_0, \ldots, \boldsymbol{x}_{N-1} \in [0, 1)^s$. QMC rules can be viewed as deterministic versions of Monte Carlo rules. The fundamental error estimate for QMC rules is the Koksma–Hlawka inequality which bounds the absolute integration error as

$$\left| \int_{[0,1]^s} f(\boldsymbol{x})\, d\boldsymbol{x} - \frac{1}{N} \sum_{n=0}^{N-1} f(\boldsymbol{x}_n) \right| \leq V(f) D_N^*(\boldsymbol{x}_0, \ldots, \boldsymbol{x}_{N-1}),$$

where $V(f)$ is the variation of $f$ in the sense of Hardy and Krause and where $D_N^*$ is the star discrepancy of the underlying sample nodes, see [6].

In the mid 1970s Harald Niederreiter started to investigate QMC methods. His first pioneering work was the paper "Quasi-Monte Carlo methods and pseudo-random numbers" [13] published in the *Bulletin of the American Mathematical Society* in 1978. Today this paper can be seen as the first systematic survey of the theoretical foundations of QMC dealing with Koksma–Hlawka type inequalities and with constructions of point sets for QMC rules such as Halton's sequence, Sobol's construction of $P_\tau$ nets and $LP_\tau$ sequences, and good lattice points in the sense of Korobov and Hlawka.

The quintessence of the Koksma–Hlawka inequality is that good QMC rules should be based on sample nodes with low discrepancy, informally often called *low-discrepancy point sets.* Today there are two main methods of constructing low-discrepancy point sets. Both constructions are intimately connected with Niederreiter, who contributed pioneering work to these topics. The first construction is the concept of lattice point sets and the second is the concept of $(t, m, s)$-nets and $(t, s)$-sequences in a base $b$.

An $N$-*element lattice point set* (cf. Section 1.2) is based on an $s$-dimensional lattice point $\boldsymbol{a} = (a_1, \ldots, a_s)$. The $n$th element of such a lattice point set is then given as

$$\boldsymbol{x}_n = \left\{ \frac{n}{N} \boldsymbol{a} \right\} \quad \text{for } n = 0, 1, \ldots, N-1,$$

where the fractional part function $\{\cdot\}$ is applied component-wise. QMC rules which are based on good lattice point sets are called the *method of good lattice points* or *lattice rules* and nowadays belong to the most popular QMC rules in practical applications. Niederreiter analyzed distribution properties and showed the existence of good lattice point sets with low discrepancy. The full power of lattice rules, however, lies in the integration of smooth one-periodic functions. One reason for this is the following relation: for $\boldsymbol{h} \in \mathbb{Z}^s$

$$\frac{1}{N} \sum_{n=0}^{N-1} \exp\left(2\pi \mathrm{i} \frac{n}{N} \boldsymbol{a} \cdot \boldsymbol{h}\right) = \begin{cases} 1 & \text{if } \boldsymbol{a} \cdot \boldsymbol{h} \equiv 0 \,(\mathrm{mod}\,N), \\ 0 & \text{if } \boldsymbol{a} \cdot \boldsymbol{h} \not\equiv 0 \,(\mathrm{mod}\,N), \end{cases}$$

where $\cdot$ denotes the usual inner product. Niederreiter studied the worst-case error $P_\alpha$ for the integration of functions $f$ which can be represented by absolutely convergent Fourier series whose Fourier coefficients $\widehat{f}(\boldsymbol{h})$ tend to zero as $\boldsymbol{h}$ moves away from the origin at a prescribed rate which is determined by the parameter $\alpha$. His most important contributions to the theory of good lattice point sets are summarized in Chapter 5 of his book *Random Number Generation and Quasi-Monte Carlo Methods* [19] which appeared in 1992. Niederreiter's most recent contributions to the theory of lattice point sets deal with the existence and construction of so-called extensible lattice point sets which have the property that the number of points in the node set may be increased while retaining the existing points (see [5, 30]).

The theory of $(t, m, s)$-*nets and* $(t, s)$-*sequences* was initiated by Niederreiter in his seminal paper "Point sets and sequences with small discrepancy" [15] published in *Monatshefte für Mathematik* in 1987. The basic idea of these concepts is that if a point set has good equidistribution properties with respect to a reasonable (finite) set of test sets, then the point set already has low star discrepancy. The definition of a $(t, m, s)$-net in base $b$ can be stated as follows.

**Definition 1.4** (Niederreiter, 1987) *Let $s, b, m, t$ be integers satisfying $s \geq 1$, $b \geq 2$ and $0 \leq t \leq m$. A set $\mathcal{P}$ consisting of $b^m$ elements in $[0, 1)^s$ is said to be a $(t, m, s)$-net in base $b$ if every so-called elementary interval of the form*

$$\prod_{j=1}^{s} \left[ \frac{a_j}{b^{d_j}}, \frac{a_j + 1}{b^{d_j}} \right)$$

*of volume $b^{t-m}$ with $d_j \in \mathbb{N}_0$ and $a_j \in \{0, 1, \ldots, b^{d_j} - 1\}$ for $j = 1, 2, \ldots, s$, contains exactly $b^t$ elements of $\mathcal{P}$.*

A $(t, s)$-sequence in base $b$ is an infinite version of a $(t, m, s)$-net.

**Definition 1.5** (Niederreiter, 1987) *Let $s, b, t$ be integers satisfying $s \geq 1$, $b \geq 2$ and $t \geq 0$. An infinite sequence $(\boldsymbol{x}_n)_{n\geq 0}$ of points in $[0, 1)^s$ is said to be a $(t, s)$-sequence in base $b$ if, for all integers $k \geq 0$ and $m > t$, the point set consisting of the $\boldsymbol{x}_n$ with $kb^m \leq n < (k + 1)b^m$ is a $(t, m, s)$-net in base $b$.*

In his work Niederreiter [15] presented a comprehensive theory of $(t, m, s)$-nets and $(t, s)$-sequences including discrepancy estimates, existence results and connections to other mathematical disciplines such as, for example, combinatorics. The fundamental discrepancy estimate for a $(t, m, s)$-net $\mathcal{P}$ in base $b$ states that

$$D_N^*(\mathcal{P}) \leq c_{s,b} b^t \frac{(\log N)^{s-1}}{N} + O_{s,b}\left( b^t \frac{(\log N)^{s-2}}{N} \right)$$

where $N = b^m$ and where $c_{s,b} > 0$ is independent of $m$ and $t$. This estimate justifies the definition of $(t, m, s)$-nets since it means that for sufficiently small $t$ one can achieve a star discrepancy of order of magnitude $O((\log N)^{s-1}/N)$. Many people working on discrepancy theory conjecture that this is the best convergence rate which can be achieved for the star discrepancy of $N$-element point sets in dimension $s$. For infinite $(t, s)$-sequences in base $b$ one can achieve a star discrepancy of order of magnitude $O((\log N)^s/N)$, which again is widely believed to be the best rate for the star discrepancy of infinite sequences in dimension $s$.

Most constructions of $(t, m, s)$-nets and $(t, s)$-sequences rely on the digital method which was introduced by Niederreiter, also in [15]. In the case of $(t, m, s)$-nets this construction requires $m \times m$ matrices $C_1, C_2, \ldots, C_s$ over a commutative ring $R$ with identity and $|R| = b$ and, in a simplified form, a bijection $\psi$ from the set of $b$-adic digits $\mathcal{Z}_b = \{0, 1, \ldots, b - 1\}$ onto $R$. For $n = 0, 1, \ldots, b^m - 1$, let $n = n_0 + n_1 b + \cdots + n_{m-1}b^{m-1}$ with all

$n_r \in \mathcal{Z}_b$. Then, for $j = 1, 2, \ldots, s$, multiply the matrix $C_j$ with the vector $\mathbf{n} = (\psi(n_0), \psi(n_1), \ldots, \psi(n_{m-1}))^\top$ whose components belong to $R$,

$$C_j \mathbf{n} = (y_{n,j,1}, y_{n,j,2}, \ldots, y_{n,j,m})^\top, \quad \text{with all } y_r \in R$$

and set $\mathbf{x}_n = (x_{n,1}, x_{n,2}, \ldots, x_{n,s})$, where

$$x_{n,j} = \frac{\psi^{-1}(y_{n,j,1})}{b} + \frac{\psi^{-1}(y_{n,j,2})}{b^2} + \cdots + \frac{\psi^{-1}(y_{n,j,m})}{b^m}.$$

The point set $\{\mathbf{x}_0, \mathbf{x}_1, \ldots, \mathbf{x}_{b^m-1}\}$ constructed this way is a $b^m$-element point set in $[0, 1)^s$ and it is therefore a $(t, m, s)$-net in base $b$ for some $t \in \{0, 1, \ldots, m\}$ which is called a digital $(t, m, s)$-net over $R$. In the case of $(t, s)$-sequences the only difference is that one uses $\infty \times \infty$ matrices.

The so-called quality parameter $t$ depends only on the chosen matrices $C_1, C_2, \ldots, C_s$. Of course $t$ should be as small as possible, in the optimal case $t = 0$. If the base $b$ is a prime power, then one chooses for $R$ the finite field $\mathbb{F}_b$ of order $b$. This makes life a bit easier and is therefore the most studied case. Then $t$ is determined by some linear independence property of the row-vectors of the generating matrices $C_1, C_2, \ldots, C_s$ which provides the link of digital nets and sequences to the theory of finite fields and linear algebra over finite fields.

Niederreiter developed several constructions of generating matrices which lead to good, often even optimal, small $t$-values. One important construction results in the now so-called *Niederreiter sequences* and is based on polynomial arithmetic over finite fields and the formal Laurent series expansion of certain rational functions over $\mathbb{F}_b$ whose Laurent coefficients are used to fill the generating matrices. If $s \le b$ this leads to an explicit construction of $(0, s)$-sequences in base $b$ which in turn implies, for $s \le b + 1$, an explicit construction of a $(0, m, s)$-net in base $b$ for every $m \ge 2$. It is known that the conditions $s \le b$ for sequences and $s \le b + 1$ for nets, respectively, are even necessary to achieve a quality parameter equal to zero. Niederreiter sequences and slight generalizations thereof recover and unify the existing constructions due to Il'ya M. Sobol' and Henri Faure.

An important subclass of $(t, m, s)$-nets which was introduced by Niederreiter in the paper "Low-discrepancy point sets obtained by digital constructions over finite fields" [18] is provided by the concept of what we call today *polynomial lattice point sets*. This name has its origin in a close relation to ordinary lattice point sets. In fact, research on polynomial lattice point sets and on ordinary lattice point sets often follows two parallel tracks with many similarities (but there are also differences).

Niederreiter's early work on $(t, m, s)$-nets and $(t, s)$-sequences is well summarized in Chapter 4 of his book, already mentioned. *Random Number Generation and Quasi-Monte Carlo Methods* [19] which appeared in 1992. Since its appearance, this book has become *the* reference book for $(t, m, s)$-nets and $(t, s)$-sequences especially, and for QMC and random number generation in general.

A disadvantage of Niederreiter sequences in dimension $s$ is that they can only achieve a $t$-value of order $O(s \log s)$ as $s$ tends to infinity. This disadvantage was overcome by the next cornerstone of Niederreiter's work in QMC, the constructions of Niederreiter–Xing sequences. In a series of papers [42, 43, 44, 50, 55] starting in 1995 and based on methods from algebraic geometry, Niederreiter developed, in collaboration with Chaoping Xing, constructions of generating matrices which achieve the currently best known quality parameters of order $O(s)$ for growing dimensions $s$. This order is known to be best possible. An introduction into this subject and an overview can be found in the book *Rational Points on Curves over Finite Fields* [54] published by Niederreiter and Xing in 2001.

In 2001 Niederreiter developed together with Gottlieb Pirsic [31] a *duality theory* for digital nets. The basic idea is that the construction of digital $(t, m, s)$-nets over $\mathbb{F}_b$ can be reduced to the construction of certain $\mathbb{F}_b$-linear subspaces of $\mathbb{F}_b^{sm}$. Using the standard inner product in $\mathbb{F}_b^{sm}$ one can define and study the dual linear subspace. If one defines a special weight on $\mathbb{F}_b^{sm}$, the so-called *Niederreiter–Rosenbloom–Tsfasman weight*, then the $t$-parameter of a digital net is closely related to the weight of the corresponding dual linear subspace. This point of view gives new possibilities for the construction of digital nets, for example, cyclic nets or hyperplane nets, and it provides a connection to the theory of linear codes. Later, in 2009, Josef Dick and Niederreiter [1] extended the duality theory for digital nets to digital sequences which became a convenient framework for the description of many constructions such as, for example, those of Niederreiter and Xing and of Niederreiter and Ferruh Özbudak [27, 28] (see also [2]).

Digital nets also have a close connection to other discrete objects such as orthogonal Latin squares and ordered orthogonal arrays. These relations were also the subject of Niederreiter's research.

Harald Niederreiter's contributions to the theory of QMC are groundbreaking. He opened new doors and developed comprehensive theories of lattice rules and of $(t, m, s)$-nets and $(t, s)$-sequences with many new ideas and facets. Today Niederreiter's work forms one of the essential pillars of QMC integration.

## Acknowledgements

G. Larcher is supported by the Austrian Science Fund (FWF): Project F5507-N26, which is a part of the Special Research Program "Quasi-Monte Carlo Methods: Theory and Applications." F. Pillichshammer is supported by the Austrian Science Fund (FWF): Project F5509-N26, which is a part of the Special Research Program "Quasi-Monte Carlo Methods: Theory and Applications." C. Xing is supported by Singapore Ministry of Education Tier 1 grant 2013-T1-002-033. A. Winterhof is supported by the Austrian Science Fund (FWF): Project F5511-N26, which is a part of the Special Research Program "Quasi-Monte Carlo Methods: Theory and Applications."

## References

[1]  J. Dick and H. Niederreiter, Duality for digital sequences. *J. Complexity* **25**, 406–414, 2009.

[2]  J. Dick and F. Pillichshammer, *Digital Nets and Sequences. Discrepancy Theory and Quasi-Monte Carlo Integration.* Cambridge University Press, Cambridge, 2010.

[3]  C. S. Ding, H. Niederreiter and C. P. Xing, Some new codes from algebraic curves, *IEEE Trans. Inf. Theory* **46**, 2638–2642, 2000.

[4]  J. von zur Gathen and D. Panario, Factoring polynomials over finite fields: a survey. Computational algebra and number theory (Milwaukee, WI, 1996). *J. Symbolic Comput.* **31**, 3–17, 2001.

[5]  F. J. Hickernell and H. Niederreiter, The existence of good extensible rank-1 lattices. *J. Complexity* **19**, 286–300, 2003.

[6]  L. Kuipers and H. Niederreiter, *Uniform Distribution of Sequences.* John Wiley, New York, 1974. Reprint, Dover Publications, Mineola, NY, 2006.

[7]  G. Larcher and H. Niederreiter, Kronecker type sequences and non-Archimedean Diophantine approximations. *Acta Arith.* **63**, 379–396, 1993.

[8]  G. Larcher and H. Niederreiter, Generalized $(t, s)$-sequences, Kronecker-type sequences, and Diophantine approximations of formal Laurent series. *Trans. Am. Math. Soc.* **347**, 2051–2073, 1995.

[9]  R. Lidl and H. Niederreiter, *Finite Fields*, second edition. Encyclopedia of Mathematics and its Applications, volume 20. Cambridge University Press, Cambridge, 1997.

[10]  H. B. Mann, The construction of orthogonal Latin squares. *Ann. Math. Stat.* **13**, 418–423, 1942.

[11]  C. Mauduit and A. Sárközy, On finite pseudorandom binary sequences. I. Measure of pseudorandomness, the Legendre symbol. *Acta Arith.* **82**, 365–377, 1997.

[12]  H. Niederreiter, An application of the Hilbert–Montgomery–Vaughan inequality to the metric theory of uniform distribution mod 1. *J. London Math. Soc.* **13**, 497–506, 1976.

[13] H. Niederreiter, Quasi-Monte Carlo methods and pseudo-random numbers. *Bull. Am. Math. Soc.* **84**, 957–1041, 1978.

[14] H. Niederreiter, Dyadic fractions with small partial quotients. *Monatsh. Math.* **101**, 309–315, 1986.

[15] H. Niederreiter, Point sets and sequences with small discrepancy. *Monatsh. Math.* **104**, 273–337, 1987.

[16] H. Niederreiter, Rational functions with partial quotients of small degree in their continued fraction expansion. *Monatsh. Math* **103**, 269–288, 1987.

[17] H. Niederreiter, A short proof for explicit formulas for discrete logarithms in finite fields. *Appl. Algebra Eng. Commun. Comput.* **1**, 55–57, 1990.

[18] H. Niederreiter, Low-discrepancy point sets obtained by digital constructions over finite fields. *Czecho. Math. J.* **42**, 143–166, 1992.

[19] H. Niederreiter, *Random Number Generation and Quasi-Monte Carlo Methods.* CBMS-NSF Series in Applied Mathematics, volume 63. SIAM, Philadelphia, PA, 1992.

[20] H. Niederreiter, A new efficient factorization algorithm for polynomials over small finite fields. *Appl. Algebra Eng. Commun. Comput.* **4**, 81–87, 1993.

[21] H. Niederreiter, Low-discrepancy sequences and non-Archimedean Diophantine approximations. *Stud. Sci. Math. Hung.* **30**, 111–122, 1995.

[22] H. Niederreiter, On the discrepancy of some hybrid sequences. *Acta Arith.* **138**, 373–398, 2009.

[23] H. Niederreiter, A discrepancy bound for hybrid sequences involving digital explicit inversive pseudorandom numbers. *Unif. Distrib. Theory* **5**, 53–63, 2010.

[24] H. Niederreiter, Further discrepancy bounds and an Erdős–Turán–Koksma inequality for hybrid sequences. *Monatsh. Math.* **161**, 193–222, 2010.

[25] H. Niederreiter, Discrepancy bounds for hybrid sequences involving matrix-method pseudorandom vectors. *Publ. Math. Debrecen* **79**, 589–603, 2011.

[26] H. Niederreiter, Improved discrepancy bounds for hybrid sequences involving Halton sequences. *Acta Arith.* **155**, 71–84, 2012.

[27] H. Niederreiter and F. Özbudak, Constructions of digital nets using global function fields. *Acta Arith.* **105**, 279–302, 2002.

[28] H. Niederreiter and F. Özbudak, Matrix-product constructions of digital nets. *Finite Fields Appl.* **10**, 464–479, 2004.

[29] H. Niederreiter and W. Philipp, Berry–Esseen bounds and a theorem of Erdős and Turán on uniform distribution mod 1. *Duke Math. J.* **40**, 633–649, 1973.

[30] H. Niederreiter and F. Pillichshammer, Construction algorithms for good extensible lattice rules. *Construct. Approx.* **30**, 361–393, 2009.

[31] H. Niederreiter and G. Pirsic, Duality for digital nets and its applications. *Acta Arith.* **97**, 173–182, 2001.

[32] H. Niederreiter and J. Rivat, On the Gowers norm of pseudorandom binary sequences. *Bull. Aust. Math. Soc.* **79**, 259–271, 2009.

[33] H. Niederreiter and K. H. Robinson, Complete mappings of finite fields. *J. Aust. Math. Soc. Ser. A* **33**(2), 197–212, 1982.

[34] H. Niederreiter and I. E. Shparlinski, On the distribution and lattice structure of nonlinear congruential pseudorandom numbers. *Finite Fields Appl.* **5**, 246–253, 1999.

[35] H. Niederreiter and I. E. Shparlinski, On the distribution of inversive congruential pseudorandom numbers in parts of the period. *Math. Comp.* **70**, 1569–1574, 2001.

[36] H. Niederreiter and I. Shparlinski, On the distribution of power residues and primitive elements in some nonlinear recurring sequences. *Bull. London Math. Soc.* **35**, 522–528, 2003.

[37] H. Niederreiter and R. F. Tichy, Solution of a problem of Knuth on complete uniform distribution of sequences. *Mathematika* **32**, 26–32, 1985.

[38] H. Niederreiter and R. F. Tichy, Metric theorems on uniform distribution and approximation theory. *Journées Arithmétiques de Besançon (Besançon, 1985),* Astérisque No. 147–148, 319–323, 346, 1987.

[39] H. Niederreiter and A. Winterhof, Incomplete character sums and polynomial interpolation of the discrete logarithm. *Finite Fields Appl.* **8**, 184–192, 2002.

[40] H. Niederreiter and A. Winterhof, Multiplicative character sums for nonlinear recurring sequences. *Acta Arith.* **111**, 299–305, 2004.

[41] H. Niederreiter and A. Winterhof, Exponential sums for nonlinear recurring sequences. *Finite Fields Appl.* **14**, 59–64, 2008.

[42] H. Niederreiter and C. P. Xing, Low-discrepancy sequences obtained from algebraic function fields over finite fields. *Acta Arith.* **72**, 281–298, 1995.

[43] H. Niederreiter and C. P. Xing, Low-discrepancy sequences and global function fields with many rational places. *Finite Fields Appl.* **2**, 241–273, 1996.

[44] H. Niederreiter and C. P. Xing, Quasirandom points and global function fields. In: S. Cohen and H. Niederreiter (eds.), *Finite Fields and Applications.* London Mathematical Society Lecture Note Series, volume 233, pp. 269–296. Cambridge University Press, Cambridge, 1996.

[45] H. Niederreiter and C. P. Xing, Explicit global function fields over the binary field with many rational places. *Acta Arith.* **75**, 383–396, 1996.

[46] H. Niederreiter and C. P. Xing, Cyclotomic function fields, Hilbert class fields and global function fields with many rational places. *Acta Arith.* **79**, 59–76, 1997.

[47] H. Niederreiter and C. P. Xing, Drinfeld modules of rank 1 and algebraic curves with many rational points II. *Acta Arith.* **81**, 81–100, 1997.

[48] H. Niederreiter and C. P. Xing, Global function fields with many rational places over the ternary field. *Acta Arith.* **83**, 65–86, 1998.

[49] H. Niederreiter and C. P. Xing, A general method of constructing global function fields with many rational places. *Algorithmic Number Theory (Portland 1998).* Lecture Notes in Computer Science, volume 1423, pp. 555–566. Springer, Berlin, 1998.

[50] H. Niederreiter and C. P. Xing, Nets, $(t, s)$-sequences, and algebraic geometry. *Random and Quasi-random Point Sets.* Lecture Notes in Statistics, volume 138, pp. 267–302. Springer, New York, 1998.

[51] H. Niederreiter and C. P. Xing, Towers of global function fields with asymptotically many rational places and an improvement on the Gilbert–Varshamov bound. *Math. Nachr.* **195**, 171–186, 1998.

[52] H. Niederreiter and C. P. Xing, Curve sequences with asymptotically many rational points. In: M. D. Fried (ed.), *Applications of Curves over Finite Fields.* Contemporary Mathematics, volume 245, pp. 3–14. American Mathematical Society, Providence, RI, 1999.

[53]  H. Niederreiter and C. P. Xing, Algebraic curves with many rational points over finite fields of characteristic 2. *Proc. Number Theory Conference (Zakopane 1997)*, pp. 359–380. de Gruyter, Berlin, 1999.

[54]  H. Niederreiter and C. P. Xing, *Rational Points on Curves over Finite Fields. Theory and Applications.* London Mathematical Society Lecture Note Series, volume 285. Cambridge University Press, Cambridge, 2001.

[55]  H. Niederreiter and C. P. Xing, Constructions of digital nets. *Acta Arith.* **102**, 189–197, 2002.

[56]  R. Shaheen and A. Winterhof, Permutations of finite fields for check digit systems. *Des. Codes Cryptogr.* **57**, 361–371, 2010.

[57]  A. Winterhof, Recent results on recursive nonlinear pseudorandom number generators. *Sequences and their Applications–SETA 2010.* Lecture Notes in Computer Science, volume 6338, pp. 113–124. Springer, Berlin, 2010.

[58]  A. Winterhof, Generalizations of complete mappings of finite fields and some applications. *J. Symbolic Comput.* **64**, 42–52, 2014.

[59]  C. P. Xing and H. Niederreiter, Drinfeld modules of rank 1 and algebraic curves with many rational points. *Monatsh. Math.* **127**, 219–241, 1999.

# 2

# Partially bent functions and their properties

*Ayça Çeşmelioğlu*
Istanbul Kemerburgaz University, Istanbul

*Wilfried Meidl and Alev Topuzoğlu*
Sabancı University, Istanbul

*Dedicated to Harald Niederreiter on the occasion of his 70th birthday.*

## Abstract

A function $f : \mathbb{F}_p^n \to \mathbb{F}_p$ is called partially bent if for all $a \in \mathbb{F}_p^n$ the derivative $D_a f(x) = f(x + a) - f(x)$ is constant or balanced, i.e., every value in $\mathbb{F}_p$ is taken on $p^{n-1}$ times. Bent functions have balanced derivatives $D_a f$ for all nonzero $a \in \mathbb{F}_p^n$, hence are partially bent. Partially bent functions may be balanced and highly nonlinear, and thus have favorable properties for cryptographic applications in stream and block ciphers. Hence they are of independent interest. Partially bent functions are also used to construct new bent functions.

The aim of this article is to provide a deeper understanding of partially bent functions. We collect their properties and describe partially bent functions with appropriate generalizations of relative difference sets and difference sets. The descriptions of bent functions as relative difference sets and of Hadamard difference sets in characteristic 2, follow from our result as special cases. We describe Hermitian matrices related to partially bent functions and interpret a secondary construction of bent functions from partially bent functions in terms of relative difference sets.

## 2.1 Introduction

Let $p$ be a prime and let $V_n$ denote an $n$-dimensional vector space over $\mathbb{F}_p$. Suppose $f$ is a function from $V_n$ to $\mathbb{F}_p$. The *Walsh transform* $\widehat{f}$ of $f$ is the complex valued function on $V_n$, defined as

$$\widehat{f}(b) = \sum_{x \in V_n} \epsilon_p^{f(x) - \langle b, x \rangle},$$

where $\epsilon_p = e^{2\pi i/p}$ and $\langle b, x \rangle$ denotes a (nondegenerate) inner product on $V_n$. Typically, either $V_n = \mathbb{F}_p^n$, and $\langle b, x \rangle$ is the conventional dot product $\langle b, x \rangle = b \cdot x$, or $V_n = \mathbb{F}_{p^n}$ and $\langle b, x \rangle = \mathrm{Tr}_n(bx)$, where $\mathrm{Tr}_n(z)$ denotes the absolute trace of $z \in \mathbb{F}_{p^n}$. Here we consider functions from $V_n$ to $\mathbb{F}_p$, so we denote the set of all such functions by $\mathcal{F}(V_n, \mathbb{F}_p)$. We recall that elements of $\mathcal{F}(V_n, \mathbb{F}_p)$ are called *Boolean* functions when $p = 2$.

If $f \in \mathcal{F}(V_n, \mathbb{F}_p)$ satisfies $|\widehat{f}(b)| = p^{n/2}$ for all $b \in V_n$, then $f$ is a *bent function*. We call a function $f \in \mathcal{F}(V_n, \mathbb{F}_p)$ *plateaued* (or *s-plateaued*), if there is a fixed integer $s$, $0 \leq s \leq n$, depending on $f$, such that $|\widehat{f}(b)| \in \{0, p^{(n+s)/2}\}$ for all $b \in V_n$. We remark that $\epsilon_p = -1$ for $p = 2$, hence $\widehat{f}(b)$ is an integer for all $b$. Consequently for an $s$-plateaued Boolean function, we always have $n \equiv s \bmod 2$. In particular, bent functions, i.e., 0-plateaued functions, can only exist for even $n$. There is no such restriction when $p$ is odd.

Let $f \in \mathcal{F}(V_n, \mathbb{F}_p)$ and $a \in V_n$. The *derivative $D_a f$ of $f$ in direction $a$* is defined as $D_a f(x) = f(x + a) - f(x)$. Using derivatives we can characterize bent functions alternatively, see [13, Theorem 2.3]. A function $f \in \mathcal{F}(V_n, \mathbb{F}_p)$ is bent if and only if $D_a f$ is balanced for all nonzero $a \in V_n$, i.e., $D_a f$ takes every value of $\mathbb{F}_p$ the same number, $p^{n-1}$ of times.

In this article we study a generalization of bent functions, which can also be defined using derivatives as follows. A function $f \in \mathcal{F}(V_n, \mathbb{F}_p)$ is called *partially bent* if the derivative $D_a f$ is either balanced or constant for any $a \in V_n$.

Partially bent Boolean functions were introduced in [2], where their properties were studied in relation to the good propagation criterion, high correlation immunity, balancedness and high nonlinearity. Partially bent functions have been shown to be favorable with respect to these important features regarding cryptographic applications in stream and block ciphers. Hence [2] confirms that they are interesting in their own right. Partially bent functions can also be used to construct bent functions. Through such constructions in [3, 4, 5, 6, 7], amongst others, the first infinite classes of not weakly regular bent functions were presented. Ternary bent functions with maximal possible algebraic degree as well as self-dual bent functions were obtained.

The aim of this article is to provide a deeper understanding of partially bent functions in several contexts. It is well known that bent functions correspond to particular (relative) difference sets, see for instance [17]. We extend such a correspondence to partially bent functions, by characterizing appropriate generalizations of (relative) difference sets. When $f$ is a bent function in $\mathcal{F}(V_n, \mathbb{F}_p)$, one can correspond a generalized Hadamard matrix $H$ to it, where $HH^* = p^n I$. We show that when $f$ is partially bent, the matrix $H$ satisfies $HH^* = p^n A$, for a Hermitian matrix $A$, and we describe the properties of $A$. We also consider the construction of bent functions by the use of partially bent functions from [4, 5, 7], and interpret it as a realization of a general method, suggested in [8], for constructing relative difference sets from *building blocks*.

This article is structured as follows. We collect basic properties of partially bent functions in Section 2.2. Several constructions of partially bent functions are given in Section 2.3. We describe the correspondence between partially bent functions and generalizations of relative difference sets in Section 2.4. Section 2.5 is on the relation of Hermitian matrices to partially bent functions. In Section 2.6 we recall the construction of bent functions from partially bent functions, presented in [3, 4, 5, 6, 7], and give an interpretation of this construction in terms of difference sets.

## 2.2 Basic properties

Let $f \in \mathcal{F}(V_n, \mathbb{F}_p)$. An element $a \in V_n$ is called a *linear structure* of $f$ if $D_a f(x) = f(x + a) - f(x)$ is constant. Obviously this implies that $f(x + a) - f(x) = f(a) - f(0)$ for all $x \in V_n$. We summarize some well-known properties of linear structures below. In order to keep the paper self-contained we also provide short proofs.

**Lemma 2.1** *Let $\Lambda$ be the set of linear structures of $f$.*

(i) *The set $\Lambda$ is a subspace of $V_n$, which is called the linear space of $f$.*
(ii) *The function $f(x) - f(0)$ is a linear transformation on $\Lambda$.*
(iii) *For $a \in \Lambda$ and $x \in V_n$ we have $f(x + a) = f(x) + f(a) - f(0) = f(x) + \langle t, a \rangle$ for some $t \in V_n$.*

*Proof.* (i) Let $a_1, a_2 \in \Lambda$, then

$$f(x + a_1 + a_2) - f(x) = f(x + a_1 + a_2) - f(x + a_1) + f(x + a_1) - f(x)$$
$$= f(a_2) - f(0) + f(a_1) - f(0) \tag{2.1}$$

is constant, hence $a_1 + a_2 \in \Lambda$.

(ii) Follows from $f(a_1 + a_2) - f(0) = f(x + a_1 + a_2) - f(x)$ and (2.1).
(iii) Follows immediately from $f(x + a) - f(x) = f(a) - f(0)$ and (ii). $\quad\square$

In what follows, we denote the set of partially bent functions from $V_n$ to $\mathbb{F}_p$ by $\mathcal{P}(V_n, \mathbb{F}_p)$. We first wish to characterize the *Walsh spectrum* of $f$ in $\mathcal{P}(V_n, \mathbb{F}_p)$, i.e., the set $\{\widehat{f}(b) \mid b \in V_n\}$. We start by collecting some well-known properties of the Walsh transform which we will use frequently in the sequel.

P1 Parseval's identity: $\sum_{b \in V_n} |\widehat{f}(b)|^2 = p^{2n}$, see [3].
P2 A function $f \in \mathcal{F}(V_n, \mathbb{F}_p)$ is balanced if and only if $\widehat{f}(0) = 0$. In particular, the derivative $D_a f$ of $f$ is balanced if and only if $\widehat{D_a f}(0) = 0$. Obviously a function $f$ is bent if $\widehat{D_a f}(0) = 0$ for all nonzero $a \in V_n$.
P3 The Walsh spectrum is invariant under *extended affine (EA) equivalence*. Recall that two functions $f, g : V_n \to \mathbb{F}_p$ are *EA-equivalent* if there exists an affine permutation $A_1$ of $V_n$, an affine map $A_2 : V_n \to \mathbb{F}_p$ and an element $a \in \mathbb{F}_p^*$ such that $g(x) = af(A_1(x)) + A_2(x)$.

Let $f \in \mathcal{F}(V_n, \mathbb{F}_p)$, and let $s$ be the dimension of $\Lambda$. In view of P3 we may always assume that $f(0) = 0$. Applying the standard Welch squaring technique, we obtain

$$|\widehat{f}(b)|^2 = \sum_{x,y \in V_n} \epsilon_p^{f(x)-f(y)-\langle b, x-y \rangle} = \sum_{y,z \in V_n} \epsilon_p^{f(y+z)-f(y)-\langle b,z \rangle} \qquad (2.2)$$
$$= \sum_{z \in V_n} \epsilon_p^{f(z)-\langle b,z \rangle} \sum_{y \in V_n} \epsilon_p^{f(y+z)-f(y)-f(z)},$$

for $b \in V_n$. The function $f$ is partially bent if and only if $g(y) = f(y+z) - f(y) - f(z)$ is balanced for all $z \notin \Lambda$. Hence we get

$$|\widehat{f}(b)|^2 = p^n \sum_{z \in \Lambda} \epsilon_p^{f(z)-\langle b,z \rangle} = \begin{cases} p^{n+s} & \text{if } f(z) - \langle b,z \rangle \equiv 0 \text{ on } \Lambda \\ 0 & \text{otherwise,} \end{cases} \qquad (2.3)$$

where in the last step we used that $f(z) - \langle b, z \rangle$ is linear on $\Lambda$.

Clearly a bent function is partially bent with $\Lambda = \{0\}$. Consider the support of $\widehat{f}$, defined as

$$\mathrm{supp}(\widehat{f}) = \{b \in V_n \mid \widehat{f}(b) \neq 0\}.$$

By (2.3) it is a certain coset of the orthogonal complement of $\Lambda$. The cardinality $|\mathrm{supp}(\widehat{f})|$ of the support of $\widehat{f}$ is $p^{n-s}$.

**Remark 2.2** A partially bent function $f \in \mathcal{F}(V_n, \mathbb{F}_p)$ with linear space $\Lambda$ is $s$-plateaued, if the dimension of $\Lambda$ is $s$. In this case we call $f$ *s-partially bent* also. We denote the set of such functions by $\mathcal{P}(V_n, \mathbb{F}_p, s)$. The set of partially bent functions is a proper subset of the set of plateaued functions.

For a construction of $s$-plateaued functions with $\dim(\Lambda) < s$ we refer to [18, Lemma 6].

The original definition of partially bent functions [2] is by a slightly different approach. For a function $f \in \mathcal{F}(V_n, \mathbb{F}_p)$ we put

$$R(f) = \{a \in V_n \mid \widehat{D_a f}(0) \neq 0\}.$$

By P2 we can describe $R(f)$ as the set of elements $a \in V_n$ for which the derivative $D_a f$ is not balanced. Note that for a partially bent function $f$ we then have $R(f) = \Lambda$. In [2] it is shown that $|\operatorname{supp}(\widehat{f})||R(f)| \geq 2^n$ for every Boolean function $f$ and partially bent functions are defined to be those Boolean functions for which the equality holds. Proposition 2.3 below shows that these properties hold in the case of odd $p$ also. Since the argument of the proof is similar to that of the Boolean case, we only give a sketch.

**Proposition 2.3** *Every function $f \in \mathcal{F}(V_n, \mathbb{F}_p)$ satisfies*

$$|\operatorname{supp}(\widehat{f})||R(f)| \geq p^n. \tag{2.4}$$

*Sketch of Proof.* By the definition of $R(f)$ and (2.2) we get

$$p^n |R(f)| \geq \sum_{a \in V_n} \widehat{D_a f}(0) = |\widehat{f}(0)|^2.$$

Since $|R(f)|$ does not change if we replace $f(x)$ by $f(x) + \langle x, t \rangle$ we have

$$p^n |R(f)| \geq |\widehat{f}(t)|^2 \tag{2.5}$$

for all $t \in V_n$. By Parseval's identity P1 we obtain

$$|\operatorname{supp}(\widehat{f})| \geq \left( \sum_{t \in V_n} |\widehat{f}(t)|^2 \right) \Big/ \sup |\widehat{f}(t)|^2 = p^{2n} \Big/ \sup |\widehat{f}(t)|^2. \tag{2.6}$$

Combining inequalities (2.5) and (2.6) completes the proof. $\qquad\square$

The following theorem extends the properties given in [2, Theorem] to arbitrary primes and shows that the two definitions of partially bent functions, stated above are equivalent.

**Theorem 2.4** *Let $f$ be a function from $V_n$ to $\mathbb{F}_p$ with linear space $\Lambda$. Then the following are equivalent.*

(i) *$f$ is partially bent.*

(ii) *There is an integer $s \geq 0$ such that $f$ is $s$-plateaued and the linear space of $f$ has dimension $s$.*

(iii) $|\text{supp}(\widehat{f})||R(f)| = p^n$.

(iv) *There exists an element* $t \in V_n$ *such that for any* $a \in V_n$, *either* $\widehat{D_a f}(0) = 0$ *or* $\widehat{D_a f}(0) = p^n \epsilon_p^{\langle t, a \rangle}$.

(v) *The function* $f_{|\Lambda^c}$ *is bent, where* $\Lambda^c$ *is an arbitrary complement of* $\Lambda$ *in* $V_n$, *and* $f_{|\Lambda^c}$ *is the restriction* $f$ *to* $\Lambda^c$.

*Proof.* (i) $\Leftrightarrow$ (ii) That (i) implies (ii) follows immediately from Equation (2.3). For the converse we use the $p$-ary version of Proposition II.1 in Canteaut *et al.* [1]. Putting $e = 1$ in [1, Proposition II.1],

$$\sum_{\alpha \in V_n} (\widehat{f}(\alpha))^4 = p^n \sum_{b \in V_n} \widehat{D_{-b} f}(0) \widehat{D_b f}(0) = p^n \sum_{b \in V_n} |\widehat{D_b f}(0)|^2$$

$$= p^n \left( \sum_{b \in \Lambda} |\widehat{D_b f}(0)|^2 + \sum_{b \in V_n \setminus \Lambda} |\widehat{D_b f}(0)|^2 \right)$$

$$= p^n \left( p^{2n} p^s + \sum_{b \in V_n \setminus \Lambda} |\widehat{D_b f}(0)|^2 \right).$$

The left hand side is

$$\sum_{\alpha \in V_n} (\widehat{f}(\alpha))^4 = |\text{supp}(\widehat{f})| p^{2(n+s)} = p^{3n+s}.$$

Therefore we have

$$\sum_{b \in V_n \setminus \Lambda} |\widehat{D_b f}(0)|^2 = 0,$$

which means $\widehat{D_b f}(0) = 0$ for all $b \in V_n \setminus \Lambda$. Hence the derivative of $f$ is either balanced or constant.

(i) $\Rightarrow$ (iii) Since we have already shown that (i) $\Leftrightarrow$ (ii), we can assume that $f$ is $s$-plateaued for some integer $s \geq 0$. Therefore we have $|\text{supp}(\widehat{f})||R(f)| = |\text{supp}(\widehat{f})||\Lambda| = p^{n-s} p^s$.

(iii) $\Rightarrow$ (i) If the equality in (2.4) holds, then $|R(f)| = p^{-n} \sup |\widehat{f}(t)|^2$ and $|\text{supp}(\widehat{f})| \sup |\widehat{f}(t)|^2 = p^{2n}$. The latter equation, together with Parseval's identity, imply that $|\widehat{f}(t)|^2 = \sup |\widehat{f}(t)|^2$ for all $t \in \text{supp}(\widehat{f})$. Hence $f$ is plateaued, i.e., $|\widehat{f}(t)|^2 = 0$ or $p^{n+s}$ for some $0 \leq s \leq n$. But then $|\text{supp}(\widehat{f})| = p^{n-s}$, implying $|R(f)| = p^s$. By Equation (2.2), we obtain $|\widehat{f}(t)|^2 = \sum_{a \in R(f)} \widehat{D_a f}(0) \epsilon_p^{\langle t, a \rangle} = p^{n+s}$ for some $t \in \text{supp}(\widehat{f})$. Therefore for all $a \in R(f)$

$$\widehat{D_a f}(0) = p^n \epsilon_p^{-\langle t, a \rangle}, \tag{2.7}$$

i.e., $D_a f(x) = \langle t, a \rangle$ is constant.

(i) and (iv) are equivalent by Equation (2.7) and the fact that $\widehat{D_a f}(0) = 0$ if and only if $D_a f$ is balanced.

(i) $\Leftrightarrow$ (v) Let $\Lambda^c$ be any (vector space) complement in $V_n$ of the linear space $\Lambda$ of $f$. Then

$$\widehat{D_a f}(0) = \sum_{x \in V_n} \epsilon_p^{f(x+a)-f(x)} = \sum_{y \in \Lambda} \sum_{z \in \Lambda^c} \epsilon_p^{f(y+z+a)-f(y+z)}$$

$$= \sum_{y \in \Lambda} \sum_{z \in \Lambda^c} \epsilon_p^{f(y)+f(z+a)-f(y)-f(z)} = |\Lambda| \sum_{z \in \Lambda^c} \epsilon_p^{f(z+a)-f(z)}.$$

Suppose that $f$ is partially bent and $a \notin \Lambda$. Then $\widehat{D_a f}(0) = 0$, hence $\sum_{z \in \Lambda^c} \epsilon_p^{f(z+a)-f(z)} = 0$. In particular, for all nonzero $a \in \Lambda^c$ the derivative $D_a f$, restricted to $\Lambda^c$ is balanced. Consequently, $f$ restricted to a complement $\Lambda^c$ of $\Lambda$ is bent.

Conversely suppose that $f$ is bent on every complement of $\Lambda$. Let $a \notin \Lambda$ and let $\Lambda^c$ be a complement of $\Lambda$ containing $a$. Since $f_{|\Lambda^c}$ is bent, we have $\sum_{z \in \Lambda^c} \epsilon_p^{f(z+a)-f(z)} = 0$, hence $\widehat{D_a f}(0) = 0$ and $D_a f$ is balanced. $\square$

## 2.3 Examples and constructions

The classical examples of partially bent functions are quadratic functions, i.e., functions of algebraic degree 2. A quadratic function $f$ from $V_n$ to $\mathbb{F}_p$ can be characterized as a function for which $D_a(f)$ is linear or constant for all $a \in V_n$. Note that then $D_a(f)$ is balanced if $a$ is not a linear structure of $f$. We remark also that $f$ is affine or constant if $D_a(f)$ is constant for all $a \in V_n$.

A quadratic function in $\mathcal{F}(\mathbb{F}_{p^n}, \mathbb{F}_p)$ can be expressed in trace form as

$$f(x) = \text{Tr}_n \left( \sum_{i=0}^{\lfloor n/2 \rfloor} a_i x^{p^i+1} \right), \; a_i \in \mathbb{F}_{p^n},$$

except for an affine term. The linear space $\Lambda$ of $f$ is then the kernel in $\mathbb{F}_{p^n}$ of the linear transformation induced by the linearized polynomial $L(x) = \sum_{i=0}^{\lfloor n/2 \rfloor} (a_i x^{p^i} + a_i^{p^{n-i}} x^{p^{n-i}})$, see [10, 12].

When $p$ is odd, a quadratic function $f \in \mathcal{F}(\mathbb{F}_p^n, \mathbb{F}_p)$ can be expressed as $f(x) = x^T A x$ for a unique symmetric matrix $A$ of rank $n-s$, for some $s \geq 0$. Such a quadratic function $f$ is partially bent with a linear space of dimension $s$, see [4, 5, 6].

In [18] the well-known Maiorana–McFarland construction of bent functions was generalized to construct $s$-plateaued Boolean functions. This construction allows the choice of the dimension of the linear space $\Lambda$ to be between 0 and $s$.

In particular, the choice $\dim(\Lambda) = s$ gives a partially bent function. We present here the analog construction for arbitrary primes $p$. For two integers $t$ and $m$, $t < m$, let $P$ be an injective linear transformation from $\mathbb{F}_p^t$ to $\mathbb{F}_p^m$. Then the function $f : \mathbb{F}_p^t \times \mathbb{F}_p^m \to \mathbb{F}_p$, given by

$$f(x, y) = P(x) \cdot y \qquad (2.8)$$

is partially bent with the $(m - t)$-dimensional linear space $\Lambda = \{(0, b) \mid b \in P^\perp\}$, where $P^\perp$ denotes the orthogonal complement of the image of $P(x)$ in $\mathbb{F}_p^m$. To confirm this statement, we have to show that $D_{(a,b)}f$ is in fact constant for $a = 0 \in \mathbb{F}_p^t$ and $b \in P^\perp$, and that $D_{(a,b)}f$ is balanced if $(a, b) \notin \Lambda$. We determine $\widehat{D_{(a,b)}f}(0)$ and employ property P2 above:

$$\widehat{D_{(a,b)}f}(0) = \sum_{x \in \mathbb{F}_p^t} \sum_{y \in \mathbb{F}_p^m} \epsilon_p^{P(x+a)\cdot(y+b) - P(x)\cdot y} = \sum_{x \in \mathbb{F}_p^t} \epsilon_p^{P(x)\cdot b} \sum_{y \in \mathbb{F}_p^m} \epsilon_p^{P(a)\cdot(y+b)}.$$

If $P(a) \neq 0$, i.e., $a \neq 0$, then the second sum vanishes, and $\widehat{D_{(a,b)}f}(0) = 0$. For $a = 0$ we have

$$\widehat{D_{(a,b)}f}(0) = p^m \sum_{x \in \mathbb{F}_p^t} \epsilon_p^{P(x)\cdot b},$$

which again vanishes if $b \notin P^\perp$. For $b \in P^\perp$ we get $\widehat{D_{(a,b)}f}(0) = p^{t+m}$. In fact, for $a = 0 \in \mathbb{F}_p^t$ and $b \in P^\perp$

$$D_{(0,b)}f(x) = P(x) \cdot (y + b) - P(x) \cdot y = P(x) \cdot b = 0.$$

For details of the construction of plateaued Boolean functions without a linear structure we refer to [18, Section VIII].

As immediately seen, the partially bent functions (2.8), obtained by the Maiorana–McFarland construction are again quadratic. Partially bent functions which are not quadratic can easily be obtained from nonquadratic bent functions. Let $f$ be a bent function from $\mathbb{F}_p^{n-s}$ to $\mathbb{F}_p$, and let $f_1 : \mathbb{F}_p^n \to \mathbb{F}_p$ be given as $f_1(x_1, \ldots, x_{n-s}, \ldots, x_n) = f(x_1, \ldots, x_{n-s})$. Denoting $(x_1, \ldots, x_{n-s})$ by $\underline{x}$ and $(x_{n-s+1}, \ldots, x_n)$ by $\bar{x}$, for $a = (a_1, \ldots, a_{n-s}, a_{n-s+1}, \ldots, a_n) = (\underline{a}, \bar{a}) \in \mathbb{F}_p^n$ we obtain $D_a f(x) = f_1(\underline{x} + \underline{a}, \bar{x} + \bar{a}) - f_1(\underline{x}, \bar{x}) = f(\underline{x} + \underline{a}) - f(\underline{x})$. Since $f$ is bent, $D_a f(x)$ is balanced if $\underline{a} \neq 0$, and constant 0 if $\underline{a} = 0$. Hence $f_1$ is $s$-partially bent with linear space $\Lambda = \{(0, \bar{a}) \mid \bar{a} \in \mathbb{F}_p^s\}$.

## 2.4 Partially bent functions and difference sets

Let $G$ be a finite (Abelian) group of order $v$. A $(k$-$)$subset $D$ of $G$ of cardinality $k$ is called a $(v, k, \lambda)$-*difference set* in $G$ if every element $g \in G$, different from the identity, can be written as $d_1 d_2^{-1}$, $d_1, d_2 \in D$, in exactly $\lambda$ different ways. It is well known that there is a correspondence between Boolean bent functions and the so-called Hadamard difference sets: a function $f \in \mathcal{F}(V_n, \mathbb{F}_2)$ is bent if and only if its support $D = \mathrm{supp}(f)$ is a

$$(2^n, 2^{n-1} \pm 2^{n/2-1}, 2^{n-1} \pm 2^{n/2-1})$$

Hadamard difference set in $V_n$, see [9, Theorem 6.2.2].

Let $G$ be a finite (Abelian) group of order $mn$, and let $N$ be a subgroup of $G$ of order $n$. A $k$-subset $R$ of $G$ is called an $(m, n, k, \lambda)$-*relative difference set in* $G$ *relative to* $N$ if the following conditions are satisfied see [16, 17]:

- every element $g \in G \setminus N$ can be represented as $r_1 r_2^{-1}$, $r_1, r_2 \in R$ in exactly $\lambda$ ways, and
- no element $n \in N$, different from identity, has such a representation.

One can also associate relative difference sets to bent functions as follows. When $p$ is an arbitrary prime, and $f \in \mathcal{F}(V_n, \mathbb{F}_p)$ is bent, then the set $R = \{(x, f(x)) \mid x \in V_n\}$ is a $(p^n, p, p^n, p^{n-1})$-relative difference set in $V_n$ relative to $\mathbb{F}_p$, see [17].

**Remark 2.5** Group rings play an important role in the study of difference sets. For background on group rings we refer the reader to [14]. One can identify a subset $R$ of a group $G$ with the group ring element $\sum_{r \in R} r \in \mathbb{C}[G]$. Then a $k$-subset $R$ of $G$ is an $(m, n, k, \lambda)$-relative difference set in $G$ relative to a subgroup $N$ if and only if $R R^{(-1)} = k + \lambda(G - N)$ in $\mathbb{C}[G]$. For details, see Sections 2.1 and 2.4 in [17].

So far we have seen how bent functions and (relative) difference sets are related. Our aim in the next sections is to extend these results to partially bent functions and appropriate generalizations of (relative) difference sets.

### 2.4.1 Partially bent relative difference sets

Let $f \in \mathcal{P}(V_n, \mathbb{F}_p, s)$, and let $\Lambda$ be its linear space. We consider the sets

$$A = \{(a, f(a)) \mid a \in \Lambda\} \quad \text{and} \quad B = \{(a, y) \mid a \in \Lambda, y \in \mathbb{F}_p\}.$$

We put $G := V_n \times \mathbb{F}_p$, and $f(0) = 0$. Obviously $|B| = p^{s+1}$, $|A| = p^s$ and $B$ is a subgroup of $G$ containing $A$ as a subgroup. We again define $R$ by

$$R = \{(x, f(x)) \mid x \in V_n\},$$

and observe that the number of solutions of the equation $f(x+a) - f(x) = b$ corresponds to the number of ways $(a, b)$ can be written as a difference of two elements in $R$, i.e., as $(x + a, f(x + a)) - (x, f(x))$. Then the following properties can be obtained easily [15].

R1  $(x, y) \in G \setminus B$ can be represented in the form $r_1 - r_2, r_1, r_2 \in R$ in exactly $\lambda$ ways, where we put $\lambda = p^{n-1}$.

R2  $(x, y) \in B \setminus A$ has no representation in the form $r_1 - r_2, r_1, r_2 \in R$.

R3  $(x, y) \in A$ can be represented in the form $r_1 - r_2, r_1, r_2 \in R$, in exactly $|R| = k$ ways.

**Remark 2.6** Using group ring notation the set $R$ satisfies

$$RR^{(-1)} = kA + \lambda(G - B) \text{ in } \mathbb{C}[G], \tag{2.9}$$

where $R$, $A$, $B$ and $G$ are as above. If $f$ is bent, i.e., $s = 0$, then $A = \{(0, 0)\}$, $B = N = \{0\} \times \mathbb{F}_p \simeq \mathbb{F}_p$, and $R$ coincides with the conventional relative difference set (relative to $N$) for bent functions.

Now we identify $\mathbb{F}_p$ with $\{0\} \times \mathbb{F}_p \subset G$, and observe that the subgroup $B$ of $G = V_n \times \mathbb{F}_p$ is the direct sum $B = A \oplus \mathbb{F}_p$. In particular, $A \cap \mathbb{F}_p = (0, 0)$. This motivates the definition of a new class of "relative difference sets" which leads to a one-to-one correspondence between partially bent functions and these new relative difference sets in $V_n \times \mathbb{F}_p$.

Let $G = H \times N$, $A$ be a subgroup of $G$ such that $A \cap N = \{(0, 0)\}$, and $B = A \oplus N$. Suppose that $|H| = m$, $|N| = n$, $|A| = l$, and therefore $l|m$ and $|B| = ln$. We call a $k$-subset $R$ of $G$ an $(m, n, l, k, \lambda)$-*partially bent relative difference set* if R1, R2, R3 hold (or equivalently if $R$ satisfies the group ring equation (2.9)).

**Lemma 2.7** *If $R$ is an $(m, n, l, m, \lambda)$-partially bent relative difference set in $G = H \times N$, then $R$ defines a function $f : H \to N$ by $f(h) = n$ if $(h, n) \in R$.*

*Proof.* We have to show that for every $h \in H$, there exists exactly one $n \in N$ such that $(h, n) \in R$. Since $N \subset B$ and $A \cap N = \{(0, 0)\}$, there is no representation of $(0, z) \in N$ as a difference of two elements of $R$ for any nonzero $z$. Hence for every $h \in H$ there is at most one $n \in N$ such that $(h, n) \in R$. On the other hand, the element $(0, 0) \in A$ can be written as a difference of two elements of $R$ in $m = |R| = |H|$ ways. Hence for every $h \in H$ there exists an element $n \in N$ such that $(h, n) \in R$. $\qquad\square$

In view of Lemma 2.7 we may denote the elements of $R$ by $(h, f(h))$.

**Proposition 2.8** *Let R be an $(m, n, l, m, \lambda)$-partially bent relative difference set in $G = H \times N$, where $B = A \oplus N$ for a subgroup A of G with $A \cap N = \{(0, 0)\}$. Suppose $f : H \to N$ is defined as in Lemma 2.7. Then*

(i) *for every $a \in H$ there exists at most one $\bar{b} \in N$ such that $(a, \bar{b}) \in A$,*
(ii) *for every $(a, \bar{b}) \in A$ the equation $D_a f(x) = \bar{b}$ is satisfied for all $x \in H$, hence $D_a f$ is constant,*
(iii) *for all $a \in H$ for which there is no $b \in H$ such that $(a, b) \in A$, the equation $D_a f(x) = b$ has $\lambda$ solutions $x \in H$, for all $b \in N$, hence $D_a f$ is balanced.*

*Proof.* We recall that the number of solutions of the equation $f(x + a) - f(x) = b$ is the number of ways $(a, b)$ can be written as a difference of two elements in $R$.

Suppose that $(a, b_1), (a, b_2) \in A$. The property R3 and $k = m$ imply that all $m$ elements $x$ of $H$ are solutions of $f(x + a) - f(x) = b_i$, $i = 1, 2$. Consequently $b_1 = b_2$, and $f(x + a) - f(x) = b$ has no solution if $b \neq b_1$. This finishes the proof of parts (i) and (ii).

Now let $a \in H$ such that $(a, b) \notin A$ for any $b \in N$. Since $B = A \oplus N$, we then also have $(a, b) \notin B$ for all $b \in N$. Hence, there are exactly $\lambda$ solutions for $f(x + a) - f(x) = b$ by R1, which shows part (iii). $\square$

Proposition 2.8(ii) and (iii) imply a one-to-one correspondence between $s$-partially bent functions and partially bent difference sets in $V_n \times \mathbb{F}_p$. Putting $s = 0$, we obtain the known relation between bent functions and relative difference sets, see [17, Proposition 2].

**Corollary 2.9** *There is a one-to-one correspondence between $s$-partially bent functions from $V_n$ to $\mathbb{F}_p$ and $(p^n, p, p^s, p^n, p^{n-1})$-partially bent relative difference sets of $V_n \times \mathbb{F}_p$.*

**Remark 2.10** For the definition of the function $f$ we used that $N$ and $A$ are subsets of $B$ and $A \cap N = \{(0, 0)\}$. That $B = A \oplus N$ is only used at the last step. Actually $B = A \oplus N$ follows from Equation (2.9) also, and hence it is sufficient to require $N, A \subset B$ and $A \cap N = \{(0, 0)\}$ in the definition of partially bent relative difference sets . One has to show that $(a, b) \in B$ implies that $(a, \bar{b}) \in A$ for some $\bar{b} \in N$. Suppose that $(a, b) \in B$, then $(a, \bar{b}) \in B$ for all $\bar{b} \in N$ (since $N \subset B$). Therefore by (2.9), either the equation $f(x + a) - f(x) = \bar{b}$ has no solution, or it holds for all $x \in H$, where the latter applies if $(a, \bar{b}) \in A$. This shows that there must exist a $\bar{b} \in N$ such that $(a, \bar{b}) \in A$.

### 2.4.2 Difference properties of partially bent Boolean functions

In this subsection we study the difference properties of the set supp($f$) = $\{x \in V_n \mid f(x) = 1\}$ for $f \in \mathcal{P}(V_n, \mathbb{F}_2, s)$. The result generalizes the correspondence between bent functions in $\mathcal{F}(V_n, \mathbb{F}_2)$, where $s = 0$, and Hadamard difference sets.

**Lemma 2.11** *Let $\Lambda$ be an $s$-dimensional vector space over $\mathbb{F}_2$ and $f : \Lambda \to \mathbb{F}_2$ be linear, and different from the zero map.*

(1) *If $f(z) = 0$, then the equation $z = z' + z''$ has*
    (i) *$2^{s-1}$ solutions $z', z''$ satisfying $f(z') = f(z'') = 0$,*
    (ii) *$2^{s-1}$ solutions $z', z''$ satisfying $f(z') = f(z'') = 1$,*
    (iii) *no solution satisfying $f(z') \neq f(z'')$.*
(2) *If $f(z) = 1$, then the equation $z = z' + z''$ has*
    (i) *$2^{s-1}$ solutions $z', z''$ satisfying $f(z') = 1$, $f(z'') = 0$,*
    (ii) *$2^{s-1}$ solutions $z', z''$ satisfying $f(z') = 0$, $f(z'') = 1$,*
    (iii) *no solution satisfying $f(z') = f(z'')$.*

*Proof.* If $f(z) = 0$ and $z = z' + z''$, then from the linearity of $f$ it follows that $f(z') = f(z'')$. For every $z'$ with $f(z') = 0$ ($f(z') = 1$) there exists exactly one $z''$ such that $z = z' + z''$. By the same argument one obtains the assertion for $z \in \Lambda$ with $f(z) = 1$. $\qquad\square$

**Theorem 2.12** *Let $f \in \mathcal{P}(V_n, \mathbb{F}_2, s)$ be partially bent with linear space $\Lambda$. Put $T = \text{supp}(f) = \{x \in V_n \mid f(x) = 1\}$.*

(1) *Suppose $f$ is the zero-function on $\Lambda$, then the following hold.*
    (i) *$|T| = 2^{n-1} \pm 2^{(n+s)/2-1}$.*
    (ii) *Any $x \in \Lambda$ can be written as $x = d - d'$, $d, d' \in T$, in $2^{n-1} \pm 2^{(n+s)/2-1}$ ways.*
    (iii) *Any $x \in V_n \setminus \Lambda$ can be written as $x = d - d'$, $d, d' \in T$, in $2^{n-2} \pm 2^{(n+s)/2-1}$ ways.*
(2) *Suppose $f$ is not the zero-function on $\Lambda$, then the following hold.*
    (i) *$|T| = 2^{n-1}$.*
    (ii) *Any $x \in V_n \setminus \Lambda$ can be written as $x = d - d'$, $d, d' \in T$, in $2^{n-2}$ ways.*
    (iii) *Any $x \in \Lambda$ with $f(x) = 0$, i.e., any $x \in \ker(f|_\Lambda)$ can be written as $x = d - d'$, $d, d' \in T$, in $2^{n-1}$ ways.*
    (iv) *If $x \in \Lambda$ with $f(x) = 1$, i.e., $x \notin \ker(f|_\Lambda)$, then $x$ cannot be written as $x = d - d'$, $d, d' \in T$.*

*Proof.* Let $n = m + s$, and $\Lambda^c$ be a complement of $\Lambda$ in $V_n$. Then by Theorem 2.4(v), the function $f$, restricted to the $m$-dimensional vector space $\Lambda^c$, is a bent function. Hence $D = \{y \in \Lambda^c \mid f(y) = 1\}$ is a $(v, k, \lambda)$-Hadamard difference set in $\Lambda^c$ with $v = 2^m$, $k = 2^{m-1} \pm 2^{m/2-1}$, $\lambda = 2^{m-2} \pm 2^{m/2-1}$. Recall that for $x \in V_n$ and $z \in \Lambda$ we have $f(x + z) = f(x) + f(z) + f(0)$. Supposing without loss of generality that $f(0) = 0$, we have $T = \{x = y + z \mid y \in \Lambda^c, z \in \Lambda, f(y) + f(z) = 1\}$. Consequently, $x = y + z$ with $y \in \Lambda^c, z \in \Lambda$, is an element of $T$ if and only if $f(y) = 1$ and $f(z) = 0$ or $f(y) = 0$ and $f(z) = 1$. For the remainder of the proof suppose that $y, y', y''$ are in $\Lambda^c$ and $z, z', z''$ are in $\Lambda$.

(1) If $f$ is the zero-function on $\Lambda$, then $x = y + z, y \in \Lambda^c, z \in \Lambda$, is in $T$ if and only if $f(y) = 1$, i.e., $y \in D$. Hence $|T| = |D||\Lambda| = 2^{m+s-1} \pm 2^{m/2+s-1} = 2^{n-1} \pm 2^{(n+s)/2-1}$. Suppose that $x \notin \Lambda$, then $x = y + z, y \in \Lambda^c, z \in \Lambda$, with $y \neq 0$. Using that $D$ is a $(v, k, \lambda)$-difference set in $\Lambda^c$ and that $\Lambda$ is an $s$-dimensional subspace of $V_n$, we obtain $\lambda$ ways of writing $y = y' + y''$ with $y', y'' \in \Lambda^c$ and $f(y') = f(y'') = 1$, and $2^s$ ways of writing $z = z' + z''$, $z', z'' \in \Lambda$. Hence we have $2^s \lambda$ ways for writing $x$ as $x = y' + z' + y'' + z''$, $y' + z', y'' + z'' \in T$, which shows (1)(iii).

Suppose that $x \in \Lambda$. Then there are $k2^s$ possibilities to write $x = y + z' + y + z''$, with $y + z', y + z'' \in T$, i.e, $y \in D$ and $z', z'' \in \Lambda$. This finishes the proof of (1).

(2) If $f$ is not the zero function on $\Lambda$, then by the linearity of $f_{|\Lambda}$, half of the elements of $\Lambda$ are mapped to 0. Let $\bar{D}$ be the complementary difference set of $D$. Since $x = y + z \in T$ if and only if $f(y) = 1$ and $f(z) = 0$ or $f(y) = 0$ and $f(z) = 1$ we have $|T| = |D|\frac{|\Lambda|}{2} + |\bar{D}|\frac{|\Lambda|}{2} = 2^m \frac{|\Lambda|}{2} = 2^{m+s-1}$, which shows part (i).

(ii) Let $x = y + z \in V_n \setminus \Lambda$, i.e., $y \neq 0$. Then we can write $y$ as

(a) $y = y' + y''$ with $y', y'' \in D$ in $\lambda = 2^{m-2} \pm 2^{m/2-1}$ ways,
(b) $y = y' + y''$ with $y', y'' \in \bar{D}$ in $\bar{\lambda} = 2^{m-2} \mp 2^{m/2-1}$ ways,
(c) $y = y' + y''$ with $y', y''$ not both in $D$ or $\bar{D}$ in $2^m - \lambda - \bar{\lambda} = 2^{m-1}$ ways.

To get a representation $x = y + z = d' - d''$ with $d' = y' + z', d'' = y'' + z'' \in T$,

- for case (a) we have to choose $z', z''$ such that $f(z') = f(z'') = 0$,
- for case (b) we have to choose $z', z''$ such that $f(z') = f(z'') = 1$,
- for case (c) the choice of $z', z''$ will satisfy $f(z') \neq f(z'')$.

First we suppose that $f(z) = 0$. Then by Lemma 2.11 we have $2^{s-1}$ choices for cases (a) and (b) and no choice for case (c). Hence we have $\lambda 2^{s-1} + \bar{\lambda} 2^{s-1} = 2^{m+s-2}$ ways of writing $x$ as $x = d' - d'', d', d'' \in T$.

Now we consider the case that $f(z) = 1$. By Lemma 2.11 we have no solution for cases (a) and (b). In case (c), we have $2^{s-1}$ possible choices for $z', z''$ for every combination of $y', y''$. Consequently we have $2^{m-1}2^{s-1}$, i.e., again $2^{m+s-2}$ ways of writing $x$ as $x = d' - d''$, $d', d'' \in T$.

To prove (iii) and (iv), let $x = y + z \in \Lambda$, i.e., $y = 0$. Consider the equation $x = d' - d'' = y' + z' + y'' + z''$. Clearly we have $2^m$ possibilities for $y' = y'' \in \Lambda^c$. If $f(y') = 0$ ($f(y') = 1$) we require $f(z') = f(z'') = 1$ ($f(z') = f(z'') = 0$) so that $d', d'' \in T$. By Lemma 2.11 this implies $f(z) = f(x) = 0$, and then we have $2^{s-1}$ choices $z', z''$ for every choice of $y'$. Consequently, we have $2^m 2^{s-1} = 2^{m+s-1}$ solutions if $f(x) = 0$ and no solution if $f(x) = 1$. $\square$

**Remark 2.13** In both cases in Theorem 2.12, all elements of $V_n$, except for those in $\Lambda$, can be written as a difference of elements of $T$ in precisely $\lambda$ ways for an integer $\lambda$. The elements in the set $\{x \in \Lambda : f(x) = 0\}$ can be written as a difference of elements of $T$ in precisely $|T|$ ways, while for the elements in the set $\{x \in \Lambda : f(x) = 1\}$ there is no such representation.

For bent functions, which trivially fall into the first class of functions, Theorem 2.12 retrieves the fact that $T$ is a Hadamard difference set.

Finally we see that a partially bent function $f$ is balanced, i.e., $|T| = 2^{n-1}$, if and only if $f$ is not the zero-function on $\Lambda$, which agrees with [2, Proposition 2].

## 2.5 Partially bent functions and Hermitian matrices

Let $f \in \mathcal{F}(V_n, \mathbb{F}_p)$. We associate to it a $p^n \times p^n$ matrix $H$, defined by $H := (\epsilon_p^{f(x-y)})_{(x,y)}$. The product of the matrix $H$ with its adjoint $H^* = (\epsilon_p^{-f(y-x)})_{(x,y)}$ is given by

$$HH^* = \left(\sum_{y \in V_n} \epsilon_p^{f(x-y)-f(z-y)}\right)_{(x,z)} = \left(\sum_{y \in V_n} \epsilon_p^{f(x-z+z-y)-f(z-y)}\right)_{(x,z)}$$

$$= \left(\sum_{y \in V_n} \epsilon_p^{D_{x-z}f(z-y)}\right)_{(x,z)} = \left(\widehat{D_{x-z}f}(0)\right)_{(x,z)}. \qquad (2.10)$$

Moreover, the sum of the entries on each row or column of the matrix $HH^*$ is $|\widehat{f}(0)|^2$ and $|\widehat{f}(b)|^2 = (HH^*)_{(1,y)} \cdot (\epsilon_p^{-b \cdot y})_{(y,b)}$. Recall that when $f$ is bent, the derivative $D_a f$ is balanced for all $a \neq 0$, hence by (2.10), we have $HH^* = p^n I$. Therefore to any bent function, one can associate a *generalized Hadamard matrix*, see [11].

Now we consider a partially bent function, the matrix $H$ that is associated to it, and determine the properties of $HH^*$. Let $f$ be in $\mathcal{P}(V_n, \mathbb{F}_p, s)$ and $H = (\epsilon_p^{f(x-y)})_{(x,y)}$. Then $HH^* = p^n A$, where

- $A$ is a Hermitian matrix, whose nonzero entries are $p$th roots of unity,
- all entries of $A$ on the main diagonal are one and each row and column contains exactly $p^s$ nonzero entries,
- the $(x, z)$-entry of $A$ is nonzero if and only if $x - z \in \Lambda$. Note that the locations of the nonzero entries may change if we use a different ordering of the elements of $V_n$.

## 2.6 Relative difference sets revisited: a construction of bent functions

We first recall a recent construction of bent functions by using partially bent functions, which was introduced in [4, 5] and further analyzed in [7]. For each $(y_1, \ldots, y_s) \in \mathbb{F}_p^s$, let $f_{(y_1,\ldots,y_s)} : \mathbb{F}_p^n \to \mathbb{F}_p$ be a bent function. Denoting $(x_1, \ldots, x_n)$ by $\underline{x}$ and $(x_{n+1}, \ldots, x_{n+s})$ by $\bar{x}$, the functions

$$g_{(y_1,\ldots,y_s)}(\underline{x}, \bar{x}) = f_{(y_1,\ldots,y_s)}(\underline{x}) + y_1 x_{n+1} + \cdots + y_s x_{n+s}$$

are $s$-partially bent functions from $\mathbb{F}_p^{n+s}$ to $\mathbb{F}_p$, see Section 2.3. Additionally, for each $b \in \mathbb{F}_p^{n+s}$, there is exactly one $g_{(y_1,\ldots,y_s)}$, satisfying $\widehat{g_{(y_1,\ldots,y_s)}}(b) \neq 0$. The function $F : \mathbb{F}_p^{n+2s} \to \mathbb{F}_p$, given by

$$F(\underline{x}, \bar{x}, y_1, \ldots, y_s) = g_{(y_1,\ldots,y_s)}(\underline{x}, \bar{x}), \tag{2.11}$$

is then a bent function.

We now interpret this construction in terms of relative difference sets, and point out that this interpretation describes a concrete realization of a general method for constructing relative difference sets from *building blocks*, suggested in [8].

For every $(y_1, \ldots, y_s) \in \mathbb{F}_p^s$, consider the set

$$R_{(y_1,\ldots,y_s)} = \{(\underline{x}, \bar{x}, g_{(y_1,\ldots,y_s)}(\underline{x}, \bar{x})) : \underline{x} \in \mathbb{F}_p^n, \bar{x} \in \mathbb{F}_p^s\},$$

which is a subset of $\mathbb{F}_p^n \times \mathbb{F}_p^s \times \mathbb{F}_p \simeq \mathbb{F}_p^{n+s+1}$. By Corollary 2.9, $R_{(y_1,\ldots,y_s)}$ is a $(p^{n+s}, p, p^s, p^{n+s}, p^{n+s-1})$-partially bent relative difference set of $\mathbb{F}_p^{n+s+1}$. We note that the classical result on bent functions and relative difference sets (cf. [17, Proposition 2]) that we described in Section 2.3, implies that

$$R = \{(\underline{x}, \bar{x}, y_1, \ldots, y_s, g_{(y_1,\ldots,y_s)}(\underline{x}, \bar{x})) : \underline{x} \in \mathbb{F}_p^n, \bar{x} \in \mathbb{F}_p^s\}$$

is the $(p^{n+2s}, p, p^{n+2s}, p^{n+2s-1})$-relative difference set corresponding to $F$ in (2.11). Identifying $(\underline{x}, \bar{x}, g_{(y_1,\dots,y_s)}(\underline{x}, \bar{x}))$ with $(\underline{x}, \bar{x}, 0, \dots, 0, g_{(y_1,\dots,y_s)}(\underline{x}, \bar{x}))$ $\in \mathbb{F}_p^{n+2s+1}$, we can interpret $R_{(y_1,\dots,y_s)}$ as a subset of $\mathbb{F}_p^{n+2s+1}$. We then obtain $R$ from $R_{(y_1,\dots,y_s)}$ as

$$R = \bigcup_{(y_1,\dots,y_s)\in\mathbb{F}_p^s} (0,\dots,0,y_1,\dots,y_s,0) + R_{(y_1,\dots,y_s)}. \tag{2.12}$$

We observe that the set $C = \{(0,\dots,0,y_1,\dots,y_s,0) \mid y_i \in \mathbb{F}_p, 1 \le i \le s\}$ is a set of coset representatives of $\mathbb{F}_p^{n+s+1}$ in $\mathbb{F}_p^{n+2s+1}$ (with the obvious interpretation of $\mathbb{F}_p^{n+s+1}$ as a subgroup of $\mathbb{F}_p^{n+2s+1}$). In fact, if we replace $C$ in (2.12) with another set of coset representatives, we again obtain a (different) $(p^{n+2s}, p, p^{n+2s}, p^{n+2s-1})$-relative difference set. In terms of the construction of a bent function from partially bent functions, changing the set of coset representatives slightly changes the partially bent functions and the order in which they are used in the construction, but not the correctness of the construction.

Our aim now is to interpret the sets $R_{(y_1,\dots,y_s)}$ in the framework given in [8]. We first recall the definition of a "building block" from [8]. A subset $R$ of a group $G$ is called a building block in $G$ if the magnitude of all nonprincipal character sums over $R$ is either 0 or $m$. Using this terminology, the sets $R_{(y_1,\dots,y_s)}$ are buildings blocks in $G = \mathbb{F}_p^{n+s+1}$.

Using building blocks in a group $G$ with a subgroup $U$, one can define a *building set in $G$ relative to $U$* as follows (see [8]). Consider a collection of $t$ building blocks with magnitude $m$ in $G$, each containing $a$ elements. Such a collection is called an $(a, m, t)$ building set in $G$ relative to $U$ if for every nonprincipal character $\chi$ of $G$, the following hold.

(1) Exactly one of the building blocks has nonzero character sum if $\chi$ is nonprincipal on $U$.
(2) If $\chi$ is principal on $U$, then character sums for all building blocks are equal to zero.

By the property that every $b \in \mathbb{F}_p^{n+s}$ is in the support of $\widehat{g_{(y_1,\dots,y_s)}}$ for exactly one $g_{(y_1,\dots,y_s)}$, the collection of the sets $R_{(y_1,\dots,y_s)}$ forms a $(p^{n+s}, p^{(n+2s)/2}, p^s)$ building set in $\mathbb{F}_p^{n+s+1}$ relative to the subgroup $U = \{0\} \times \{0\} \times \dots \times \{0\} \times \mathbb{F}_p$ of $\mathbb{F}_p^{n+s+1}$. By Lemma 2.1, a $(p^{n+2s}, p, p^{n+2s}, p^{n+2s-1})$-relative difference set in $\mathbb{F}_p^{n+2s+1}$ relative to the subgroup $U \simeq \mathbb{F}_p$ is then obtained as (2.12).

## Acknowledgement

Wilfried Meidl and Alev Topuzoğlu would like to express their gratitude to Harald Niederreiter for the essential role he played in their academic lives, not only as an excellent mentor but also as a great friend.

A. Çeşmelioğlu is supported by Tübitak BİDEB 2219 Scholarship Programme. W. Meidl and A. Topuzoğlu are supported by Tübitak Project no.111T234.

## References

[1]  A. Canteaut, C. Carlet, P. Charpin and C. Fontaine, On cryptographic properties of the coset of $R(1, m)$. *IEEE Trans. Inf. Theory* **47**, 1494–1513, 2001.

[2]  C. Carlet, Partially bent functions. *Des. Codes Cryptogr.* **3**, 135–145, 1993.

[3]  A. Çeşmelioğlu and W. Meidl, Bent functions of maximal degree. *IEEE Trans. Inf. Theory* **58**, 1186–1190, 2012.

[4]  A. Çeşmelioğlu and W. Meidl, A construction of bent functions from plateaued functions. *Des. Codes Cryptogr.* **66**, 231–242, 2013.

[5]  A. Çeşmelioğlu, G. McGuire and W. Meidl, A construction of weakly and non-weakly regular bent functions. *J. Comb. Theory Ser. A* **119**, 420–429, 2012.

[6]  A. Çeşmelioğlu, W. Meidl and A. Pott, On the dual of (non)-weakly regular bent functions and self-dual bent functions. *Adv. Math. Commun.* **7**, 425–440, 2013.

[7]  A. Çeşmelioğlu, W. Meidl and A. Pott, Generalized Maiorana–McFarland class and normality of $p$-ary bent functions. *Finite Fields Appl.* **24**, 105–117, 2013.

[8]  J. Davis and J. Jedwab, A unifying construction for difference sets. *J. Comb. Theory Ser. A* **80**, 13–78, 1997.

[9]  J. F. Dillon, Elementary Hadamard difference sets. PhD Dissertation, University of Maryland, 1974.

[10]  T. Helleseth and A. Kholosha, Monomial and quadratic bent functions over the finite fields of odd characteristic. *IEEE Trans. Inf. Theory* **52**, 2018–2032, 2006.

[11]  P. V. Kumar, R. A. Scholtz and L. R. Welch, Generalized bent functions and their properties. *J. Comb. Theory Ser. A* **40**, 90–107, 1985.

[12]  W. Meidl and A. Topuzoğlu, Quadratic functions with prescribed spectra. *Des. Codes Cryptogr.* **66**, 257–273, 2013.

[13]  K. Nyberg, Perfect nonlinear S-boxes. *Advances in Cryptology, EUROCRYPT '91 (Brighton, 1991).* Lecture Notes in Computer Science, volume 547, pp. 378–386. Springer, Berlin, 1991.

[14]  D. S. Passman, *The Algebraic Structure of Group Rings.* Wiley, New York, 1977.

[15]  A. Pott, private communication.

[16]  A. Pott, *Finite Geometry and Character Theory.* Springer-Verlag, Berlin, 1995.

[17]  Y. Tan, A. Pott and T. Feng, Strongly regular graphs associated with ternary bent functions. *J. Comb. Theory Ser. A* **117**, 668–682, 2010.

[18]  Y. Zheng and X. M. Zhang, On plateaued functions. *IEEE Trans. Inf. Theory* **47**, 1215–1223, 2001.

# 3

# Applications of geometric discrepancy in numerical analysis and statistics

*Josef Dick*
The University of New South Wales, Sydney

*Dedicated to Harald Niederreiter on the occasion of his 70th birthday.*

## Abstract

In this paper we discuss various connections between geometric discrepancy measures, such as discrepancy with respect to convex sets (and convex sets with smooth boundary in particular), and applications to numerical analysis and statistics, such as point distributions on the sphere, the acceptance-rejection algorithm and certain Markov chain Monte Carlo algorithms.

## 3.1 Introduction

The local discrepancy function of a point set $P_{N,s} = \{\boldsymbol{x}_0, \boldsymbol{x}_1, \ldots, \boldsymbol{x}_{N-1}\} \subset [0, 1]^s$ measures the difference of the empirical distribution from the uniform distribution with respect to a collection $\mathcal{A} \subset \mathcal{P}([0, 1]^s)$ of test sets, where $\mathcal{P}([0, 1]^s)$ denotes the power set of $[0, 1]^s$, that is

$$\Delta_{P_{N,s}}(A) = \frac{1}{N} \sum_{n=0}^{N-1} 1_A(\boldsymbol{x}_n) - \lambda_s(A),$$

where $A \in \mathcal{A}$, $1_A$ is the indicator function of the set $A$ and $\lambda_s$ denotes the $s$-dimensional Lebesgue measure. The supremum of $|\Delta_{P_{N,s}}(A)|$ over all sets $A \in \mathcal{A}$ is called the star discrepancy of $P_{N,s}$ (with respect to the test sets $\mathcal{A}$)

$$D^*_{\mathcal{A}}(P_{N,s}) = \sup_{A \in \mathcal{A}} |\Delta_{P_{N,s}}(A)|.$$

Depending on the choice of test sets, one has different types of convergence behavior. One well-studied example of test sets is that of boxes anchored at the origin $\mathcal{B} = \{[\mathbf{0}, t) : t \in [\mathbf{0}, \mathbf{1}]\}$, where $\mathbf{0} = (0, 0, \ldots, 0)$, $t = (t_1, t_2, \ldots, t_s) \in [0, 1]^s$ and $[\mathbf{0}, t) = \prod_{j=1}^{s}[0, t_j)$. In this case upper and lower bounds are known, as well as explicit constructions of point sets which match the best known upper bounds. Variations of anchored boxes, like boxes anchored at different places, boxes which are not anchored or boxes on the torus $(\mathbb{R}/\mathbb{Z})^s$ are all similar and many results are also known in these cases [7, 17, 21, 31, 38, 45]. If the test sets are allowed to be more general, for example, if they are all convex sets or all convex sets with smooth boundary, then the situation is more complicated. Upper and lower bounds are known, but the upper bounds are often based on probabilistic arguments and are therefore not constructive.

In this paper we provide examples of applications which naturally yield problems in discrepancy theory. The case of anchored boxes is the best understood example of these and provides a connection of low-discrepancy point sets to applications in numerical integration. The connection of discrepancy measures to applications, when the discrepancy is defined with respect to other types of test sets, is less well known. We relate point distributions on the sphere, points transformed via inversion to different distributions and spaces, the acceptance-rejection algorithm and the Markov chain Monte Carlo algorithm to various discrepancy measures of point sets in the unit cube. This provides a motivation for studying discrepancy with respect to various test sets in the cube.

## 3.2  Numerical integration in the unit cube

We explain how numerical integration in the unit cube using equal weight quadrature rules leads one to discrepancy with respect to anchored boxes $\mathcal{B}$. A simple explanation in one dimension is the following. Assume that $f : [0, 1] \to \mathbb{R}$ is absolutely continuous and let $P_{N,1} = \{x_0, x_1, \ldots, x_{N-1}\} \subset [0, 1]$. We have

$$f(x) = f(1) - \int_0^1 1_{[0,t)}(x) f'(t)\, \mathrm{d}t$$

and

$$\int_0^1 f(x)\, \mathrm{d}x = f(1) - \int_0^1 \int_0^1 1_{[0,t)}(x)\, \mathrm{d}x f'(t)\, \mathrm{d}t = f(1) - \int_0^1 f'(t) t\, \mathrm{d}t.$$

Therefore

$$
\int_0^1 f(x)\,dx - \frac{1}{N}\sum_{n=0}^{N-1} f(x_n)
$$

$$
= \int_0^1 f'(t)\left[\frac{1}{N}\sum_{n=0}^{N-1} 1_{[0,t)}(x_n) - \int_0^1 1_{[0,t)}(x)\,dx\right] dt
$$

$$
= \int_0^1 f'(t)\Delta_{P_{N,1}}(t)\,dt,
$$

where $\Delta_{P_{N,1}}$ is the local discrepancy function given by

$$
\Delta_{P_{N,1}}(t) = \frac{1}{N}\sum_{n=0}^{N-1} 1_{[0,t)}(x_n) - t.
$$

Using Hölder's inequality we therefore get

$$
\left|\int_0^1 f(x)\,dx - \frac{1}{N}\sum_{n=0}^{N-1} f(x_n)\right| \le \left(\int_0^1 |f'(t)|^p\,dt\right)^{1/p}
$$

$$
\times \left(\int_0^1 |\Delta_{P_{N,1}}(t)|^q\,dt\right)^{1/q}, \qquad (3.1)
$$

for Hölder conjugates $1 \le p, q \le \infty$, with the obvious modifications for $p$ or $q = \infty$. Inequality (3.1) is a variant of an inequality due to Koksma [30].

From these considerations, one obtains the $L^q$ discrepancy as a quality criterion for the point set $P_{N,1}$:

$$
L^q_{\mathcal{B}}(P_{N,1}) = \left(\int_{[0,1]} |\Delta_{P_{N,1}}(t)|^q\,dt\right)^{1/q} \qquad \text{for } 1 \le q \le \infty,
$$

again with the obvious modifications for $q = \infty$. As above, $\mathcal{B}$ is the set of boxes anchored at the origin.

There is a natural generalization of the above approach to dimensions $s > 1$ by using partial derivatives of $f$. This leads to discrepancy measures with respect to anchored boxes. Let a point set $P_{N,s} = \{x_0, x_1, \ldots, x_{N-1}\} \subset [0,1]^s$ be given and let $t = (t_1, \ldots, t_s) \in [0,1]^s$. Then we define the local discrepancy function by

$$
\Delta_{P_{N,s}}(t) = \frac{1}{N}\sum_{n=0}^{N-1} 1_{[0,t]}(x_n) - \prod_{j=1}^{s} t_j.
$$

where $[\mathbf{0}, \mathbf{t}] = \prod_{j=1}^{s}[0, t_j]$. Again, by taking the $L^q$ norm of the local discrepancy function, we obtain the $L^q$ discrepancy with respect to anchored boxes $\mathcal{B}$ given by

$$L_{\mathcal{B}}^q(P_{N,s}) = \left( \int_{[0,1]^s} |\Delta_{P_{N,s}}(\mathbf{t})|^q \, d\mathbf{t} \right)^{1/q} \qquad \text{for } 1 \le q \le \infty,$$

with obvious modifications for $q = \infty$.

There is a generalization of (3.1), which is due to Hlawka [28]. Let $f : [0, 1]^s \to \mathbb{R}$ and $P_{N,s} = \{\mathbf{x}_0, \mathbf{x}_1, \ldots, \mathbf{x}_{N-1}\} \subset [0, 1]^s$, then

$$\left| \int_{[0,1]^s} f(\mathbf{x}) \, d\mathbf{x} - \frac{1}{N} \sum_{n=0}^{N-1} f(\mathbf{x}_n) \right| \le \left( \int_{[0,1]^s} \left| \frac{\partial^s f}{\partial \mathbf{t}}(\mathbf{t}) \right|^p \, d\mathbf{t} \right)^{1/p} L_{\mathcal{B}}^q(P_{N,s}),$$

(3.2)

where $\frac{\partial^{|u|} f}{\partial t_u}(\mathbf{t}_u, \mathbf{1}) = 0$ for all $u \subsetneq \{1, 2, \ldots, s\}$.

Several important variants of (3.2) are known, see for instance Dick and Pillichshammer [17], Hickernell [27] and Sloan and Woźniakowski [57] (but these are not discussed here in further detail).

In the following we write $A(N, s) \ll_s B(N, s)$ if there is a constant $c_s > 0$ which depends only on $s$ (but not on $N$) such that $A(N, s) \le c_s B(N, s)$ for all $N$, with an analogous meaning for $\gg_s$. We write $A(N, s) \asymp B(N, s)$ if $A(N, s) \ll B(N, s)$ and $A(N, s) \gg B(N, s)$. If the implied constant depends on some coefficients (say $q$ and $s$), then we write $\asymp_{q,s}$.

The $L^q$ discrepancy has been intensively studied and many precise results are known. Lower bounds by Roth [51] and Schmidt [54] and upper bounds via explicit constructions by Chen and Skriganov [13] and Skriganov [56] show that for all $1 < q < \infty$ we have

$$L_{\mathcal{B}}^q(P_{N,s}) \asymp_{q,s} \frac{(\log N)^{\frac{s-1}{2}}}{N}.$$

An explicit construction of sequences which achieve the optimal rate of convergence for $q = 2$ has been introduced by Dick and Pillichshammer [18]. The point sets introduced by Skriganov [56] depend on $q$, i.e., as $q$ increases one has to use different point sets. In [16] a construction was introduced for which the choice of the point set $P_{N,s}$ does not depend on $q$.

The endpoint cases $q = 1$ and $q = \infty$ are still open. The following lower bounds are due to Halász [24] for $q = 1$ and Bilyk and Lacey [8] and Bilyk et al. [9] for $q = \infty$:

$$L_{\mathcal{B}}^1(P_{N,s}) \gg_s \frac{\log N}{N},$$

$$L_{\mathcal{B}}^{\infty}(P_{N,s}) \gg_s \frac{(\log N)^{\frac{s-1}{2}+\eta}}{N}.$$

Explicit constructions of point sets are known in each case, see Chen and Skriganov [13], Dick [16], Dick and Pillichshammer [18], Faure [22], Halton [25], Hammersley [26], Niederreiter [44], Niederreiter and Xing [47, 48, 63], Skriganov [56], Sobol' [58] and others, which show that

$$L_{\mathcal{B}}^{\infty}(P_{N,s}) \ll_s \frac{(\log N)^{s-1}}{N},$$

$$L_{\mathcal{B}}^{1}(P_{N,s}) \ll_s \frac{(\log N)^{\frac{s-1}{2}}}{N}.$$

In the following we consider generalizations of discrepancy measures, which we then relate to problems from numerical analysis and statistics.

### 3.2.1 Generalizations of the discrepancy with respect to anchored boxes

The $L^{\infty}$ discrepancy is often called the star discrepancy and is denoted by $D^*$. Consider the star discrepancy with respect to anchored boxes $\mathcal{B}$

$$D_{\mathcal{B}}^*(P_{N,s}) = \sup_{t \in [0,1]^s} \left| \frac{1}{N} \sum_{n=0}^{N-1} 1_{[\mathbf{0},t)}(\mathbf{x}_n) - \prod_{j=1}^{s} t_j \right|.$$

The supremum over the boxes $[\mathbf{0}, t)$ can be replaced by other test sets. This will yield other discrepancy criteria. For instance, the isotropic discrepancy is defined with respect to the collection $\mathcal{C}$ of convex subsets of $[0,1]^s$. The local isotropic discrepancy is in this case defined by

$$\Delta_{P_{N,s}}(C) = \frac{1}{N} \sum_{n=0}^{N-1} 1_C(\mathbf{x}_n) - \lambda_s(C),$$

where $C \in \mathcal{C}$ is a convex set and $\lambda_s$ is the $s$-dimensional Lebesgue measure. The isotropic discrepancy is then defined by

$$D_{\mathcal{C}}^*(P_{N,s}) = \sup_{C \in \mathcal{C}} \left| \Delta_{P_{N,s}}(C) \right|.$$

The connection to numerical integration is not as clear in this case as for the case of anchored boxes.

Again, a number of results are known about the isotropic discrepancy due to Beck [6], Hlawka [28], Laczkovich [33], Mück and Philipp [41], Niederreiter [42, 43], Niederreiter and Wills [46], Schmidt [53], Stute [61] and Zaremba [64]. In dimension 2 it is known that

$$\frac{1}{N^{2/3}} \ll D_{\mathcal{C}}^*(P_{N,2}) \ll \frac{(\log N)^4}{N^{-2/3}},$$

whereas in dimension $s > 2$ we have

$$\frac{1}{N^{\frac{2}{s+1}}} \ll_s D_{\mathcal{C}}^*(P_{N,s}) \leq (D^*(P_{N,s}))^{1/s}.$$

Using known constructions of point sets with small star discrepancy, one obtains the upper bound

$$D_{\mathcal{C}}^*(P_{N,s}) \leq C_s \frac{\log N}{N^{1/s}}.$$

The construction of the point sets is explicit in this case. However, there is a gap between the upper and lower bound and the precise rate of convergence remains unknown.

In the next section we consider numerical integration over the unit sphere.

## 3.3 Numerical integration over the unit sphere

Spherical caps are a natural class of test sets on the unit sphere $\mathbb{S}^s = \{x \in \mathbb{R}^{s+1} : \|x\| = 1\}$, where $\|\cdot\|$ denotes the Euclidean norm. These are defined in the following way. The spherical cap with center $x \in \mathbb{S}$ and height $-1 \leq t \leq 1$ is defined by

$$C(x, t) = \{z \in \mathbb{S}^s : \langle z, x \rangle > t\}.$$

Let $\mathcal{S} = \{C(x, t) : x \in \mathbb{S}^s, -1 \leq t \leq 1\}$ denote the set of spherical caps. The spherical cap $L^q$ discrepancy is now given by

$$L_{\mathcal{S}}^q(P_{N,s}) = \left( \int_{-1}^{1} \int_{\mathbb{S}^s} \left| \frac{1}{N} \sum_{n=0}^{N-1} 1_{C(x,t)}(x_n) - \sigma_s(C(x, t)) \right|^q \, d\sigma_s(x) \, dt \right)^{1/q},$$

where $\sigma_s$ is the normalized surface Lebesgue measure on the sphere $\mathbb{S}^s$. As in the case of discrepancy with respect to anchored boxes in $[0, 1]^s$, the spherical cap discrepancy is related to numerical integration. The Koksma–Hlawka type inequality is of the form

$$\left| \int_{\mathbb{S}_s} f(x) \, d\sigma_s(x) - \frac{1}{N} \sum_{n=0}^{N-1} f(x_n) \right| \leq L_{\mathcal{S}}^q(P_{N,s}) \|f\|_p,$$

where $\|f\|_p$ is a suitable function norm, see Brauchart and Dick [11].

Bounds on the spherical cap discrepancy have been established by Beck [5], Schmidt [52] and Stolarsky [60]. It is known that

$$\frac{1}{N^{1/2+1/2s}} \ll_{q,s} L^q_{\mathbb{S}}(P_{N,s}) \ll_{q,s} \frac{\sqrt{\log N}}{N^{1/2+1/2s}}.$$

However, the upper bound is based on probabilistic arguments and there are no known explicit constructions satisfying this upper bound.

A number of nonoptimal results follow from the work of Grabner and Tichy [23], Lubotzky *et al.* [36, 37] and Aistleitner *et al.* [3].

We briefly discuss the explicit constructions of points on the sphere $\mathbb{S}^2$ from Aistleitner *et al.* [3]. The idea there is to use a transformation from the square $[0, 1]^2$ to the sphere $\mathbb{S}^2$ which preserves the measure. The so-called Lambert transform $\Phi : [0, 1]^2 \to \mathbb{S}^2$ given by

$$\Phi(x, y) = \left( 2\cos(2\pi x)\sqrt{y - y^2}, 2\sin(2\pi x)\sqrt{y - y^2}, 1 - 2y \right)$$

has this property, i.e., for any Lebesgue measurable set $J \subseteq [0, 1]^2$ we have $\lambda_2(J) = \sigma_2(\Phi(J))$, where $\Phi(J) = \{\Phi(x) : x \in J\}$.

In order to obtain points on the sphere $\mathbb{S}^2$, we proceed in the following way. We map the points $\{x_0, x_1, \ldots, x_{N-1}\} \in [0, 1]^2$ to $\Phi(x_0), \Phi(x_1), \ldots, \Phi(x_{N-1}) \in \mathbb{S}^2$. It was shown by Aistleitner *et al.* [3] that if $x_0, x_1, \ldots, x_{N-1}$ have low discrepancy with respect to anchored boxes in the square, then $\Phi(x_0), \Phi(x_1), \ldots, \Phi(x_{N-1})$ have low spherical cap discrepancy. Figure 3.1 shows some numerical results for a digital net mapped to the sphere $\mathbb{S}^2$. The result indicates that these point sets achieve the optimal rate of convergence of the spherical cap $L^2$ discrepancy.

First note that the optimal rate of convergence for boxes in the square differs from the optimal rate of convergence for spherical caps on the sphere. This is not surprising when considering the inverse sets of spherical caps $B(x, t) = \Phi^{-1}(C(x, t)) = \{z \in [0, 1]^2 : \Phi(z) \in C(x, t)\}$. These sets have particular shapes which change as $x$ and $t$ vary. They are not convex, however, they can be broken up into a small number of convex parts and parts whose complement with respect to some rectangle is convex. Their boundary is smooth except for the pole caps where $x$ is either the north pole $(1, 0, 0)$ or the south pole $(-1, 0, 0)$, in which case $B(x, t)$ is a rectangle. The curvature of the boundary is unbounded, which can be seen when $x$ moves to one of the poles where the smooth boundary curve of $B(x, t)$ turns into a rectangle. Thus the sets $B(x, t)$ do not have any discernible features. However, for most sets $B(x, t)$, the boundary is smooth and has bounded curvature. Thus, for the most part, the

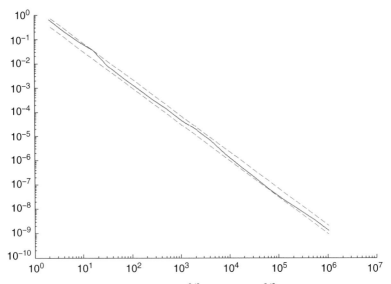

Figure 3.1 The dashed lines show $N^{-3/2}$ and $(9/4)N^{-3/2}$, and the curve shows the squared spherical cap $L^2$ discrepancy, where the quadrature points are a digital net mapped to the sphere.

sets $B(x, t)$ can be described by convex sets with smooth boundary of bounded curvature.

In the following we briefly discuss discrepancy in the cube with respect to convex test sets with smooth boundary, since the problem of the discrepancy of points mapped to the sphere using the Lambert transform is, to a large degree, related to this discrepancy.

### 3.3.1 Discrepancy with respect to convex sets with smooth boundary

We consider now the star discrepancy in the torus $(\mathbb{R} \setminus \mathbb{Z})^2$ with respect to the convex test sets $\mathcal{E}$, whose boundary curve is twice continuously differentiable with minimal curvature divided by maximal curvature bounded away from 0. It was shown by Beck and Chen [7] that

$$\frac{1}{\sqrt{\log N}N^{3/4}} \ll L_{\mathcal{E}}^{\infty}(P_{N,2}) \ll \frac{\sqrt{\log N}}{N^{3/4}}.$$

A generalization (using slightly different assumptions) to arbitrary dimension can be found in Drmota [20], which shows that

$$\frac{1}{N^{1/2+1/2s}} \ll_s L_{\mathcal{E}}^{\infty}(P_{N,s}) \ll_s \frac{\sqrt{\log N}}{N^{1/2+1/2s}}.$$

The discrepancy bounds for this case are very similar to the discrepancy bounds for the spherical cap discrepancy on $\mathbb{S}^s$. As in the sphere case, no explicit constructions of point sets achieving the upper bound on the discrepancy with respect to convex sets with smooth boundary are known. Indeed, a solution to one of these problems may also yield a solution to the other problem. The numerical results for the spherical cap $L^2$ discrepancy indicate that classical low-discrepancy constructions such as digital nets and Fibonacci lattices are optimal. Hence the question arises whether this is also true for the discrepancy in the square with respect to convex sets with smooth boundary, as studied by Beck and Chen [7].

## 3.4 Inverse transformation and test sets

Assume now we want to approximate the integral

$$\int_G f(x)\psi(x)\,dx,$$

where $\psi$ is a probability density function on $G \subseteq \mathbb{R}^s$. In some cases there is a mapping $\Phi : [0,1]^s \to G$ which is measure preserving in the following sense. For every Lebesgue measurable set $A \subseteq [0,1]^s$ we have

$$\lambda_s(A) = \int_{\Phi(A)} \psi(x)\,dx.$$

See the inverse Rosenblatt transform [50], or the monographs by Devroye [15] and Hörmann *et al.* [29]. In the previous section we saw an example of this situation, however such problems come up in other contexts as well (see for instance Kuo *et al.* [32] for an example in the context of quasi-Monte Carlo integration). Now assume that we want to study discrepancy with respect to boxes in $G$. Let $P = \{x_0, x_1, \dots, x_{N-1}\} \subset [0,1]^s$. Then we can define the discrepancy

$$\sup_{t\in\mathbb{R}^s}\left|\frac{1}{N}\sum_{n=0}^{N-1}1_{G\cap(-\infty,t]}(\Phi(x_n)) - \int_{G\cap(-\infty,t]}\psi(x)\,dx\right|.$$

This discrepancy can be translated to a discrepancy in the unit cube

$$\sup_{A\in\mathcal{A}}\left|\frac{1}{N}\sum_{n=0}^{N-1}1_A(x_n) - \lambda_s(A)\right|,$$

where $\mathcal{A}$ consists of all sets $A_t = \{x \in [0, 1]^s : \Phi(x) \in G \cap [-\infty, t)\}$ for all $t \in [0, 1]^s$. In general, the sets $A_t$ are not boxes anymore and therefore one wants to have point sets $x_0, x_1, \ldots, x_{N-1} \in [0, 1]^s$ which have small discrepancy with respect to the test sets $\mathcal{A}$ rather than boxes. Such a problem was also observed for instance by L'Ecuyer *et al.* [34] in their array RQMC method. In some cases, one can do stratified sampling, that is, divide the cube into subcubes with side length $2^{-k}$ and randomly place a point in each box. Then one can show a convergence rate of order $N^{-1/2-1/(2s)}$. However, for instance, L'Ecuyer *et al.* [34] observed better rates of convergence when using low-discrepancy point sets. Kuo *et al.* [32] observed that the choice of transformation $\Phi$ influences the rate of convergence, but it is a priori not clear what choice of $\Phi$ yields the best results. For these types of applications it would be interesting to have point sets which achieve good convergence rates for various types of test sets.

In this context, we mention one construction of explicit point sets in dimension 2 which considers more general test sets, namely the construction by Bilyk *et al.* [10] where discrepancy with respect to certain rotated boxes is considered.

## 3.5 Acceptance-rejection sampler

In statistical sampling one often wants to sample from a target distribution. The standard procedure for obtaining samples from a given distribution is to invert the cumulative distribution function (cdf), which can be used to map points from the cube $[0, 1]^s$ to the required domain. However, this is often not possible (or it is numerically expensive), in which case one has to resort to other methods.

To illustrate the procedure we consider a one-dimensional example. In higher dimensions one can use the Rosenblatt transformation [50]. See also Chelson's result [12] for discrepancy bounds in higher dimensions using the Rosenblatt transformation in higher dimensions (this result can also be found in Spanier and Maize [59, Theorem 4.1]).

As an example, consider the unnormalized density function $\psi(x) = x^2 + \sin(\pi x)$ for $x \in [0, 2]$ (see the work of Barekat and Caflisch [4] and the discussion in Aistleitner and Dick [1] for a motivation for this type of example). The normalization constant is $\int_0^2 (x^2 + \sin(\pi x))\, dx = 8/3$. The cdf is given by

$$\Psi(t) = \frac{3}{8} \int_0^t \psi(x)\, dx = \frac{3}{8} \int_0^t (x^2 + \sin(\pi x))\, dx = \frac{t^3}{8} + \frac{3(1 - \cos(\pi t))}{8\pi}.$$

In order to be able to sample directly from the density $3\psi/8$, one would have to invert $\Psi$. Since this cannot be done exactly, we resort to other methods like the acceptance-rejection algorithm.

The acceptance rejection sampler proceeds in the following way.

**Algorithm 3.1** *Let* $\psi : [0, 1]^s \to \mathbb{R}$ *be a (unnormalized) density function.*

(1) *Choose* $L > 0$ *such that* $\frac{\psi(x)}{L} \leq 1$ *for all* $x \in [0, 1]^s$.
(2) *Generate a point set* $P_{M,s+1} = \{x_0, x_1, \ldots, x_{M-1}\} \subset [0, 1]^{s+1}$. *Assume that* $x_n = (x_{n,1}, x_{n,2}, \ldots, x_{n,s+1})$.
(3) *Choose*

$$I = \{0 \leq n < M : \psi(x_{n,1}, x_{n,2}, \ldots, x_{n,s}) \leq Lx_{n,s+1}\}.$$

(4) *Return the point set* $Q = \{(x_{n,1}, x_{n,2}, \ldots, x_{n,s}) : n \in I\}$.

It is well known that if $P_{M,s+1} = \{x_0, x_1, \ldots, x_{M-1}\}$ is chosen i.i.d. uniformly distributed, then the point set $Q$ has distribution with law $\widetilde{\psi}$ (where $\widetilde{\psi} = (\int_{[0,1]^s} \psi(x)\, dx)^{-1}\psi$ is the normalized density function). For a proof, see for instance Robert and Casella [49, Section 2.3].

Numerical tests have been performed by Morokoff and Caflisch [39], Moskowitz and Caflisch [40] and Wang [62], where the random point set $P_{M,s+1}$ in Algorithm 3.1 is replaced by a low-discrepancy point set with the intention to obtain samples which have better distribution properties. The difference between random point sets and deterministic point sets in the acceptance-rejection algorithm is illustrated in Figure 3.2.

Let $Q = \{y_0, y_1, \ldots, y_{N-1}\}$ be the set of points generated by Algorithm 3.1. To study the performance of this type of algorithm, we introduce the discrepancy

$$D^*(Q) = \sup_{t \in [0,1]^s} \left| \frac{1}{N} \sum_{n=0}^{N-1} 1_{[0,t)}(y_n) - \frac{1}{C} \int_{[0,t)} \psi(x)\, dx \right|,$$

where $C = \int_{[0,1]} \psi(x)\, dx$.

Some simple numerical tests confirm that low-discrepancy point sets in Algorithm 3.1 can improve the performance of the acceptance-rejection algorithm. For instance, Zhu and Dick [65] considered the following example. Let an unnormalized target density be given by

$$\psi(x) = \frac{3}{4} - \left(x - \frac{1}{2}\right)^2, \quad x \in [0, 1].$$

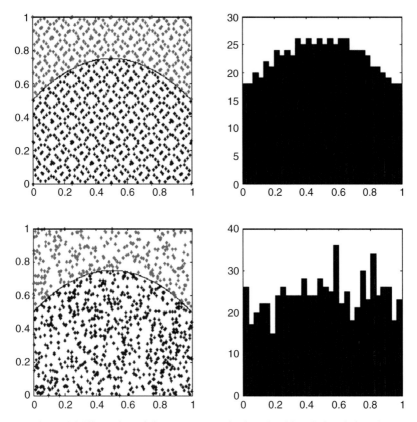

Figure 3.2 Illustration of the acceptance-rejection algorithm. Points below the
curved line are accepted and then projected onto the $x$-axis. The top row shows the
acceptance-rejection algorithm using a deterministic point set $P_{M,s+1}$, whereas
the bottom row shows the acceptance-rejection sampler using random samples
$P_{M,s+1}$. In both cases the number of points is $M = 2^7$. (See color plate.)

Figure 3.3 shows the discrepancy of the point set when the proposal points are
a digital net.

The discrepancy $D^*(Q)$ can be written in terms of the discrepancy of the
proposal points $P_{M,s+1}$. To do so, let

$$A = \{x = (x_1, x_2, \ldots, x_{s+1}) \in [0, 1]^{s+1} : f(x_1, x_2, \ldots, x_s) \le Lx_{s+1}\}.$$

Then we have

$$D^*(Q) = \sup_{t \in [0,1]^{s+1}} \left| \frac{1}{M} \sum_{n=0}^{M-1} 1_{A \cap [0,t]}(x_n) - \lambda_{s+1}(A \cap [0, t]) \right|.$$

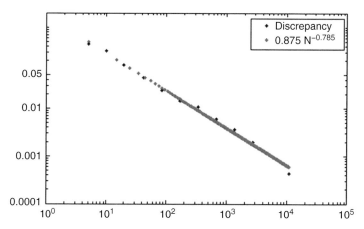

Figure 3.3 Numerical result of the acceptance-rejection algorithm using low-discrepancy point sets. The convergence rate is approximately of order $N^{-0.8}$, which is better than the rate one would expect when using random samples (which is $N^{-0.5}$). (See color plate.)

Thus the discrepancy of the samples obtained from Algorithm 3.1 coincides with the discrepancy of the points $P_{M,s+1}$ with respect to the test sets $A \cap [0, t)$. If $\psi$ is smooth and concave, then the set $A \cap [0, t]$ is convex and has a smooth boundary, except at the intersection points of the boundaries of $A$ and $[0, t]$ and the intersection of the faces of the box. Thus bounds on the discrepancy with respect to convex sets with smooth boundary may also be of help in this problem.

Some results about the discrepancy $D^*(Q)$ are known from Zhu and Dick [65]. If $P_{M,s+1}$ is a low-discrepancy point set, then for any concave unnormalized density $\psi$ we have $D^*(Q) \ll N^{-\frac{1}{s+1}}$. On the other hand, for every point set $P_{M,s+1}$, there is a concave function $\psi$ such that $D^*(Q) \gg N^{-\frac{2}{s+2}}$. Thus, in general, the upper bound cannot be significantly improved without using further assumptions. Smoothness of $\psi$ would be an assumption where one would hope to be able to get better results.

## 3.6 Markov chain Monte Carlo and completely uniformly distributed sequences

Let $G \subseteq \mathbb{R}^d$ be measurable and let $\varphi : G \times [0, 1]^s \rightarrow G$ be a measurable function. We call $G$ the "state space" and $\varphi$ the "update function," respectively. Then one obtains a Markov chain $(x_n)_{n \geq 0}$ by choosing a starting point $x_0 \in G$, generating a sequence of random numbers $u_0, u_1, \ldots \in [0, 1]^s$ and setting $x_n = \varphi(x_{n-1}; u_{n-1})$. Chen *et al.* [14] studied Markov chains and conditions under which the Markov chain consistently samples a target distribution $\psi$, that is, for every continuous function defined on $G$ we have

$$\lim_{N \to \infty} \frac{1}{N} \sum_{n=0}^{N-1} f(x_n) = \int_G f(x) \psi(x) \, dx. \tag{3.3}$$

Chen *et al.* [14] showed that if the random numbers $u_0, u_1, \ldots$ are completely uniformly distributed (and some further assumptions on the update function are satisfied), then (3.3) holds. Complete uniform distribution is a condition on a sequence of numbers $u_0, u_1, u_2, \ldots \in [0, 1]$ which ensures statistical independence of successive terms in a specific sense. The definition is based on discrepancy and works as follows. Let $\boldsymbol{u}_0^{(s)} = (u_0, u_1, \ldots, u_{s-1})$, $\boldsymbol{u}_1^{(s)} = (u_s, u_{s+1}, \ldots, u_{2s-1})$, and so on. Let $\mathcal{U}_{N,s} = \{\boldsymbol{u}_0^{(s)}, \boldsymbol{u}_1^{(s)}, \ldots, \boldsymbol{u}_{N-1}^{(s)}\}$. Then the sequence $u_0, u_1, u_2, \ldots$ is completely uniformly distributed (CUD) if for all dimensions $s \geq 1$ we have

$$\lim_{N \to \infty} D_{\mathcal{B}}^*(\mathcal{U}_{N,s}) = 0.$$

Explicit constructions of sequences which are completely uniformly distributed have been established by Levin [35] and Shparlinski [55]. For instance, Levin [35] showed error bounds of the form $D_{\mathcal{B}}^*(P_{N,s}) \ll N^{-1}(\log N)^{s+\varepsilon}$. However, in the application arising in Chen *et al.* [14], one needs $s \asymp \log N$. In this case we have

$$N^{-1}(\log N)^{s+\varepsilon} \approx N^{-1}(\log N)^{\log N} = N^{-1+\log \log N}.$$

Thus these results do not guarantee convergence of the discrepancy as $N$ tends to $\infty$. Instead, a different approach is required. Chen *et al.* [14] showed the existence of a sequence of numbers such that for all $s \geq 1$ we have

$$D_{\mathcal{B}}^*(\mathcal{U}_{N,s}) \leq C \sqrt{s \frac{\log N}{N}}.$$

For $s \asymp \log N$ one therefore has $D_{\mathcal{B}}^*(\mathcal{U}_{N,s}) \leq C \frac{\log N}{\sqrt{N}}$. Aistleitner and Weimar [2] obtained an improvement where $D_{\mathcal{B}}^*(\mathcal{U}_{N,s}) \leq C \sqrt{s \frac{\log \log N}{N}}$. However, no explicit construction of such a sequence is known. Such sequences could be useful in applications of Markov chain Monte Carlo methods.

## 3.7 Uniformly ergodic Markov chains and push-back discrepancy

The result in Chen *et al.* [14] yields consistency for certain Markov chains for completely uniformly distributed sequences, but does not yield any convergence rates. This question was addressed by Dick *et al.* [19]. Therein, the discrepancy of the sample points in the state space $G$ with respect to the target distribution was related to the discrepancy of the driver sequence $u_0, u_1, u_2, \ldots \in [0, 1]$ with respect to the uniform measure. Again, through the update function, the test sets defined in $G$ are distorted in the unit cube, as described in Section 3.4. In general, one therefore does not have boxes as test sets anymore. Additionally, one also needs the statistical independence of the driver sequence measured in terms of complete uniform distribution. This yields a generalized definition of completely uniformly distributed point sets, where one does not have boxes as test sets. We call the underlying discrepancy a "push-back discrepancy," since it is derived from the discrepancy in the state space by inverting the update function. We provide some details in the following.

Let $\mathcal{U}_{N,s} = \{u_0, u_1, \ldots, u_{N-1}\} \subset [0, 1]^s$ denote the points which drive the Markov chain via the update function. Let $\varphi : G \times [0, 1]^s \to G$ denote again the update function, so that $x_n = \varphi(x_{n-1}; u_{n-1})$. The $n$ times iterated update function is denoted by $\varphi_n$, that is, we have $x_n = \varphi_n(x_0; u_0, u_1, \ldots, u_{n-1})$. Then we define the sets

$$C_{n,x_0}(A) = \{z \in [0, 1]^{ns} : \varphi_n(x_0; z) \in A\},$$

for $A \in \mathcal{B}(G)$, the Borel $\sigma$ algebra of $G$, and $n \in \mathbb{N}$. The local discrepancy function of the driver point set $\mathcal{U}_{N,s}$ is then given by

$$\Delta_{N,\mathcal{U}_{N,s},\psi,\varphi}(A) = \frac{1}{N} \sum_{n=1}^{N} \left[ 1_{(u_0,\ldots,u_{n-1}) \in C_{n,\psi}(A)} - \lambda_{ns}(C_{n,\psi}(A)) \right],$$

and the discrepancy of the driver sequence is given by

$$D^*_{\mathcal{A},\psi,\varphi}(\mathcal{U}_{N,s}) = \sup_{A \in \mathcal{A}} \left| \Delta_{N,\mathcal{U}_{N,s},\psi,\varphi}(A) \right|.$$

We call $D^*_{\mathcal{A},\psi,\varphi}(\mathcal{U}_{N,s})$ the push-back discrepancy of $\mathcal{U}_{N,s}$.

The push-back discrepancy combines two principles: the discrepancy in the cube with respect to general test sets and the principle of complete uniform distribution. As for the discrepancy with respect to general test sets, there are no explicit constructions of point sets with small push-back discrepancy known. However, Dick *et al.* [19] showed that there exist points $u_0, u_1, \ldots, u_{N-1} \in [0, 1]^s$ such that

$$D^*_{\mathcal{A},\psi,\varphi}(\mathcal{U}_n) \ll \sqrt{\frac{\log N}{N}}.$$

Such a point set would have direct applications in Markov chain Monte Carlo methods.

## Acknowledgment

The author is supported by an ARC Queen Elizabeth II Fellowship and an ARC Discovery Project. The help of Houying Zhu and the remarks of the reviewer are gratefully acknowledged. The author is also grateful to Ch. Aistleitner for pointing out Chelson's result and to J. Spanier for providing another reference for Chelson's result. The comments of the editors F. Pillichshammer and A. Winterhof are greatly appreciated.

## References

[1]  C. Aistleitner and J. Dick, Functions of bounded variation, signed mea-sures, and a general Koksma–Hlawka inequality. Submitted 2014, available at http://arxiv.org/pdf/1406.0230.pdf.

[2]  C. Aistleitner and M. Weimar, Probabilistic star discrepancy bounds for dou-ble infinite random matrices. In: J. Dick, F. Y. Kuo, G. W. Peters and I. H. Sloan (eds.), *Monte Carlo and Quasi-Monte Carlo Methods 2012*, pp. 271–287. Springer-Verlag, Heidelberg, 2013.

[3]  C. Aistleitner, J. S. Brauchart and J. Dick, Point sets on the sphere $\mathbb{S}^2$ with small spherical cap discrepancy. *Discrete Comput. Geom.* **48**, 990–1024, 2012.

[4]  F. Barekat and R. Caflisch, Simulation with fluctuating and singular rates. Preprint, available at http://arxiv.org/pdf/1310.4555.pdf.

[5]  J. Beck, Sums of distances between points on a sphere an application of the the-ory of irregularities of distribution to discrete geometry. *Mathematika* **31**, 33–41, 1984.

[6]  J. Beck, On the discrepancy of convex plane sets. *Monatsh. Math.* **105**, 91–106, 1988.

[7]  J. Beck and W. W. L. Chen, *Irregularities of Distribution*. Cambridge Tracts in Mathematics, volume 89. Cambridge University Press, Cambridge, 2008.

[8]  D. Bilyk and M. T. Lacey, On the small ball inequality in three dimensions. *Duke Math. J.* **143**, 81–115, 2008.

[9]  D. Bilyk, M. T. Lacey and A. Vagharshakyan, On the small ball inequality in all dimensions. *J. Funct. Anal.* **254**, 2470–2502, 2008.

[10]  D. Bilyk, X. Ma, J. Pipher and C. Spencer, Directional discrepancy in two dimensions. *Bull. London Math. Soc.* **43**, 1151–1166, 2011.

[11]  J. S. Brauchart and J. Dick, A simple proof of Stolarsky's invariance principle. *Proc. Am. Math. Soc.* **141**, 2085–2096, 2013.

[12]  P. O. Chelson, Quasi-random techniques for Monte Carlo methods. PhD Disser-tation, Claremont Graduate School, Claremont, CA, 1976.

[13] W. W. L. Chen and M. M. Skriganov, Explicit constructions in the classical mean squares problem in irregularities of point distribution. *J. Reine Angew. Math.* **545**, 67–95, 2002.

[14] S. Chen, J. Dick and A. B. Owen, Consistency of Markov chain quasi-Monte Carlo on continuous state spaces. *Ann. Stat.* **39**, 673–701, 2011.

[15] L. Devroye, *Non-Uniform Random Variate Generation.* Springer, New York, 1986.

[16] J. Dick, Discrepancy bounds for infinite-dimensional order two digital sequences over $\mathbb{F}_2$. *J. Number Theory* **136**, 204–232, 2014.

[17] J. Dick and F. Pillichshammer, *Digital Nets and Sequences. Discrepancy Theory and Quasi-Monte Carlo Integration.* Cambridge University Press, Cambridge, 2010.

[18] J. Dick and F. Pillichshammer, Optimal $\mathcal{L}_2$ discrepancy bounds for higher order digital sequences over the finite field $\mathbb{F}_2$. *Acta Arith.* **162**, 65–99, 2014.

[19] J. Dick, D. Rudolf and H. Zhu, Discrepancy bounds for uniformly ergodic Markov chain quasi-Monte Carlo. Submitted, 2013.

[20] M. Drmota, Irregularities of distribution and convex sets. *Österreichisch-Ungarisch-Slowakisches Kolloquium über Zahlentheorie (Maria Trost, 1992).* Grazer Math. Ber., 318, pp. 9–16. Karl-Franzens-Univ. Graz, Graz, 1993.

[21] M. Drmota and R. F. Tichy, *Sequences, Discrepancies and Applications.* Lecture Notes in Mathematics, volume 1651. Springer-Verlag, Berlin, 1997.

[22] H. Faure, Discrépance de suites associées à un système de numération (en dimension *s*). *Acta Arith.* **41**, 337–351, 1982.

[23] P. J. Grabner and R. F. Tichy, Spherical designs, discrepancy and numerical integration. *Math. Comp.* **60**, 327–336, 1993.

[24] G. Halász, On Roth's method in the theory of irregularities of point distributions. *Recent Progress in Analytic Number Theory (Durham, 1979)*, volume 2, pp. 79–94. Academic Press, London, 1981.

[25] J. H. Halton, On the efficiency of certain quasi-random sequences of points in evaluating multi-dimensional integrals. *Numer. Math.* **2**, 84–90, 1960.

[26] J. M. Hammersley, Monte Carlo methods for solving multivariable problems. *Ann. N. Y. Acad. Sci.* **86**, 844–874, 1960.

[27] F. J. Hickernell, A generalized discrepancy and quadrature error bound. *Math. Comp.* **67**, 299–322, 1998.

[28] E. Hlawka, Funktionen von beschränkter Variation in der Theorie der Gleichverteilung (German). *Ann. Mat. Pura Appl.* **54**, 325–333, 1961.

[29] W. Hörmann, J. Leydold and G. Derflinger, Automatic nonuniform random variate generation. *Statistics and Computing.* Springer-Verlag, Berlin, 2004.

[30] J. F. Koksma, Een algemeene stelling uit de theorie der gelijkmatige verdeeling modulo 1, *Mathematica.* **11**, 7–11, 1942.

[31] L. Kuipers and H. Niederreiter, *Uniform Distribution of Sequences.* Dover, New York, 1974.

[32] F. Y. Kuo, W. T. M. Dunsmuir, I. H. Sloan, M. P. Wand and R. S. Womersley, Quasi-Monte Carlo for highly structured generalised response models. *Methodol. Comput. Appl. Probab.* **10**, 239–275, 2008.

[33] M. Laczkovich, Discrepancy estimates for sets with small boundary. *Stud. Sci. Math. Hung.* **30**, 105–109, 1995.

[34] P. L'Ecuyer, Ch. Lécot and A. L'Archevêque-Gaudet, On array-RQMC for Markov chains: mapping alternatives and convergence rates. In: P. L'Ecuyer and A. B. Owen (eds.), *Monte Carlo and quasi-Monte Carlo Methods 2008*, pp. 485–500. Springer, Berlin, 2009.

[35] M. B. Levin, Discrepancy estimates of completely uniformly distributed and pseudorandom number sequences. *Int. Math. Res. Notices*, 1231–1251, 1999.

[36] A. Lubotzky, R. Phillips and P. Sarnak, Hecke operators and distributing points on the sphere. I. Frontiers of the mathematical sciences 1985 (New York, 1985). *Commum. Pure Appl. Math.* **39**, S149–S186, 1986.

[37] A. Lubotzky, R. Phillips and P. Sarnak, Hecke operators and distributing points on $\mathbb{S}^2$. II. *Commum. Pure Appl. Math.* **40**, 401–420, 1987.

[38] J. Matoušek, *Geometric Discrepancy. An illustrated guide*. Algorithms and Combinatorics, volume 18. Springer-Verlag, Berlin, 1999.

[39] W. J. Morokoff and R. E. Caflisch, Quasi-Monte Carlo integration. *J. Comput. Phys.* **122**, 218–230, 1995.

[40] B. Moskowitz and R. E. Caflisch, Smoothness and dimension reduction in quasi-Monte Carlo methods. *Math. Comput. Modell.* **23**, 37–54, 1996.

[41] R. Mück and W. Philipp, Distances of probability measures and uniform distribution mod 1. *Math. Z.* **142**, 195–202, 1975.

[42] H. Niederreiter, Discrepancy and convex programming. *Ann. Mat. Pura Appl.* **93**, 89–97, 1972.

[43] H. Niederreiter, Methods for estimating discrepancy. *Applications of Number Theory to Numerical Analysis (Proc. Symp., Univ. Montreal, Montreal, 1971)*, pp. 203–236. Academic Press, New York, 1972.

[44] H. Niederreiter, Low-discrepancy and low-dispersion sequences. *J. Number Theory* **30**, 51–70, 1988.

[45] H. Niederreiter, *Random Number Generation and Quasi-Monte Carlo Methods*. CBMS-NSF Regional Conference Series in Applied Mathematics, volume 63. Society for Industrial and Applied Mathematics (SIAM), Philadelphia, PA, 1992.

[46] H. Niederreiter and J. M. Wills, Diskrepanz und Distanz von Maßen bezüglich konvexer und Jordanscher Mengen (German). *Math. Z.* **144**, 125–134, 1975.

[47] H. Niederreiter and C. P. Xing, Low-discrepancy sequences and global function fields with many rational places. *Finite Fields Appl.* **2**, 241–273, 1996.

[48] H. Niederreiter and C. P. Xing, *Rational Points on Curves over Finite Fields: Theory and Applications*. London Mathematical Society Lecture Note Series, volume 285. Cambridge University Press, Cambridge, 2001.

[49] C. Robert and G. Casella, *Monte Carlo Statistical Methods*, second edition. Springer-Verlag, New York, 2004.

[50] M. Rosenblatt, Remarks on a multivariate transformation. *Ann. Math. Stat.* **23**, 470–472, 1952.

[51] K. F. Roth, On irregularities of distribution. *Mathematika* **1**, 73–79, 1954.

[52] W. M. Schmidt, Irregularities of distribution IV. *Invent. Math.* **7**, 55–82, 1969.

[53] W. M. Schmidt, On irregularities of distribution IX. *Acta Arith.* **27**, 385–396, 1975.

[54] W. M. Schmidt, Irregularities of distribution X. *Number Theory and Algebra*, pp. 311–329. Academic Press, New York, 1977.

[55] I. E. Shparlinski, On a completely uniform distribution. *Comput. Math. Math. Phys.* **19**, 249–253, 1979.

[56] M. M. Skriganov, Harmonic analysis on totally disconnected groups and irregularities of point distributions. *J. Reine Angew. Math.* **600**, 25–49, 2006.

[57] I. H. Sloan and H. Woźniakowski, When are quasi-Monte Carlo algorithms efficient for high-dimensional integrals? *J. Complexity* **14**, 1–33, 1998.

[58] I. M. Sobol', Distribution of points in a cube and approximate evaluation of integrals (Russian). *Ž. Vyčisl. Mat. Mat. Fiz.* **7**, 784–802, 1967.

[59] J. Spanier and E. Maize, Quasi-random methods for estimating integrals using relatively small samples. *SIAM Rev.* **36**, 18–44, 1994.

[60] K. B. Stolarsky, Sums of distances between points on a sphere II. *Proc. Am. Math. Soc.* **41**, 575–582, 1973.

[61] W. Stute, Convergence rates for the isotrope discrepancy. *Ann. Probab.* **5**, 707–723, 1977.

[62] X. Wang, Improving the rejection sampling method in quasi-Monte Carlo methods. *J. Comput. Appl. Math.* **114**, 231–246, 2000.

[63] C. P. Xing and H. Niederreiter, A construction of low-discrepancy sequences using global function fields. *Acta Arith.* **73**, 87–102, 1995.

[64] S. C. Zaremba, La discrépance isotrope et l'intégration numérique. *Ann. Mat. Pura Appl.* **87**, 125–135, 1970.

[65] H. Zhu and J. Dick, A discrepancy bound for a deterministic acceptance-rejection sampler. *Electron. J. Stat.* **8**, 678–707, 2014.

# 4

# Discrepancy bounds for low-dimensional point sets

*Henri Faure*
Aix Marseille University, Marseille

*Peter Kritzer*
Johannes Kepler University Linz

*Dedicated to H. Niederreiter on the occasion of his 70th birthday.*

## Abstract

The class of $(t, m, s)$-nets and $(t, s)$-sequences, introduced in their most general form by Niederreiter, are important examples of point sets and sequences that are commonly used in quasi-Monte Carlo algorithms for integration and approximation. Low-dimensional versions of $(t, m, s)$-nets and $(t, s)$-sequences, such as Hammersley point sets and van der Corput sequences, form important subclasses, as they are interesting mathematical objects from a theoretical point of view, and simultaneously serve as examples that make it easier to understand the structural properties of $(t, m, s)$-nets and $(t, s)$-sequences in arbitrary dimension. For these reasons, a considerable number of papers have been written on the properties of low-dimensional nets and sequences.

In this paper, we summarize recent results on the distribution properties of low-dimensional examples of $(t, m, s)$-nets and $(t, s)$-sequences, state a new result regarding lower star discrepancy bounds, and formulate some open questions.

## 4.1 Introduction

In many applications of mathematics, such as in finance or computer graphics, one needs to approximate numerically the value of an integral $I_s(f) = \int_{[0,1]^s} f(x) \, dx$ of a function $f$ defined on $[0, 1]^s$. One way of dealing with this problem is to use a *quasi-Monte Carlo* (QMC) integration rule, which is an equal weight quadrature rule of the form $Q_{N,s}(f) := N^{-1} \sum_{n=0}^{N-1} f(x_n)$, where the integration nodes $x_0, \ldots, x_{N-1}$ are deterministically chosen points in the unit cube. It is well known in the theory of QMC methods that a useful property of the points $x_0, \ldots, x_{N-1}$ is that they are very evenly distributed in the integration domain $[0, 1]^s$. In this context, we frequently refer to the collection of the points $x_0, \ldots, x_{N-1}$ as a point set, by which we mean a multi-set, i.e., points may occur repeatedly. In general, the term "point set" also includes infinite sets, i.e., infinite sequences.

In order to measure the uniformity of distribution of a given point set $P$ with points $x_0, \ldots, x_{N-1}$ in $[0, 1)^s$, one frequently studies the star discrepancy $D^*$, which is defined by

$$D^*(N, P) = \sup_J |A(J, N, P) - N\lambda(J)| := \sup_J |\Delta(J, N, P)|,$$

where the supremum is extended over all intervals $J \subseteq [0, 1)^s$ of the form $J = \prod_{j=1}^{s}[0, \alpha_j)$, $0 < \alpha_j \leq 1$, $A(J, N, P)$ denotes the number of $i$ with $x_i \in J$, and $\lambda$ is the Lebesgue measure ($\Delta(J, N, P)$ is the so-called local discrepancy function).

In one dimension, where very precise results exist, we will also need the notion of (extreme) discrepancy $D$ obtained by taking the supremum of $|\Delta(J, N, P)|$ over all intervals $J$ (not necessarily anchored in the origin).

When considering an infinite sequence $S = (x_n)_{n \geq 0}$ of points in $[0, 1)^s$, we denote by $D^*(N, S)$ (respectively $D(N, S)$) the star discrepancy (respectively the discrepancy) of the first $N$ elements of $S$. In the case of a finite point set $P$ with $N$ points, we frequently write $D^*(P)$ (respectively $D(P)$) if there is no possible confusion regarding the number of points.

**Relation between sequences and finite point sets** A general principle (also valid in arbitrary dimension) states the link between one-dimensional sequences and two-dimensional point sets deduced from them [10, 34]. Let $S = (x_n)_{n \geq 0}$ be an infinite sequence taking its values in $[0, 1]$ and let $P$ be the two-dimensional point set $P = \left\{ \left(x_n, \frac{n}{N}\right) ; 0 \leq n < N \right\} \subset [0, 1]^2$. Then

$$\max_{1 \leq M \leq N} D^*(M, S) \leq D^*(N, P) \leq \max_{1 \leq M \leq N} D^*(M, S) + 1. \qquad (4.1)$$

These inequalities will be used to deduce results for two-dimensional point sets from results for one-dimensional sequences and vice versa, see Section 4.2.

The link between numerical integration and uniformly distributed point sets is provided by the Koksma–Hlawka inequality, which bounds the integration error of a QMC rule by means of the discrepancy of the node sets,

$$\left| I_s(f) - \frac{1}{N} \sum_{n=0}^{N-1} f(x_n) \right| \leq V(f) D^*(\{x_0, x_1, \ldots, x_{N-1}\})/N,$$

where $V(f)$ is the variation of $f$ in the sense of Hardy and Krause (see, e.g., [29] for further information). However, it should be noted that, considering the huge bounds for the discrepancy of low-discrepancy sequences in the usual ranges of $N$, this inequality is unsatisfactory and is not really meaningful for practical applications (see among others [19, 20]), at least for high dimensions $s$.

For overviews of QMC methods and their applications, uniform distribution of point sets, and their relations, we refer to the monographs [5, 6, 29, 32, 34, 43].

There are two families of low-discrepancy sequences widely used in QMC methods: Halton sequences [23] and their generalizations, and the so-called $(t, s)$-sequences. In this paper we only consider the second family along with $(t, m, s)$-nets, their associated finite point sets of cardinality $b^m$, which we now define in detail. The concepts of (digital) $(t, m, s)$-nets and (digital) $(t, s)$-sequences provide very efficient methods to construct point sets with small star discrepancy. These notions go back to ideas of Sobol' [44], Faure [9] and Niederreiter [33], and extensive information on this topic is presented by Niederreiter [34] (see also [5, 35] for a recent overview). We first give the general definition of a $(t, m, s)$-net.

**Definition 4.1** *Let $b \geq 2$, $s \geq 1$, and $0 \leq t \leq m$ be integers. Then a point set $P$ consisting of $b^m$ points in $[0, 1)^s$ forms a $(t, m, s)$-net in base $b$ if every subinterval $J = \prod_{j=1}^{s} [a_j b^{-d_j}, (a_j + 1)b^{-d_j})$ of $[0, 1)^s$, with integers $d_j \geq 0$ and integers $0 \leq a_j < b^{d_j}$ for $1 \leq j \leq s$ and of volume $b^{t-m}$, contains exactly $b^t$ points of $P$.*

Observe that a $(t, m, s)$-net is extremely well distributed if the quality parameter $t$ is small.

**Example 4.2** *A very special, though equally prominent example of a $(0, m, s)$-net in base $b$ is the Hammersley net $\mathcal{H}_{b,m}$ in dimension $s = 2$, consisting of $b^m$ points of the form*

$$\boldsymbol{x}_n = \left(\phi_b(n), \frac{n}{b^m}\right), \quad 0 \leq n \leq b^m - 1,$$

*where $\phi_b$ is the radical inverse function, defined as $\phi_b(n) = \sum_{k=0}^{m-1} \frac{n_k}{b^{k+1}}$ for an integer $n$ with base $b$ representation $n = n_0 + n_1 b + \cdots + n_{m-1} b^{m-1}$.*

Sometimes we do not consider all $(t, m, s)$-nets in their full generality, but restrict ourselves to studying a special construction, namely digital

$(t, m, s)$-nets over a finite field. To this end, as usual, for $b \geq 2$ we set $\mathbb{Z}_b := \mathbb{Z}/b\mathbb{Z}$ the residue class ring modulo $b$ equipped with addition and multiplication modulo $b$. To keep the notation simple, we shall sometimes associate the elements of $\mathbb{Z}_b$ with the set $\{0, 1, \ldots, b - 1\}$. Of course, if $p$ is prime $\mathbb{Z}_p$ is isomorphic to the field with $p$ elements and we do not explicitly distinguish between $\mathbb{Z}_p$ and this field.

For prime $p$, a digital $(t, m, s)$-net over $\mathbb{Z}_p$, which is a special type of $(t, m, s)$-net in base $p$, is defined as follows (for a more general definition of $(t, m, s)$-nets over commutative rings, see [34]).

For the construction of a digital $(t, m, s)$-net choose $s$ $(m \times m)$-matrices $C_1, \ldots, C_s$ over $\mathbb{Z}_p$ with the following property. For each choice of nonnegative integers $d_1, \ldots, d_s$ with $d_1 + \cdots + d_s = m - t$, the system of the

first $d_1$ rows of $C_1$ together with the
first $d_2$ rows of $C_2$ together with the
$$\vdots$$
first $d_s$ rows of $C_s$

is linearly independent over $\mathbb{Z}_p$. For a fixed $n \in \{0, \ldots, p^m - 1\}$, let $n$ have base $p$ representation $n = n_0 + n_1 p + \cdots + n_{m-1} p^{m-1}$. For $j \in \{1, \ldots, s\}$, multiply the matrix $C_j$ by the vector of digits of $n$ in $\mathbb{Z}_p^m$, which gives

$$C_j \cdot (n_0, \ldots, n_{m-1})^\top =: (y_1^{(j)}(n), \ldots, y_m^{(j)}(n))^\top \in \mathbb{Z}_p^m.$$

Then we set

$$x_n^{(j)} := \sum_{k=1}^{m} \frac{y_k^{(j)}(n)}{p^k}.$$

Finally, let $\boldsymbol{x}_n := (x_n^{(1)}, \ldots, x_n^{(s)})$. The point set consisting of the points $\boldsymbol{x}_0, \boldsymbol{x}_1, \ldots, \boldsymbol{x}_{p^m - 1}$ is called a digital $(t, m, s)$-net over $\mathbb{Z}_p$ with generating matrices $C_1, \ldots, C_s$.

Note that a digital $(t, m, s)$-net over $\mathbb{Z}_p$ is a $(t, m, s)$-net in base $p$ in the sense of Definition 4.1 (see [33]).

**Remark 4.3** Let $P$ be a digital $(t, m, s)$-net over $\mathbb{Z}_p$ with generating matrices $C_1, \ldots, C_s$, and let $D$ be a nonsingular $(m \times m)$-matrix over $\mathbb{Z}_p$. Then the digital net $Q$ over $\mathbb{Z}_p$ that is generated by $C_1 \cdot D, C_2 \cdot D, \ldots, C_s \cdot D$ is, up to the order of the points, the same point set as $P$, since multiplication of the generating matrices by $D$ from the right can be interpreted as a reordering of the indices $n \in \{0, \ldots, p^m - 1\}$ of the points. In particular, $D^*(P) = D^*(Q)$.

**Example 4.4** *If $b$ is prime, then $\mathcal{H}_{b,m}$ is a digital $(0, m, 2)$-net over $\mathbb{Z}_b$ with generating matrices*

$$
C_1 = \begin{pmatrix} 1 & 0 & \cdots & 0 \\ 0 & \ddots & \ddots & \vdots \\ \vdots & \ddots & \ddots & 0 \\ 0 & \cdots & 0 & 1 \end{pmatrix}, \quad C_2 = \begin{pmatrix} 0 & \cdots & 0 & 1 \\ \vdots & \cdot^{\cdot} & \cdot^{\cdot} & 0 \\ 0 & \cdot^{\cdot} & \cdot^{\cdot} & \vdots \\ 1 & 0 & \cdots & 0 \end{pmatrix}.
$$

The definition of a $(t, s)$-sequence is based on $(t, m, s)$-nets and is given in the following.

**Definition 4.5** *Let $b \geq 2$, $s \geq 1$, and $t \geq 0$ be integers. A sequence $(\boldsymbol{x}_n)_{n \geq 0}$ in $[0, 1)^s$ is a $(t, s)$-sequence in base $b$ if for all $l \geq 0$ and $m > t$ the point set consisting of the points $\boldsymbol{x}_{lb^m}, \ldots, \boldsymbol{x}_{(l+1)b^m - 1}$ is a $(t, m, s)$-net in base $b$.*

Again, a $(t, s)$-sequence is particularly well distributed if the quality parameter $t$ is small.

**Example 4.6** *Probably the best known example of a $(0, s)$-sequence in base $b$ is the van der Corput sequence $S_b^{id}$ in dimension $s = 1$, with points of the form*

$$
S_b^{id}(n) = (\phi_b(n)), \quad n \geq 0,
$$

*where the radical inverse function $\phi_b$ is defined as above (see Example 4.2).*

A digital $(t, s)$-sequence over $\mathbb{Z}_p$ ($p$ prime), which is a special type of $(t, s)$-sequence in base $p$ (cf. [34]), is constructed as follows. Choose $s$ ($\infty \times \infty$)-matrices $C_1, \ldots, C_s$ over $\mathbb{Z}_p$ such that for any $m > t$ the left upper ($m \times m$)-submatrices of $C_1, \ldots, C_s$ generate a digital $(t, m, s)$-net over $\mathbb{Z}_p$. For $n \geq 0$, let $n$ have base $p$ representation $n = n_0 + n_1 p + \cdots$. For $j \in \{1, \ldots, s\}$, multiply the matrix $C_j$ by the vector of digits of $n$, which yields

$$
C_j \cdot (n_0, n_1, \ldots)^{\top} =: (y_1^{(j)}(n), y_2^{(j)}(n), \ldots)^{\top},
$$

and set

$$
x_n^{(j)} := \sum_{k=1}^{\infty} \frac{y_k^{(j)}(n)}{p^k}.
$$

Then the sequence consisting of the points $x_0, x_1, \ldots$ with $x_n :=$ $(x_n^{(1)}, \ldots, x_n^{(s)})$ is called a digital $(t, s)$-sequence over $\mathbb{Z}_p$ and $C_1, \ldots, C_s$ are its generating matrices.

A technical requirement on a digital $(t, s)$-sequence is that, for each $n \geq 0$ and $1 \leq j \leq s$, we have $y_k^{(j)}(n) < p - 1$ for infinitely many $k$ (cf. [34]).

Later, in order to include new important constructions, Niederreiter and Xing [36, 37] and Tezuka [45] introduced a new definition of $(t, s)$-sequences in arbitrary base $b \geq 2$ using the so-called truncation operator.

**Definition 4.7** *Let $x = \sum_{i=1}^{\infty} x_i b^{-i}$ be a b-adic expansion of $x \in [0, 1]$, where it is allowed that $x_i = b - 1$ for all but finitely many $i$. For every integer $m \geq 1$, the m-truncation of $x$ is defined by $[x]_{b,m} = \sum_{i=1}^{m} x_i b^{-i}$ (depending on $x$ via its expansion). For $x \in I^s$, the notation $[x]_{b,m}$ means that m-truncation is applied to each coordinate of $x$.*

**Definition 4.8** *An s-dimensional sequence $(x_n)_{n\geq 0}$, with prescribed b-adic expansions for each coordinate, is a $(t, s)$-sequence in base $b$ (in the broad sense) if the subset $\{[x_n]_{b,m}; lb^m \leq n < (l + 1)b^m\}$ is a $(t, m, s)$-net in base $b$ for all integers $l \geq 0$ and $m > t$.*

The former $(t, s)$-sequences are now called $(t, s)$-sequences *in the narrow sense* and the latter (in the sense of Definition 4.8) simply $(t, s)$-sequences or sometimes $(t, s)$-sequences *in the broad sense* (cf. Niederreiter and Xing [37, Definition 2 and Remark 1]). Note that $(t, s)$-sequences in the narrow sense form a subclass of $(t, s)$-sequences in the broad sense. These new definitions will be used in the following for one-dimensional sequences.

**Example 4.9** *If $b$ is prime, then $S_b^{id}$ is a digital $(0, 1)$-sequence over $\mathbb{Z}_b$ with generating matrix*

$$
C_1 = \begin{pmatrix}
1 & 0 & 0 & 0 & \cdots \\
0 & 1 & 0 & \ddots & \\
0 & 0 & \ddots & \ddots & \\
0 & \ddots & \ddots & \ddots & \\
\vdots & & & &
\end{pmatrix}.
$$

The notion of point sets that are *digitally permuted* offers a generalization of the notions of nets and sequences. These are constructed by a basic variation in the point generating procedures outlined above. To be more precise, consider first the one-dimensional case.

Let $\mathfrak{S}_b$ be the set of all permutations of $\mathbb{Z}_b$. Choose a sequence $\Sigma = (\sigma_k)_{k \geq 0}$ of permutations $\sigma_k \in \mathfrak{S}_b$ and define the sequence $S_b^\Sigma$, namely the *generalized van der Corput sequence associated with* $\Sigma$ [8], by

$$S_b^\Sigma(n) := \sum_{k=0}^{\infty} \frac{\sigma_k(n_k)}{b^{k+1}},$$

for $n \geq 0$ with base $b$ representation $n = n_0 + n_1 b + \cdots$.

Notice that the sequences $S_b^\Sigma$ are $(0, 1)$-sequences in the broad sense [14, Proposition 3.1], the truncation from Definition 4.7 being required to prove this property when $\sigma_k(0) = b - 1$ for all sufficiently large $k$.

If, for all $k \geq 0$, $\sigma_k = \sigma$ is constant then we write $S_b^\Sigma = S_b^\sigma$. And if for all $k \geq 0$ we set $\sigma_k = \mathrm{id}$ (the identity), we recover the van der Corput sequence $S_b^{\mathrm{id}}$ in Example 4.6 above.

Now, in the $s$-dimensional case, choose $s$ sequences of permutations $\Sigma^{(1)}, \ldots, \Sigma^{(s)}$ and from an existing point set with points $x_n = (x_n^{(1)}, \ldots, x_n^{(s)})$, define the *digitally permuted point set* with points $\tilde{x}_n = (\tilde{x}_n^{(1)}, \ldots, \tilde{x}_n^{(s)})$, as

$$\tilde{x}_n^{(j)} := \sum_{k=1}^{\infty} \frac{\sigma_k^{(j)}(x_{n,k}^{(j)})}{b^k},$$

where the $x_{n,k}^{(j)}$ are the base $b$ digits of $x_n^{(j)}$.

If the permutations $\sigma$ are of the form $\sigma(x) = fx + g \pmod{b}$ with $f \neq 0$, we speak of a *linear digit scrambling* of the point set, and if $f = 1$, we speak of a *digital shift* $g$ of the point set. Linear digit scramblings are widely used in QMC methods to improve the distribution of point sets derived from Halton and $(t, s)$-sequences and hence can improve approximate computations of integrals (see among others [19, 20]).

Regarding the star discrepancy of $(t, m, s)$-nets and $(t, s)$-sequences, it is known from general results of Niederreiter [33] that for any $(t, m, s)$-net $P$ in base $b$

$$D^*(P) \leq b^t C(s, b) (\log b^m)^{s-1} + \mathcal{O}\left((\log b^m)^{s-2}\right), \qquad (4.2)$$

with $C(s, b)$ and the implied constant in the $\mathcal{O}$-notation independent of $b^m$. Furthermore, for the star discrepancy of the first $N \geq b^t$ points of a $(t, s)$-sequence $S$ in base $b$,

$$D^*(N, S) \leq b^t D(s, b) (\log N)^s + \mathcal{O}\left((\log N)^{s-1}\right), \qquad (4.3)$$

with $D(s, b)$ and the implied constant in the $\mathcal{O}$-notation independent of $N$. Values of $C(s, b)$ and $D(s, b)$ for which (4.2) and (4.3) hold were given explicitly in [33], and later improved on in [18, 21, 26].

Concerning lower discrepancy bounds, a famous theorem of Schmidt [42], improved further with respect to the constant by Béjian [1], states that for any sequence $S$ in [0, 1) and infinitely many $N$,

$$D(N, S) \geq 0.12 \log N,$$

hence achieving the exact order in $N$ for the discrepancy of sequences in one dimension. A similar result of Béjian for the star discrepancy, which states that $D^*(N, S) \geq 0.06 \log N$, was recently improved by Larcher [30]. Furthermore, from [29, Example 2.2], it follows that for any two-dimensional point set $P$ of $N$ points, $D^*(N, P) \geq 0.03 \log N$ (see also [29, Corollary 2.2]).

For $s \geq 3$, another famous theorem due to Roth [41] was recently improved [2] to

$$D^*(N, P) \geq c(s)(\log N)^{\frac{s-1}{2} + \delta_s},$$

where $c(s)$ is a constant only depending on $s$ and not on the point set or its cardinality, and where $\delta_s$ is an unknown constant in [0, 1/2).

In this paper, we review recent discrepancy estimates for low-dimensional $(t, m, s)$-nets and $(t, s)$-sequences (i.e., for $s = 1$ and $s = 2$). The reason why low-dimensional examples of nets and sequences have attracted much interest, particularly in the field of uniform distribution theory, is that special instances (for example, the Hammersley net or the van der Corput sequence) have nice mathematical properties, and that a sound understanding of low-dimensional point sets helps in dealing with those point sets in arbitrary dimension.

The rest of the paper is structured as follows. There are three main sections, Section 4.2 on upper discrepancy bounds for low-dimensional $(t, s)$-sequences and related sequences, Section 4.3 on upper discrepancy bounds for low-dimensional $(t, m, s)$-nets, and Section 4.4 on lower discrepancy bounds for both low-dimensional infinite sequences and low-dimensional finite point sets.

Section 4.2 is divided into several parts. After an introductory part, a few preliminary remarks, and a section on general upper bounds for one-dimensional sequences (Section 4.2.3), Sections 4.2.4 and 4.2.5 focus on special classes of one-dimensional sequences and contain very precise results on the discrepancy of these sequences. We conclude Section 4.2 by some remarks on $(t, 1)$-sequences (Section 4.2.6) and two-dimensional sequences (Section 4.2.7).

Section 4.3 is organized as follows. In an introductory section we review selected earlier results, and then discuss three recent streams of research: one uses Walsh functions to derive discrepancy bounds for certain two-dimensional nets (Section 4.3.2), another uses counting arguments to derive upper discrepancy bounds (Section 4.3.3), and the third relates suitably chosen one-dimensional sequences to generalized Hammersley nets (Section 4.3.4).

Finally, Section 4.4 contains one section reviewing lower discrepancy bounds for nets (Section 4.4.1), a section where we derive a new result (Section 4.4.2), and a section discussing lower discrepancy bounds for $(t, s)$-sequences (Section 4.4.3).

We conclude the paper in Section 4.5, where we also state some open questions.

## 4.2 Upper discrepancy bounds for low-dimensional sequences

### 4.2.1 Introductory remarks

We leave aside the family of $(n\alpha)$ sequences (see, e.g., [6, 29]), the other great family of one-dimensional low-discrepancy sequences, and focus on $(0, 1)$-sequences with a short insight into $(t, 1)$-sequences (Section 4.2.6), in relation to $(t, 2)$-sequences (Section 4.2.7).

The family of $(0, 1)$-sequences (in the broad sense) contains two large sub-families as shown in [14, Proposition 3.1]: the family of $S_b^\Sigma$ sequences introduced in [8] and the family of digital $(0, 1)$-sequences in (prime, for simplicity) base $b$, denoted by $X_b^C$ in the following. In this case, we assume that the generating matrix $C$ has the property that for any integer $m \geq 1$ every left upper $(m \times m)$-submatrix is nonsingular. Note that such $X_b^C$ sequences can require the truncation operator (see Definition 4.7) since we may have the digits $y_k(n) = b - 1$ for all but finitely many $k$ in the definition. An important special case is the case of nonsingular upper triangular (NUT) matrices $C$ for which the summation over $k$ in the definition is finite, so that these (so-called) *NUT digital $(0, 1)$-sequences* do not need the truncation.

Quite recently, Faure and Pillichshammer [22] introduced a mixed construction containing both the above families. Such sequences are denoted $X_b^{\Sigma,C}$ and called *NUT $(0, 1)$-sequences* over $\mathbb{Z}_b$ where $b \geq 2$ is an arbitrary base. They are obtained by putting arbitrary permutations from the sequence $\Sigma$ in place of the diagonal entries of the NUT matrix $C$. More precisely we have the following definition.

**Definition 4.10** *For any integer $b \geq 2$, let $\Sigma = (\sigma_k)_{k \geq 0}$ be a sequence of permutations $\sigma_k \in \mathfrak{S}_b$ and let $C = (c_r^k)_{r \geq 0, k \geq r+1}$ be a strict upper triangular matrix with entries in $\mathbb{Z}_b$ (i.e., an upper triangular matrix with zero diagonal* *entries). Then, for all integers $n \geq 0$, the nth element of the sequence $X_b^{\Sigma, C}$ is defined by*

$$X_b^{\Sigma, C}(n) = \sum_{r=1}^{\infty} \frac{x^{(r)}(n)}{b^r} \quad \text{where} \quad x^{(r)}(n) = \sigma_r(n_r) + \sum_{k=r+1}^{\infty} c_r^k n_k \pmod{b},$$

*in which n has base b representation with digits $n_k$ as in Example 4.2.*

If all entries above the diagonal of $C$ are zero, we recover $S_b^{\Sigma}$ sequences. If the permutations $\sigma_k$ are linear digit scramblings with shift $g = 0$, we recover classical NUT digital $(0, 1)$-sequences (with arbitrary base $b$). Sequences $S_b^{\Sigma}$ and sequences $X_b^C$ have been extensively studied [8, 12] but, since their generalization leads formally to the same formulas, we will only present these formulas for the case of sequences $X_b^{\Sigma, C}$.

### 4.2.2 Prerequisites

We first introduce two more notions of discrepancy in one dimension:

$$D^+(N, X) = \sup_{0 \leq \alpha \leq 1} \Delta([0, \alpha), N, X)$$

$$D^-(N, X) = \sup_{0 \leq \alpha \leq 1} (-\Delta([0, \alpha), N, X)).$$

The discrepancies $D^+$ and $D^-$ are linked to $D$ and $D^*$ by the relations

$$D(N, X) = D^+(N, X) + D^-(N, X)$$

$$D^*(N, X) = \max(D^+(N, X), D^-(N, X)).$$

Then we need to define the so-called $\varphi$-functions first introduced in the study of van der Corput sequences [8]. For any $\sigma \in \mathfrak{S}_b$ (the set of all permutations of $\mathbb{Z}_b$), set

$$\mathcal{Z}_b^{\sigma} = (\sigma(0)/b, \sigma(1)/b, \ldots, \sigma(b-1)/b).$$

For $h \in \{0, 1, \ldots, b-1\}$ and $x \in [(k-1)/b, k/b)$ where $k \in \{1, \ldots, b\}$, define

$$\varphi_{b,h}^{\sigma}(x) = \begin{cases} A([0, h/b); k; \mathcal{Z}_b^{\sigma}) - hx & \text{if } 0 \leq h \leq \sigma(k-1) \\ (b-h)x - A([h/b, 1); k; \mathcal{Z}_b^{\sigma}) & \text{if } \sigma(k-1) < h < b. \end{cases}$$

Further, the functions $\varphi_{b,h}^{\sigma}$ are extended to the reals by periodicity. Based on $\varphi_{b,h}^{\sigma}$ we now define

$$\psi_b^{\sigma,+} = \max_{0 \leq h < b} \varphi_{b,h}^{\sigma}, \quad \psi_b^{\sigma,-} = \max_{0 \leq h < b} (-\varphi_{b,h}^{\sigma}) \quad \text{and} \quad \psi_b^{\sigma} = \psi_b^{\sigma,+} + \psi_b^{\sigma,-},$$

which appear in the formulas for the discrepancies $D^+$, $D^-$ and $D$.

Moreover, we need further definitions to deal with (digital or not) NUT $(0, 1)$-sequences. The symbol $\uplus$ is used to denote the translation (or shift) of a given permutation $\sigma \in \mathfrak{S}_b$ by an element $t \in \mathbb{Z}_b$ in the following sense: $(\sigma \uplus t)(i) := \sigma(i) + t \pmod{b}$ for all $i \in \mathbb{Z}_b$, and for any integer $r \geq 0$ we introduce the quantity

$$\theta_r(N) := \sum_{k=r+1}^{\infty} c_r^k a_k(N) \pmod{b},$$

where the $c_r^k$ are the entries of the matrix $C$ and the $a_k(N)$ are the digits of $N-1$ in its $b$-adic expansion. Note that $a_k(N) = 0$ for all $k \geq n$ if $1 \leq N \leq b^n$, thus $\theta_r(N) = 0$ for all $r \geq n-1$ in this case. This quantity determines the translated permutations that appear in the formulas for $D^+$, $D^-$ in Theorem 4.15.

Lastly, for any $r \geq 0$ we define the permutation $\delta_r$ by $\delta_r(i) := c_r^r i \pmod{b}$ for all $i \in \mathbb{Z}_b$ and the swapping permutation $\tau(i) = b - 1 - i = (b-1)i + b - 1 \pmod{b}$ for all $i \in \mathbb{Z}_b$. The name of the permutation $\tau$ comes from the fact that $\tau$ swaps the functions $\psi_b^{\sigma,+}$ and $\psi_b^{\sigma,-}$ and hence is useful to minimize $D^* = \max(D^+, D^-)$ in the asymptotic behavior of the star discrepancy (see below).

### 4.2.3 A general upper bound and two counter-examples

It is known [8, 13] that the original van der Corput sequences are the worst distributed with respect to the star discrepancy, to the $L_2$ discrepancy and to the diaphony among all $S_b^\Sigma$ sequences and among all NUT digital sequences $X_b^C$. This is also true for all $(0, 1)$-sequences (in the broad sense, see Definition 4.8) according to Theorem 4.11 below. However, and surprisingly, this property is no longer true for the discrepancy $D$ among all $(0, 1)$-sequences (in the broad sense) according to our two counter-examples below.

**Theorem 4.11** (Faure, Kritzer) *The original van der Corput sequences in arbitrary base $b \geq 2$ are the worst distributed with respect to the star discrepancy among all $(0, 1)$-sequences $X_b$ (in the broad sense), that is,*

$$D^*(N, X_b) \leq D^*(N, S_b^{id}) = D(N, S_b^{id}).$$

**Remark 4.12** The main idea of the proof was first used by Dick and Kritzer [4] in the context of two-dimensional Hammersley point sets. Then Kritzer [25] proved Theorem 4.11 for $(0, 1)$-sequences in the narrow sense using the result for Hammersley point sets. Finally, Faure [14, Theorem 5.1] proved the theorem "in the broad sense" using the sequence $S_b^\tau$, whose functions $\psi_b^{\tau,+}$ and

$\psi_b^{\tau,-}$ are exchanged with functions $\psi_b^{\mathrm{id},-}$ and $\psi_b^{\mathrm{id},+}$ associated with $S_b^{\mathrm{id}}$. The good control of discrepancy by means of $\psi$-functions allows a shorter proof. In the broad sense, we can say that there are two worst sequences with respect to $D^*$, namely $S_b^{\mathrm{id}}$ and $S_b^{\tau}$, while in the narrow sense there is only one, namely $S_b^{\mathrm{id}}$, since $S_b^{\tau}$ is not a sequence in the narrow sense.

Now, we give the first counter-example where we show that the original van der Corput sequence in base 2 is not the worst distributed sequence with respect to $D$ among digital $(0, 1)$-sequences in base 2 [14, Theorem 5.2].

**Theorem 4.13** (Faure) *Let $C_0$ be the generating matrix in base 2 for which all entries are zero except on the diagonal and in the first column (where they are equal to 1). Then, for any $N \geq 1$,*

$$2D(N, S_2^{id}) - \frac{5}{2} \leq D(N, X_2^{C_0}) \leq 2D(N, S_2^{id}),$$

$$D^*(N, S_2^{id}) - \frac{3}{2} \leq D^*(N, X_2^{C_0}) \leq D^*(N, S_2^{id}) = D(N, S_2^{id}).$$

*Moreover the sequence $X_2^{C_0}$ is the worst distributed among all $(0, 1)$-sequences in base 2 (in the broad sense) with respect to $D$.*

For base $b \geq 3$, we have not found a digital $(0, 1)$-sequence with discrepancy $D$ twice greater than that of $S_b^{\mathrm{id}}$. The study of the sequence $X_b^{C_0}$ is, in this case, more complicated and does not give the factor 2 as for $b = 2$. However, we have a simple construction inspired by the proof of Theorem 4.13 which gives the same result [14, Theorem 5.3].

**Theorem 4.14** (Faure) *Let $b \geq 3$ be an integer. Let us define the sequence $X_b^{id\tau} = (x_n)_{n \geq 0}$ by $x_{bk} = S_b^{id}(bk)$, $x_{bk+1} = S_b^{\tau}(bk)$ and $x_{bk+l} = S_b^{id}(bk + l - 1)$ if $2 \leq l \leq b - 1$, for all $k \geq 0$. Then, the sequence $X_b^{id\tau}$ is a $(0, 1)$-sequence (not digital and not in the narrow sense), for which*

$$2D(N, S_b^{id}) - 2(b - 1) \leq D(N, X_b^{id\tau}) \leq 2D(N, S_b^{id}),$$

$$D^*(N, S_b^{id}) - (b - 1) \leq D^*(N, X_b^{id\tau}) \leq D^*(N, S_b^{id}) = D(N, S_b^{id}).$$

*Moreover the sequence $X_b^{id\tau}$ is the worst distributed among all $(0, 1)$-sequences in base $b$ (in the broad sense) with respect to $D$.*

### 4.2.4 Exact formulas for the discrepancies of NUT $(0, 1)$-sequences

We now turn to results on NUT $(0, 1)$-sequences $X_b^{\Sigma, C}$ as defined in Section 4.2.1, Definition 4.10. Here, we restrict ourselves to the discrepancies $D^+$, $D^-$ (and so $D^*$) and $D$. Similar formulas exist for the $L_2$ discrepancy and the diaphony (see [22]).

**Theorem 4.15** (Faure, Pillichshammer) *With the notation introduced in Sections 4.2.1 and 4.2.2, we have, for all integers $b \geq 2$ and $N \geq 1$,*

$$D^+(N, X_b^{\Sigma, C}) = \sum_{j=1}^{\infty} \psi_b^{\sigma_{j-1} \uplus \theta_{j-1}(N), +} \left( \frac{N}{b^j} \right),$$

$$D^-(N, X_b^{\Sigma, C}) = \sum_{j=1}^{\infty} \psi_b^{\sigma_{j-1} \uplus \theta_{j-1}(N), -} \left( \frac{N}{b^j} \right),$$

$$D(N, X_b^{\Sigma, C}) = \sum_{j=1}^{\infty} \psi_b^{\sigma_{j-1}} \left( \frac{N}{b^j} \right).$$

For sequences $X_b^{\Sigma, C}$, the formula for $D$ depends only on the permutations $\sigma_r$ and not on the entries (above the diagonal) of $C$. For sequences $X_b^C$ in [13] the formula for $D$ depends only on the permutations $\delta_r$ (associated with the diagonal entries of $C$) and not on the entries above the diagonal of $C$. This remarkable feature shows that NUT $(0,1)$-sequences having the same sequence of permutations $\Sigma$ (or $\Delta = (\delta_r)_{r \geq 0}$ for $X_b^C$ sequences) have the same extreme discrepancy $D$. In this case, studies on the asymptotic behavior of $D$ for generalized van der Corput sequences $S_b^{\Sigma}$ (see [8, 11, 15]) apply, especially to NUT digital $(0,1)$-sequences. The same remark is also valid for the diaphony, see [13, 22].

On the other hand, the formulas for $D^+$, $D^-$, and hence for $D^*$, involve the quantity $\theta_{j-1}(N)$ which depends on $N$ via its $b$-adic expansion and on the generating NUT matrix $C$ via its entries above the diagonal; this dependence is a big handicap for the precise study of the asymptotic behavior of the star discrepancy of NUT $(0,1)$-sequences. As far as we know, the only result available is that of Pillichshammer for $X_2^C$ [40], where $C$ is the matrix for which all entries are equal to 1. Indeed, Pillichshammer showed the following result, which can be deduced from the study of digital $(0, m, 2)$-nets in base 2 by Larcher and Pillichshammer [31] (see Section 4.3.2).

For an infinite sequence $X$ in $[0, 1)$, set

$$\rho(X) := \limsup_{N \to \infty} \frac{D(N, X)}{\log N} \quad \text{and} \quad \rho^*(X) := \limsup_{N \to \infty} \frac{D^*(N, X)}{\log N}.$$

The result of Pillichshammer is as follows.

**Theorem 4.16** (Pillichshammer) *For the star discrepancy of the sequence $X_2^C$ described above it is true that*

$$0.2885\ldots = \frac{1}{5 \log 2} \leq \rho^*(X_2^C) \leq \frac{5099}{22528 \log 2} = 0.3265\ldots.$$

### 4.2.5 Upper bounds for one-dimensional low-discrepancy sequences

**Reminders on the asymptotic behavior of $S_b^\Sigma$ sequences** We only recall here two main theorems going back to 1981 useful for our short review of results on upper bounds for one-dimensional low-discrepancy sequences [8, Théorèmes 2 and 3].

**Theorem 4.17** (Faure) *Let $\sigma \in \mathfrak{S}_b$ and let $\Sigma = (\sigma)$ be constant (so that $S_b^\Sigma = S_b^\sigma$). Then*

$$\rho(S_b^\sigma) = \frac{\alpha_b^\sigma}{2 \log b}$$

*where*

$$\alpha_b^\sigma = \inf_{n \geq 1} \frac{1}{n} \sup_{x \in [0,1]} \sum_{j=1}^{n} \psi_b^\sigma \left( \frac{x}{b^j} \right),$$

$$D(N, S_b^\sigma) \leq \frac{\alpha_b^\sigma}{\log b} \log N + \alpha_b^\sigma + 2 \quad \text{for all } N \geq 1.$$

The previous result concerns the extreme discrepancy. We now turn to the star discrepancy, where best results are obtained with special sequences of permutations $\Sigma$ using the swapping permutation $\tau$ introduced at the end of Section 4.2.2 and recalled below.

**Theorem 4.18** (Faure) *Let $\tau \in \mathfrak{S}_b$ be the permutation defined by $\tau(k) = b-k-1$ for all $k \in \mathbb{Z}_b$ and let $\mathcal{A}$ be the subset of $\mathbb{N}_0$ defined by $\mathcal{A} = \bigcup_{H=1}^{\infty} \mathcal{A}_H$*

with $\mathcal{A}_H = \{H(H-1), \ldots, H^2 - 1\}$. *For any permutation* $\sigma \in \mathfrak{S}_b$, *let* $\overline{\sigma} := \tau \circ \sigma$ *and let*

$$\Sigma_{\mathcal{A}}^{\sigma} = (\sigma_r)_{r \geq 0} := (\sigma, \overline{\sigma}, \sigma, \sigma, \overline{\sigma}, \overline{\sigma}, \sigma, \sigma, \sigma, \overline{\sigma}, \overline{\sigma}, \overline{\sigma}, \ldots)$$

*be the sequence of permutations defined by* $\sigma_r = \sigma$ *if* $r \in \mathcal{A}$ *and* $\sigma_r = \overline{\sigma}$ *if* $r \notin \mathcal{A}$. *Then*

$$\rho^*(S_b^{\Sigma_{\mathcal{A}}^{\sigma}}) = \frac{\alpha_b^{\sigma,+} + \alpha_b^{\sigma,-}}{2 \log b}, \tag{4.4}$$

*where*

$$\alpha_b^{\sigma,+} = \inf_{n \geq 1} \frac{1}{n} \sup_{x \in [0,1]} \sum_{j=1}^n \psi_b^{\sigma,+}\left(\frac{x}{b^j}\right),$$

$$\alpha_b^{\sigma,-} = \inf_{n \geq 1} \frac{1}{n} \sup_{x \in [0,1]} \sum_{j=1}^n \psi_b^{\sigma,-}\left(\frac{x}{b^j}\right).$$

**Updated review of results on one-dimensional low discrepancy sequences**
For reasonably small $b$, the constants $\alpha_b^{\sigma,+}$ and $\alpha_b^{\sigma,-}$ are not difficult to compute and for the identity id, in which case $\psi_b^{\mathrm{id},-} = 0$, it is even possible to find them explicitly. We have

$$\frac{\alpha_b^{\mathrm{id}}}{\log b} = \frac{b-1}{4 \log b} \qquad \text{if } b \text{ is odd,}$$

$$\frac{\alpha_b^{\mathrm{id}}}{\log b} = \frac{b^2}{4(b+1) \log b} \qquad \text{if } b \text{ is even.}$$

These constants are the worst possible leading constants for the discrepancies $D$ and $D^*$ of $S_b^{\Sigma}$ sequences since for any sequence of permutations $\Sigma$, we have [8, Section 5.5.4]

$$D(N, S_b^{\Sigma}) \leq D(N, S_b^{\mathrm{id}}) = D^*(N, S_b^{\mathrm{id}}) \qquad \text{for all } N \geq 1$$

(i.e., the original van der Corput sequence $S_b^{\mathrm{id}}$ is the worst distributed among the $S_b^{\Sigma}$ sequences with respect to the discrepancy and the star discrepancy).
  Concerning the sequences in Theorem 4.18, we obtain

$$\rho(S_b^{\Sigma_{\mathcal{A}}^{\mathrm{id}}}) = \rho^*(S_b^{\Sigma_{\mathcal{A}}^{\mathrm{id}}}) = \frac{\alpha_b^{I,+}}{2 \log b} = \frac{b-1}{8 \log b} \qquad \text{if } b \text{ is odd,}$$

$$\rho(S_b^{\Sigma_{\mathcal{A}}^{\mathrm{id}}}) = \rho^*(S_b^{\Sigma_{\mathcal{A}}^{\mathrm{id}}}) = \frac{\alpha_b^{I,+}}{2 \log b} = \frac{b^2}{8(b+1) \log b} \qquad \text{if } b \text{ is even.}$$

In these formulas, base $b = 2$ is interesting: in this case the swapping permutation $\tau$ reads as $\tau(k) = k + 1 \pmod 2$ and so $\tau$ is the digital shift(mod 2). Hence in 2007, Kritzer *et al.* [28, Section 5] rediscovered Theorem 4.18 in the special case of base 2 and the corresponding constant $1/(6 \log 2)$ above.

Notice that in general we have $\alpha_b^\sigma \le \alpha_b^{\sigma,+} + \alpha_b^{\sigma,-}$. Hence we cannot deduce any relation between a general lower bound on $\rho^*(S_b^{\Sigma^\sigma_S})$, where $S$ is an arbitrary subset of $\mathbb{N}_0$, and the upper bound on $\rho^*(S_b^{\Sigma^\sigma_A})$ from (4.4). Only if for the permutation $\sigma$ we have either $\psi_b^{\sigma,+} = 0$ or $\psi_b^{\sigma,-} = 0$, in which case we get $\alpha_b^\sigma = \alpha_b^{\sigma,+} + \alpha_b^{\sigma,-}$, do we obtain that $\alpha_b^\sigma/(2\log b)$ is the best possible lower bound for the star discrepancy of sequences $S_b^\Sigma$ with $\Sigma \in \{\sigma, \overline{\sigma}\}^{\mathbb{N}_0}$. In other words, we have

$$\inf_{\Sigma \in \{\sigma, \tau \circ \sigma\}^{\mathbb{N}_0}} \rho^*(S_b^\Sigma) = \frac{\alpha_b^\sigma}{2\log b}$$

for any $\sigma \in \mathfrak{S}_b$ such that $D^*(S_b^\sigma) = D(S_b^\sigma)$. Recall that for any $\Sigma = (\sigma_r)_{r \ge 0}$ we have $D^*(S_b^\Sigma) = D(S_b^\Sigma)$ if and only if $\psi_b^{\sigma_r,+} = 0$ for all $r \ge 0$ or $\psi_b^{\sigma_r,-} = 0$ for all $r \ge 0$. See [8, Corollaire 2, p. 160]) and [22, Section 5.2] for more details.

The smallest extreme discrepancies currently known, a new record, were obtained recently by Ostromoukhov [38] in bases 84 and 60, after a lot of computations using the method worked out in [8, Section 5]. There exist permutations $\sigma_0 \in \mathfrak{S}_{84}$ and $\sigma_1 \in \mathfrak{S}_{60}$ such that

$$\rho(S_{84}^{\Sigma_A^{\sigma_0}}) = \frac{130}{83\log 84} = 0.3534\ldots$$

$$\rho^*(S_{60}^{\Sigma_A^{\sigma_1}}) = \frac{32209}{35400\log 60} = 0.2222\ldots,$$

improving preceding results in [11, Theorem 1.2] and [8, Théorème 5] with permutations $\sigma_2 \in \mathfrak{S}_{36}$ and $\sigma_3 \in \mathfrak{S}_{12}$ that give $\rho(S_{36}^{\sigma_2}) = 0.3667\ldots$ and $\rho^*(S_{12}^{\Sigma_A^{\sigma_3}}) = 0.2235\ldots$. It is interesting to note that $\psi_{12}^{\sigma_1,-} \ne 0$ whereas $\psi_{60}^{\sigma_0,-} = 0$. This last property is quite remarkable with regard to the multitude of permutations involved in the computational search for $(60, \sigma_0)$ among all pairs $(b, \sigma)$.

### 4.2.6 A general upper bound for $(t, 1)$-sequences

As announced at the beginning of Section 4.2.1, we now turn to a slight generalization of Theorem 4.11, see [21, Section 2].

**Theorem 4.19** (Faure, Lemieux)  *For any base $b$, the original van der Corput sequences are the worst distributed with respect to the star discrepancy among all $(t, 1)$-sequences $X_b^t$ (in the broad sense), i.e., for all integers $N \geq 1$,*

$$D^*(b^t N, X_b^t) \leq b^t D^*(N, S_b^{id}) = b^t D(N, S_b^{id}).$$

As a corollary, we obtain that for any $(t, 1)$-sequence $X_b^t$ (in the broad sense) and for any integer $N \geq 1$,

$$D^*(b^t N, X_b^t) \leq b^t \left( \frac{b-1}{4 \log b} \log N + \frac{b-1}{4} + 2 \right) \qquad \text{if } b \text{ is odd,}$$

$$D^*(b^t N, X_b^t) \leq b^t \left( \frac{b^2}{4(b+1) \log b} \log N + \frac{b^2}{4(b+1)} + 2 \right) \qquad \text{if } b \text{ is even.}$$

These upper bounds in one dimension complete the upper bounds in dimension $s \geq 2$ recently obtained by Faure and Kritzer, see Section 4.2.7 below.

### 4.2.7  Two-dimensional sequences

Regarding two-dimensional sequences, the currently best known general upper discrepancy bound was first shown for digital sequences in base 2 in [39], and then for arbitrary $(t, 2)$-sequences in [4].

**Theorem 4.20** (Dick, Kritzer)  *For the star discrepancy of the first $N$ points of an arbitrary $(t, 2)$-sequence $X_b^t$ in base $b$, it is true that*

$$D^*(N, X_b^t) \leq \begin{cases} \frac{b^t}{16} \frac{b^2(b-1)^2}{(b^2-1)(\log b)^2} (\log N)^2 + \mathcal{O}(\log N) & \text{if } b \text{ is even} \\[2ex] \frac{b^t}{16} \frac{(b-1)^2}{(\log b)^2} (\log N)^2 + \mathcal{O}(\log N) & \text{if } b \text{ is odd,} \end{cases}$$

*where the implied constants in the $\mathcal{O}$-notation do not depend on $N$.*

**Remark 4.21**  Theorem 4.20 has just been generalized to arbitrary dimensions $s \geq 2$ by Faure and Kritzer [18, Theorem 2 and Corollary 2]. The proofs follow the approach which consists first in studying $(t, m, s)$-nets for which formulas are proved with the help of a recursion based on preliminary counting lemmas, then going from $(t, m, s)$-nets to results for $(t, s)$-sequences from a classical relation already used by Sobol', Faure and Niederreiter (see [33, Lemma 4.1]). There is another approach dealing directly with $(t, s)$-sequences and using an adaptation of Atanassov's method for Halton sequences [21], but the bounds obtained in this way are greater by a factor of about 2 (even though the constants hidden in the $\mathcal{O}$-notation are smaller in this approach). Another interesting remark is that Corollary 2 in [18] and Theorem 4.20 are also true in one dimension thanks to Theorem 4.19.

## 4.3 Upper discrepancy bounds for low-dimensional nets

Let us now turn to upper discrepancy bounds for low-dimensional finite point sets.

### 4.3.1 Introductory remarks

For the case $s = 1$, there exists a closed formula for the discrepancy of a finite point set in $[0, 1)$, so one usually does not explicitly study the discrepancy of $(t, m, 1)$-nets. Indeed, assume that $x_0, \ldots, x_{N-1} \in [0, 1)$ such that, without loss of generality, $x_0 \le x_1 \le \cdots \le x_{N-1}$. Then

$$D^*(\{x_0, \ldots, x_{N-1}\}) = \frac{1}{2} + N \max_{0 \le n \le N-1} \left| x_n - \frac{2n+1}{2N} \right|. \tag{4.5}$$

Due to this formula, the derivation of discrepancy bounds for finite one-dimensional point sets is obsolete. However, if one wants to study infinite point sets $S$ in $[0, 1)$ and derive bounds on $D^*(N, S)$ that hold for all or at least infinitely many $N$, this is still a nontrivial problem, see Section 4.2.

Upper discrepancy bounds for $(t, m, s)$-nets were, in a very general form, shown in [33, 34], where the formula (4.2) was proved. In particular, Niederreiter also presented discrepancy bounds for dimension $s = 2$ [33, 34], as follows.

**Theorem 4.22** (Niederreiter) *The star discrepancy of a $(t, m, 2)$-net $P$ in base $b$ satisfies*

$$D^*(P) \le \left\lfloor \frac{b-1}{2}(m-t) + \frac{3}{2} \right\rfloor b^t. \tag{4.6}$$

Regarding further, special discrepancy bounds for $(t, m, 2)$-nets, there exist results that can be grouped into two larger streams of research. On the one hand, approaches by Faure used explicit formulas for the local discrepancy function $\Delta$, see [16, Lemma 3], which make it possible to deduce discrepancy bounds for low-dimensional point sets from the study of low-dimensional sequences. On the other hand, initiated by Larcher and Pillichshammer in 2003 [31], a series of papers dealt with tightening discrepancy bounds for low-dimensional nets by using Walsh series.

### 4.3.2  Using Walsh functions to obtain bounds for $(0, m, 2)$-nets in base 2

Regarding the results of Larcher and Pillichshammer [31], these are relevant for parts of the remainder of this paper (see Theorem 4.30 in Section 4.4.2), so let us briefly recall them here. In [31], the discrepancy of digital $(0, m, 2)$-nets over $\mathbb{Z}_2$ was analyzed by the means of Walsh functions. Indeed, let $P$ be a digital $(0, m, 2)$-net over $\mathbb{Z}_2$. Without loss of generality (see Remark 4.3), we assume that $P$ is generated by matrices

$$
C_1 = \begin{pmatrix} 1 & 0 & \cdots & 0 \\ 0 & \ddots & \ddots & \vdots \\ \vdots & \ddots & \ddots & 0 \\ 0 & \cdots & 0 & 1 \end{pmatrix}, \quad
C_2 = \begin{pmatrix} c_{1,1} & c_{1,2} & \cdots & c_{1,m} \\ c_{2,1} & c_{2,2} & \cdots & c_{2,m} \\ \vdots & \vdots & \vdots & \vdots \\ c_{m,1} & c_{m,2} & \cdots & c_{m,m} \end{pmatrix} = \begin{pmatrix} c_1 \\ c_2 \\ \vdots \\ c_m \end{pmatrix}.
$$

If we would like to estimate $D^*(P)$, then it is often sufficient to restrict our study to the local discrepancy function $\Delta(J, P)$ only for certain well-chosen intervals $J$. In [31], the authors considered the local discrepancy function $\Delta$ of a digital net, evaluated at $(\eta, \beta) \in [0, 1)^2$ for $m$-bit numbers $\eta$ and $\beta$, i.e., numbers with base 2 representation

$$
\eta = \frac{\eta_1}{2} + \frac{\eta_2}{2^2} + \cdots + \frac{\eta_m}{2^m}, \quad \beta = \frac{\beta_1}{2} + \frac{\beta_2}{2^2} + \cdots + \frac{\beta_m}{2^m}.
$$

As pointed out in [31],

$$
\left| D^*(P) - 2^m \max_{\eta, \beta \ m\text{-bit}} |\Delta([0, \eta) \times [0, \beta), P)| \right| \leq 2 - 1/2^m,
$$

which means that in many cases one can resort to studying the discrepancy of $P$ only for $m$-bit numbers, without significant deviations from the precise value of the discrepancy.

We introduce some further notation. We write $\boldsymbol{\eta} := (\eta_1, \ldots, \eta_m)^\top$ and $\boldsymbol{\beta} := (\beta_1, \ldots, \beta_m)^\top$. Let

$$
\boldsymbol{\gamma} = \boldsymbol{\gamma}(\eta, \beta) = (\gamma_1, \gamma_2, \ldots, \gamma_m)^\top = C_2 \cdot \boldsymbol{\eta} + \boldsymbol{\beta},
$$

and let $\boldsymbol{\gamma}(u)$ denote the vector consisting of the first $u$ components of $\boldsymbol{\gamma}$. Furthermore, we write, for $1 \leq u \leq m$,

$$
C_2'(u) = \begin{pmatrix} c_{1,m-u+1} & c_{2,m-u+1} & \cdots & c_{u,m-u+1} \\ c_{1,m-u+2} & c_{2,m-u+2} & \cdots & c_{u,m-u+1} \\ \vdots & \vdots & \vdots & \vdots \\ c_{1,m} & c_{2,m} & \cdots & c_{u,m} \end{pmatrix}^{-1},
$$

which exists according to the $(0, m, 2)$-net property of $P$.

In the following, we denote by $(\cdot|\cdot)$ the usual inner product. Moreover, we denote by $\|x\|$ the distance to the nearest integer of a real number $x$. Notice that $\|x\| = \psi_2^{\mathrm{id}}(x)$, the $\psi$-function in base 2 associated with identity in Section 4.2.2.

In [31], the following formula for the local discrepancy $\Delta(\eta, \beta)$ was shown,

$$\Delta(\eta, \beta) = \sum_{u=0}^{m-1} \left\|2^u \beta\right\| (-1)^{(c_{u+1}|\eta)} (-1)^{(\gamma(u)|C_2'(u)(c_{u+1,m-u+1},\ldots,c_{u+1,m})^\top)}$$
$$\times \frac{(-1)^{\eta_{m-u}} - (-1)^{\eta_{m+1-j(u)}}}{2}, \tag{4.7}$$

where

$$m(u) = \begin{cases} 0 & \text{if } u = 0 \\ 0 & \text{if } (\gamma(u)|C_2'(u)e_1) = 1 \\ \max\{1 \le j \le u : (\gamma(u)|C_2'(u)e_i) = 0, i = 1,\ldots, j\} & \text{otherwise,} \end{cases}$$

and $j(u) = u - m(u)$. Here $e_i$ denotes the $i$th unit vector in $\mathbb{Z}_2^u$. Furthermore, we set $\eta_{m+1} := 0$ and $(\gamma(u)|C_2'(u)(c_{u+1,m-u+1},\ldots,c_{u+1,m})^\top) := 0$ for $u = 0$.

Formula (4.7) is a powerful tool. Indeed, by means of (4.7), it was shown in [31] that the following improvement of Theorem 4.22 holds.

**Theorem 4.23** (Larcher, Pillichshammer) *The star discrepancy of a digital $(0, m, 2)$-net $P$ over $\mathbb{Z}_2$ satisfies*

$$D^*(P) \le m/3 + 19/9. \tag{4.8}$$

This result was further sharpened, and extended to arbitrary two-dimensional nets, in the following Theorem 4.24 in [4].

### 4.3.3 Using a counting argument to obtain bounds for $(t, m, 2)$-nets

Surprisingly, the proof of Theorem 4.24 does not need any of the technical tools mentioned above, but is based on a counting argument.

**Theorem 4.24** (Dick, Kritzer) *The star discrepancy of an arbitrary $(t, m, 2)$-net $P$ in base $b$ satisfies*

$$D^*(P) \le b^t D^*(\mathcal{H}_{b,m-t}) + b^t. \tag{4.9}$$

The bound in Theorem 4.24 is practically useful, as $D^*(\mathcal{H}_{b,m})$ can be computed exactly by using formulas provided by DeClerck [3]. In particular, this implies that (4.9) is indeed an improvement of (4.8) for the case $t = 0$, $b = 2$, and also an improvement of (4.6) for arbitrary choices of $t$ and $b$. Furthermore, the bound in (4.9) can be seen as an extension of the corollary to Theorem 4.25 below and is, up to the constant $b^t$, sharp.

If we focus on the case $t = 0$, we know by Equation (4.9) that the Hammersley point set is basically the $(0, m, 2)$-net in base $b$ with the highest star discrepancy. This observation sparked interest in the question whether there are examples of other $(0, m, 2)$-nets with a significantly lower star discrepancy than $\mathcal{H}_{b,m}$. This question can be answered positively and, not surprisingly, digitally permuted Hammersley nets play a crucial role in this context, since they are connected to digitally permuted $(0, 1)$-sequences (see Section 4.3.4).

### 4.3.4 Using results on $S_b^\Sigma$ sequences to obtain bounds for Hammersley nets

We first define what we mean by digitally permuted Hammersley nets associated with $S_b^\Sigma$ sequences. In order to match the traditional definition of arbitrary (shifted or not) Hammersley point sets whose points are "$m$-bits," we restrict the infinite sequence of permutations $\Sigma$ to permutations such that $\sigma_r(0) = 0$ for all $r \geq m$, for instance $\sigma := (\sigma_0, \ldots, \sigma_{m-1}, \mathrm{id}, \mathrm{id}, \mathrm{id}, \ldots)$. Then the *generalized two-dimensional Hammersley point set in base $b$* consisting of $b^m$ points associated with $\sigma$ is defined by

$$\mathcal{H}_{b,m}^\sigma := \left\{ \left( S_b^\sigma(n), \frac{n}{b^m} \right); 0 \leq n \leq b^m - 1 \right\}.$$

Notice that the behavior of $\mathcal{H}_{b,m}^\sigma$ only depends on the finite sequence $(\sigma_0, \ldots, \sigma_{m-1})$ which we identify with $\sigma$ from now on (see [16, Definition 2] for more details). If we choose in the above definition $\sigma_j = \mathrm{id}$ for all $j$, then we obtain the classical two-dimensional Hammersley point set in base $b$, $\mathcal{H}_{b,m}^{\mathrm{id}} = \mathcal{H}_{b,m}$.

The main results concerning the star discrepancy of two-dimensional Hammersley point sets are of two kinds: some give exact formulas including complementary terms [3, 24, 31] and others give formulas for the leading terms within an error not computable, usually lower than a small additive constant [4, 10, 15, 26, 28]. For the sake of brevity, we will only refer to these latter results in the generalizations we are going to give in the following. Theorems 4.25 to 4.29 below stem from [16].

First, we make the tight link between $\mathcal{H}^{\sigma}_{b,m}$ and $S^{\sigma}_b$ via $\psi$-functions more precise.

**Theorem 4.25** (Faure) *For any integer $m \geq 1$ and any $\sigma = (\sigma_0, \ldots, \sigma_{m-1})$ we have, with some $c_m \in [0, 2]$,*

$$D^*(\mathcal{H}^{\sigma}_{b,m})$$

$$= \max \left( \max_{1 \leq n \leq b^m} \sum_{j=1}^{m} \psi_b^{\sigma_{j-1},+} \left( \frac{n}{b^j} \right) , \max_{1 \leq n \leq b^m} \sum_{j=1}^{m} \psi_b^{\sigma_{j-1},-} \left( \frac{n}{b^j} \right) \right) + c_m.$$

This formula, valid for an arbitrary base $b$, is the analog of [28, Lemma 1] in base 2. It permits all extensions to bases $b$ of results on base 2 belonging to the second kind mentioned above.

As a first corollary, we obtain that $\mathcal{H}^{\mathbf{id}}_{b,m}$ is the worst Hammersley net among generalized ones within a constant less than 2: for any integer $m \geq 1$ and any $\sigma = (\sigma_0, \ldots, \sigma_{m-1})$ we have, with some $c_m \in [-2, 2]$,

$$D^*(\mathcal{H}^{\sigma}_{b,m}) \leq D^*(\mathcal{H}^{\mathbf{id}}_{b,m}) + c_m. \tag{4.10}$$

**Swapping with the identical permutation** First let us consider the following sequence

$$i\tau = (\overbrace{\mathrm{id}, \ldots, \mathrm{id}}^{\frac{m}{2}}, \overbrace{\tau, \ldots \tau}^{\frac{m}{2}}) \quad \text{if } m \text{ is even}$$

$$i\tau = (\overbrace{\mathrm{id}, \ldots, \mathrm{id}}^{\frac{m-1}{2}}, \overbrace{\tau, \ldots \tau}^{\frac{m+1}{2}}) \quad \text{if } m \text{ is odd},$$

like Kritzer did in base 2 [27]. Applying Theorem 4.25, we can easily extend his result [27, Theorem 3.2 and Proposition 3.1] to arbitrary bases.

**Theorem 4.26** (Faure) *For any integer $m \geq 1$ we have, with some $c_m \in [0, 3]$,*

$$\text{if } b \text{ is odd} \quad D^*(\mathcal{H}^{i\tau}_{b,m}) = \begin{cases} \dfrac{b-1}{8} m + c_m & \text{if } m \text{ is even} \\ \dfrac{b-1}{8}(m+1) + c_m & \text{if } m \text{ is odd}, \end{cases}$$

$$\text{if } b \text{ is even} \quad D^*(\mathcal{H}^{i\tau}_{b,m}) = \begin{cases} \dfrac{b^2}{8(b+1)} m + c_m & \text{if } m \text{ is even} \\ \dfrac{b^2}{8(b+1)}(m+1) + c_m & \text{if } m \text{ is odd}. \end{cases}$$

The interval for $c_m$ could be reduced. Of course, we recover the result of Kritzer [27] and Kritzer *et al.* [28], with the same sequence $i\tau$, in the case of $b = 2$. The best constant is obtained for $b = 3$ with $1/(4 \log 3) = 0.227\ldots$, whereas for $b = 2$ we only have $1/(6 \log 2) = 0.240\ldots$.

Now, we show that the choice of the sequence $i\tau$ in the set $\{id, \tau\}^m$ of sequences $\sigma = (\sigma_0, \ldots, \sigma_{m-1})$ is best possible in the sense that the leading terms in Theorem 4.26 cannot be made smaller whatever the $\sigma_{j-1} \in \{id, \tau\}$, $1 \le j \le m$, are.

**Theorem 4.27** (Faure) *For any integer $m \ge 1$ and any $\sigma \in \{id, \tau\}^m$ we have*

$$\lim_{m \to \infty} \frac{D^*(\mathcal{H}^\sigma_{b,m})}{\log b^m} \ge \frac{b-1}{8 \log b} \qquad \text{if } b \text{ is odd}$$

$$\lim_{m \to \infty} \frac{D^*(\mathcal{H}^\sigma_{b,m})}{\log b^m} \ge \frac{b^2}{8(b+1) \log b} \qquad \text{if } b \text{ is even.}$$

In base 2, Theorem 4.27 has been shown in [28] by other arguments involving more computations. Another question raised and solved in base 2 in [28] is the following.

Is the star discrepancy $D^*(\mathcal{H}^\sigma_{b,m})$ independent of the distribution of id and $\tau$ in the sequence $\sigma = (\sigma_0, \ldots, \sigma_{m-1}) \in \{id, \tau\}^m$ and does only depend on the number of id and $\tau$?

In arbitrary base, the answer is *No* like in base 2, with the same counter-example as in [28], the sequence (id, $\tau$, id, $\tau$, ..., id, $\tau$).

**Theorem 4.28** (Faure) *For any even integer $m \ge 2$, let $\widetilde{i\tau} = (id, \tau, id, \tau, \ldots, id, \tau) \in \{id, \tau\}^m$. Then, with some $c_m \in [0, 3]$, we have*

$$D^*(\mathcal{H}^{\widetilde{i\tau}}_{b,m}) = \begin{cases} \dfrac{(b-1)(b+2)}{8(b+1)} m + c_m & \text{if } b \text{ is odd} \\[2ex] \dfrac{b^3}{8(b^2+1)} m + c_m & \text{if } b \text{ is even.} \end{cases}$$

*These constants are greater than $(b-1)/8$ and $b^2/(8(b+1))$, hence the answer to the question above is No.*

For $b = 2$ we recover the result of [28, end of Section 4] with the constant $1/5$. This result has been known for a long time since Halton and Zaremba [24] obtained, in 1969, exact formulas for $D^*(\mathcal{H}^{id}_{2,m})$ and $D^*(\mathcal{H}^{\widetilde{i\tau}}_{2,m})$ after a lot of technical computations (see [31, Sections 1 and 4] for comments).

**Swapping with an arbitrary permutation** In this section, we fix an arbitrary permutation $\sigma$ of $\{0, 1, \ldots, b - 1\}$ and consider sequences produced by swapping $\sigma$ with $\tau$, as we did in Theorem 4.26, to obtain sequences $\boldsymbol{\sigma} = (\sigma_0, \ldots, \sigma_{m-1}) \in \{\sigma, \overline{\sigma}\}^m$. The situation is not so clear as with the identity and we will only consider sequences

$$\sigma\overline{\sigma} = (\overbrace{\sigma, \ldots, \sigma}^{\frac{m}{2}}, \overbrace{\overline{\sigma}, \ldots \overline{\sigma}}^{\frac{m}{2}}) \qquad \text{if } m \text{ is even}$$

$$\sigma\overline{\sigma} = (\overbrace{\sigma, \ldots, \sigma}^{\frac{m-1}{2}}, \overbrace{\overline{\sigma}, \ldots \overline{\sigma}}^{\frac{m+1}{2}}) \qquad \text{if } m \text{ is odd.}$$

This choice allows us to improve on the discrepancy, but until now we have not been able to prove it is the best, like as the identity.

**Theorem 4.29** (Faure) *For any integer $m \geq 1$ we have, with some $c_m \in [-1, 4]$,*

$$D^*(\mathcal{H}_{b,m}^{\sigma\overline{\sigma}}) = \begin{cases} \dfrac{\alpha_b^{\sigma,+} + \alpha_b^{\sigma,-}}{2} m + c_m & \text{if } m \text{ is even} \\ \dfrac{\alpha_b^{\sigma,+} + \alpha_b^{\sigma,-}}{2} (m + 1) + c_m & \text{if } m \text{ is odd.} \end{cases}$$

Of course, we recover the constants of Theorem 4.26 when $\sigma = $ id. Obtained as a by-product of Theorem 4.18, the best result coming from [8, Théorème 5], with $b = 12$, $\sigma_3$ and constant $0.2235\ldots$, was recently improved by Ostromoukhov [38] with $b = 60$, $\sigma_1$ and constant $0.2222\ldots$, a bit better than $b = 3$ and $\sigma = $ id with $0.227\ldots$ (see the comments following Theorem 4.14). Even if such improvements seem small, they concern the leading constants in discrepancy formulas and we think it is more important to improve on these constants than to search for exact formulas or to reduce the complementary terms $c_m$ in estimations.

## 4.4 Lower discrepancy bounds for low-dimensional point sets

### 4.4.1 Lower discrepancy bounds for nets

We now survey lower bounds on the star discrepancy of low-dimensional nets. As in Section 4.3, we focus on the case of dimension $s = 2$, as the case $s = 1$ is essentially answered by Equation (4.5).

Generally speaking, it is a lot more challenging to provide tight lower discrepancy bounds than it is to provide upper bounds. The only general result that holds for all $(t, m, 2)$-nets is the aforementioned bound of Béjian [1] (see the end of Section 4.1), from which it results that any finite point set $P$ of $N$ points in $[0, 1)^2$ satisfies

$$D^*(N, P) \geq 0.03 \log N. \qquad (4.11)$$

In view of the bounds (4.8), (4.9), (4.10), many researchers have tried to sharpen lower bounds for at least some well-chosen subclasses of $(t, m, 2)$-nets, and there has been a major focus on $(0, m, 2)$-nets.

For example, it was shown in [28] that a digitally shifted Hammersley net $\mathcal{H}_{2,m}$ in base 2 always has a discrepancy such that $D^*(\mathcal{H}_{2,m}) \geq m/6 + c$, where $c$ is a constant independent of $m$. Faure [15] showed a similar result in this vein, proving that it is true for any digitally permuted Hammersley net $\mathcal{H}_{b,m}^\sigma$ in base $b$ that

$$D^*(\mathcal{H}_{b,m}^\sigma) \geq \left(3/4 - \sqrt{3b - 1}/(2b)\right) m + c,$$

where $c$ is some positive constant that does not depend on $m$.

### 4.4.2 A new result on the discrepancy of digital $(0, m, 2)$-nets

In this section, we show a new lower discrepancy bound that holds for certain digital $(0, m, 2)$-nets over $\mathbb{Z}_2$. For $m \in \mathbb{N}$, we write $m_0 := \lfloor \frac{m}{2} \rfloor$. Within this section, we consider digital $(0, m, 2)$-nets over $\mathbb{Z}_2$ generated by

$$C_1 = I_{m,m}, \qquad C_2 = \left(\begin{array}{c|c} A & B \\ \hline C & D \end{array}\right), \qquad (4.12)$$

where $I_{m,m}$ is the $(m \times m)$ identity matrix, and where $A$ is a nonsingular $(m_0 \times m_0)$-matrix over $\mathbb{Z}_2$. We show the following new result.

**Theorem 4.30** *Let $P$ be a digital $(0, m, 2)$-net generated by two generating matrices as in (4.12). Then it is true that*

$$D^*(P) \geq m/12 + c,$$

*where $c$ is a constant independent of $m$.*

**Remark 4.31** We remark that, even though the bound in Theorem 4.30 is weaker than the lower bound for digitally shifted Hammersley nets from [28] mentioned before, the result in Theorem 4.30 covers a relatively large class of digital $(0, m, 2)$-nets.

*Proof.* We use the approach of Larcher and Pillichshammer [31] that was summarized in Section 4.3 and the same notation as there. We make a specific choice of two base 2 $m$-bit numbers $\eta$ and $\beta$. Let

$$\beta^{(0)} = (b_{m_0}, \underbrace{0, 0, \ldots\ldots\ldots, 0}_{m-m_0 \text{ components}}),$$

where

$$b_{m_0} = \begin{cases} \underbrace{1, 0, 1, 0, \ldots, 1, 0, 1, 0}_{m_0 \text{ components}} & \text{if } m_0 \text{ is even} \\ \underbrace{1, 0, 1, 0, \ldots 1, 0, 1}_{m_0 \text{ components}} & \text{if } m_0 \text{ is odd.} \end{cases}$$

For the following, denote by $\mathbf{0}_{k,l}$ the $k \times l$ zero matrix. We define an $m$-bit number $\eta^{(0)}$ such that

$$\left( \begin{array}{c|c} A & B \\ \mathbf{0}_{m-m_0,m_0} & I_{m-m_0,m-m_0} \end{array} \right) \cdot \eta^{(0)} = \begin{pmatrix} \delta_1 \\ \delta_2 \\ \vdots \\ \delta_m \end{pmatrix} = \boldsymbol{\delta},$$

where the first and the last $m_0$ components, respectively, of $\boldsymbol{\delta}$ satisfy

$$\delta_{m-u} \oplus 1 = \delta_{u+1} = \beta^{(0)}_{u+1} \quad \text{for } 0 \le u \le m_0 - 1,$$

and where $\oplus$ denotes addition modulo 2. Note that $\eta^{(0)}$ can certainly be chosen in such a way due to the assumptions made on the matrix $A$ in Equation (4.12).

Using the notation in Section 4.3, note that the choice of $\eta^{(0)}$ and $\beta^{(0)}$ guarantees that $\boldsymbol{\gamma}(u) = \mathbf{0}$ for $1 \le u \le m_0 - 1$, which implies that $m(u) = u$ and $j(u) = 0$ for $0 \le u \le m_0 - 1$. Note furthermore, that $\left\| 2^u \beta^{(0)} \right\| = 0$ for all $u \ge m_0$. Hence the discrepancy function $\Delta$ evaluated at $(\eta^{(0)}, \beta^{(0)})$ simplifies to

$$\Delta(\eta^{(0)}, \beta^{(0)}) = \sum_{u=0}^{m_0-1} \left\| 2^u \beta^{(0)} \right\| (-1)^{\beta^{(0)}_{u+1}} \frac{(-1)^{\beta^{(0)}_{u+1}+1} - 1}{2}.$$

However,

$$(-1)^{\beta^{(0)}_{u+1}} \frac{(-1)^{\beta^{(0)}_{u+1}+1} - 1}{2} = \begin{cases} -1 & \text{if } \beta^{(0)}_{u+1} = 0 \\ 0 & \text{if } \beta^{(0)}_{u+1} = 1. \end{cases}$$

Thus,

$$\left| \Delta(\eta^{(0)}, \beta^{(0)}) \right| = \sum_{\substack{u=1 \\ u \text{ odd}}}^{m_0-1} \left\| 2^u \beta^{(0)} \right\|.$$

Note that, for the case of $m_0$ even,

$$\sum_{\substack{u=1 \\ u \text{ odd}}}^{m_0-1} \left\| 2^u \beta^{(0)} \right\| = \sum_{\substack{u=1 \\ u \text{ odd}}}^{m_0-1} \left\| 2^u \sum_{k=1}^{m_0/2} \frac{1}{2^{2k-1}} \right\| = \sum_{\substack{u=1 \\ u \text{ odd}}}^{m_0-1} \sum_{k=1}^{(m_0-1-u)/2} \frac{1}{2^{2k}}$$

$$= \sum_{\substack{u=1 \\ u \text{ odd}}}^{m_0-1} \frac{1}{3}\left(1 - \frac{1}{2^{m_0-1-u}}\right)$$

$$= \sum_{k=1}^{m_0/2} \frac{1}{3}\left(1 - \frac{1}{2^{m_0-1-(2k-1)}}\right)$$

$$= \frac{m_0}{6} + \frac{4}{9}\left(\frac{1}{2^{m_0}} - 1\right), \quad (4.13)$$

and, for the case of $m_0$ odd, we can show in a similar way as for the derivation of (4.13),

$$\sum_{\substack{u=1 \\ u \text{ odd}}}^{m_0-1} \left\| 2^u \beta^{(0)} \right\| = \frac{m_0}{6} + \frac{1}{9}\left(\frac{1}{2^{m_0}} - 1\right). \quad (4.14)$$

Now note that $m_0$ is of order $m/2$, and hence we conclude from (4.13) and (4.14) that

$$\left| \Delta(\eta^{(0)}, \beta^{(0)}) \right| \geq m/12 + c,$$

where $c$ is some constant independent of $m$. The result follows. $\qquad \square$

### 4.4.3 Lower discrepancy bounds for sequences

**Lower discrepancy bounds for one-dimensional sequences** In this part, we give an application of Theorem 4.15 and show best possible lower bounds on the star discrepancy of NUT $(0, 1)$-sequences. This study is motivated by a best possible lower bound on the star discrepancy of digitally shifted van der Corput sequences in base 2 shown in [28] and the question of whether this bound remains true also for digitally shifted NUT digital sequences in base 2.

Theorems 4.15 and 4.18 are two main ingredients of the following result, shown in [22], which leads to best possible lower bounds for large sub-families of NUT $(0, 1)$-sequences to be stated afterwards.

**Theorem 4.32** (Faure, Pillichshammer) *For any integer $b \geq 2$, let $\sigma \in \mathfrak{S}_b$ and let $C$ be a strict upper triangular matrix with entries in $\mathbb{Z}_b$. Then, for any subset $S$ of $\mathbb{N}_0$, we have*

$$D^*(N, X_b^{\Sigma_S^\sigma, C}) \geq \frac{1}{2} D(N, S_b^\sigma) \quad \text{and hence} \quad \rho^*(X_b^{\Sigma_S^\sigma, C}) \geq \frac{\alpha_b^\sigma}{2 \log b},$$

where $\Sigma_S^\sigma = (\sigma_r)_{r \geq 0}$ with $\sigma_r = \sigma$ if $r \in S$ and $\sigma_r = \tau \circ \sigma$ if $r \notin S$, and $X_b^{\Sigma_S^\sigma, C}$ is a NUT $(0, 1)$-sequence.

We start with a best possible lower bound for NUT $(0, 1)$-sequences $X_b^{\Sigma, C}$ associated with sequences of permutations $\Sigma \in \{\sigma, \tau \circ \sigma\}^{\mathbb{N}_0}$ for which $\sigma$ gives permuted van der Corput sequences with $D = D^*$.

**Corollary 4.33** (Faure, Pillichshammer) *Let $\mathcal{C}_{\mathrm{SUT}}$ be the set of all strict upper triangular matrices and let $\sigma \in \mathfrak{S}_b$ such that $D^*(S_b^\sigma) = D(S_b^\sigma)$. Then*

$$\inf_{\substack{\Sigma \in \{\sigma, \tau \circ \sigma\}^{\mathbb{N}_0} \\ C \in \mathcal{C}_{\mathrm{SUT}}}} \rho^*(X_b^{\Sigma, C}) = \frac{\alpha_b^\sigma}{2 \log b}.$$

Besides the identity id, it is not difficult to find permutations satisfying the condition $D^*(S_b^\sigma) = D(S_b^\sigma)$. Further, a systematic computer search performed by F. Pausinger (IST Austria, personal communication) has given 26, 58, 340, and 1496 such permutations in bases 6, 7, 8, and 9, respectively.

The case of identity in Corollary 4.33 is of special interest because $\alpha_b^{\mathrm{id}}$ is explicitly known for any integer $b \geq 2$.

**Corollary 4.34** (Faure, Pillichshammer) *With the notation of Corollary 4.33, we obtain*

$$\inf_{b \geq 2} \inf_{\substack{\Sigma \in \{id, \tau\}^{\mathbb{N}_0} \\ C \in \mathcal{C}_{\mathrm{SUT}}}} \rho^*(X_b^{\Sigma, C}) = \frac{1}{4 \log 3} = 0.2275 \ldots.$$

This result can be seen as the analog for NUT $(0, 1)$-sequences of the best possible lower bound for the star discrepancy of $(n\alpha)$ sequences obtained by Dupain and Sós [7], with $\rho^*((n\sqrt{2})) = 0.2836 \ldots$. We see that NUT $(0, 1)$-sequences yield a much smaller value of $\rho^*$.

Finally, we consider digitally permuted NUT digital sequences by means of linear digit scramblings. Such sequences, denoted $Z_b^{\Pi, C}$, are defined as follows. Let $\mathcal{C}_{\mathrm{NUT}}^1$ be the set of NUT matrices $C$ such that all the diagonal entries $c_r^r = 1$ and let $\Pi = (\pi_r)_{r \geq 0} \in \mathfrak{S}_b^{\mathbb{N}_0}$ be a sequence of linear digit scramblings. Then, for any $n \geq 0$,

$$Z_b^{\Pi,C}(n) = \sum_{r=0}^{\infty} \frac{\pi_r(x_{n,r})}{b^{r+1}} \quad \text{with} \quad x_{n,r} = \sum_{k=r}^{\infty} c_r^k n_k \pmod{b},$$

where the $n_k$ are the base $b$ digits of $n$. We have the following analog of Corollary 4.34.

**Corollary 4.35** (Faure, Pillichshammer) *Let* $Z_b^{\Pi,C}$ *be a linearly digit scrambled NUT digital* $(0, 1)$*-sequence associated with* $C \in \mathcal{C}_{\mathrm{NUT}}^1$ *and* $\Pi = (\pi_r)_{r \geq 0} \in \{id, \tau\}^{\mathbb{N}_0}$. *Then we have*

$$\inf_{\substack{b \geq 2 \\ C \in \mathcal{C}_{\mathrm{NUT}}^1}} \inf_{\Pi \in \{id, \tau\}^{\mathbb{N}_0}} \rho^*(Z_b^{\Pi,C}) = \frac{1}{4 \log 3} = 0.2275\ldots.$$

Notice that id and $\tau$ are the only linear digit scramblings satisfying $D^*(S_b^{\pi}) = D(S_b^{\pi})$.

In the case $b = 2$, Corollary 4.35 permits us to answer the question asked at the beginning: "Is it true that the constant $1/(6 \log 2)$ is best possible for any *digitally shifted NUT digital sequence in base 2*, as it is the case for any digitally shifted van der Corput sequence according to [28, Corollary 4]?" Taking into account that, in base 2, $\tau$ is the nonzero shift and the diagonal entries of $C$ are all equal to 1, this question was answered in the affirmative in [22].

**Corollary 4.36** (Faure, Pillichshammer) *We have*

$$\inf_{\substack{\Delta \in \mathbb{Z}_2^{\mathbb{N}_0} \\ C \in \mathcal{C}_{\mathrm{NUT}}}} \rho^*(Z_2^{\Delta,C}) = \frac{1}{6 \log 2}.$$

For more information on the context of Theorem 4.32 and its corollaries we refer to [22, Section 5], where an overview of this topic is given.

**Lower discrepancy bounds for two-dimensional sequences** Regarding lower bounds for $(t, 2)$-sequences, only very little is known, except for one example by Faure and Chaix [17], who were able to obtain the exact order of the star discrepancy for a $(0, 2)$-sequence $S_{\mathrm{Sob}}$ in base 2 first introduced by Sobol' [44].

**Theorem 4.37** (Faure, Chaix) *The digital* $(0, 2)$*-sequence in base 2 generated by the identity matrix and the Pascal matrix* mod 2, *denoted* $S_{\mathrm{Sob}}$, *satisfies the inequality*

$$\frac{1}{24(\log 2)^2} \leq \limsup_{N\to\infty} \frac{D^*(N, S_{\text{Sob}})}{(\log N)^2}.$$

In combination with Theorem 4.20, this is the only case of a low-discrepancy sequence in dimension greater than one for which the exact order of discrepancy is known. Moreover, based on thorough numerical experiments that allowed them to find the subsequence leading to their lower bound, Faure and Chaix stated the conjecture that the inequality above should actually be an equality, i.e., $\limsup_{N\to\infty} \frac{D^*(N,S_{\text{Sob}})}{(\log N)^2} = \frac{1}{24(\log 2)^2}$.

## 4.5 Conclusion

In this survey, we have illustrated that there has been a considerable history of results on discrepancy bounds for low-dimensional $(t, m, s)$-nets, $(t, s)$-sequences, and related point sets. We have summarized 20 theorems, a large part of them dealing with two-dimensional nets, an equally large part dealing with one-dimensional sequences, and further results on two-dimensional sequences. All results on one-dimensional sequences in Section 4.2 stem from the initial study [8] on generalized van der Corput sequences, Theorem 4.15 being the foremost new generalization for these sequences. Regarding two-dimensional nets in Section 4.3, we discussed several different approaches; two theorems are obtained by counting arguments (Theorems 4.22 and 4.24), two others result from Walsh function analysis of discrepancy (Theorems 4.23 and 4.30), and the remaining five theorems stem from the study of generalized van der Corput sequences (Theorems 4.25–4.29). While these results deal with the precise study of a special class of nets (namely generalized Hammersley nets), the previous results concern arbitrary $(t, m, 2)$-nets (Theorems 4.22 and 4.24) or digital $(0, m, 2)$-nets (Theorem 4.23 and Theorem 4.30, which is the only previously unpublished result of this paper).

Finally, two main results deal with two-dimensional sequences: Theorem 4.20 on upper bounds for arbitrary $(t, 2)$-sequences has been extended recently to arbitrary $(t, s)$-sequences [18], and Theorem 4.37 on lower bounds remains the only exception for which the exact order is attained in dimension $s > 1$.

Most of the results mentioned in this paper have been obtained by methods of number theory and algebra, and the precise analysis of the discrepancy of the point sets, even though they are "only" one-dimensional or two-dimensional, is very challenging. The recent results [18] on the discrepancy of $(t, m, s)$-nets and $(t, s)$-sequences in arbitrary dimension $s$, which are partly obtained by an inductive argument on the dimension $s$, demonstrate that it may be crucial to have excellent discrepancy bounds for low-dimensional examples, as the better the low-dimensional starting point, the better the results obtained inductively can be expected to be.

We end this paper by stating two selected open problems that would be interesting to solve in the near future.

- Find other two-dimensional sequences like that in Theorem 4.37 having the "correct" order of star discrepancy. Natural candidates are $(0, 2)$-sequences in arbitrary bases and two-dimensional Halton sequences, for instance in bases 2 and 3. This open problem seems to be a very challenging task.
- Find an exact formula for the discrepancy function of one-dimensional digital sequences or two-dimensional digital nets in base $b$, i.e., extend Equation (4.7) from [31] to other bases $b > 2$ . Such a formula has been extended to arbitrary bases for digital $(0, 1)$-sequences generated by NUT matrices [12], but until now, no analog exists for generating matrices having nonzero entries below the diagonal. In relation to this question, we refer to Theorem 4.13, where a generating matrix having nonzero entries below the diagonal leads to a surprising result. Further investigations on such matrices could help to make progress in the understanding of digital nets and sequences and their distribution properties.

## Acknowledgements

The authors would like to thank G. Larcher and F. Pillichshammer for suggestions and remarks, and the referee for valuable comments improving the consistency of this article.

P. Kritzer gratefully acknowledges the support of the Austrian Science Fund (FWF), Projects P23389-N18 and F5506-N26, which is part of the Special Research Program "Quasi-Monte Carlo Methods: Theory and Applications."

## References

[1]  R. Béjian, Minoration de la discrépance d'une suite quelconque sur $T$. *Acta Arith.* **41**, 185–202, 1982.

[2]  D. Bilyk, M.T. Lacey and A. Vagharshakyan. On the small ball inequality in all dimensions. *J. Funct. Anal.* **254**, 2470–2502, 2008.

[3]  L. DeClerck, A method for exact calculation of the star discrepancy of plane sets applied to the sequences of Hammersley. *Monatsh. Math.* **101**, 261–278, 1986.

[4]   J. Dick and P. Kritzer, A best possible upper bound on the star discrepancy of $(t, m, 2)$-nets. *Monte Carlo Methods Appl.* **12**, 1–17, 2006.

[5]   J. Dick and F. Pillichshammer, *Digital Nets and Sequences. Discrepancy Theory and Quasi-Monte Carlo Integration.* Cambridge University Press, Cambridge, 2010.

[6]   M. Drmota and R.F. Tichy, *Sequences, Discrepancies and Applications.* Lecture Notes in Mathematics, volume 1651. Springer, Berlin, 1997

[7]   Y. Dupain and V. Sós, On the discrepancy of $(n\alpha)$ sequences. Topics in Classical Number Theory, Colloq. Budapest 1981, volume I, *Colloq. Math. Soc. Janos Bolyai* **34**, 355–387, 1984.

[8]   H. Faure, Discrépance de suites associées à un système de numération (en dimension un). *Bull. Soc. Math. France* **109**, 143–182, 1981.

[9]   H. Faure, Discrépance de suites associées à un système de numération (en dimension $s$). *Acta Arith.* **41**, 337–351, 1982.

[10]  H. Faure, On the star-discrepancy of generalized Hammersley sequences in two dimensions. *Monatsh. Math.* **101**, 291–300, 1986.

[11]  H. Faure, Good permutations for extreme discrepancy. *J. Number Theory* **42**, 47–56, 1992.

[12]  H. Faure, Discrepancy and diaphony of digital $(0, 1)$-sequences in prime bases. *Acta Arith.* **117**, 125–148, 2005.

[13]  H. Faure, Irregularities of distribution of digital $(0, 1)$-sequences in prime bases. *Integers* **5**, A07 (electronic), 2005.

[14]  H. Faure, Van der Corput sequences towards $(0, 1)$-sequences in base $b$. *J. Théor. Nombres Bordeaux* **19**, 125–140, 2007.

[15]  H. Faure, Improvements on low discrepancy one-dimensional sequences and two-dimensional point sets. In: A. Keller, S. Heinrich and H. Niederreiter (eds.), *Monte Carlo and Quasi-Monte Carlo Methods 2006*, pp. 327–341. Springer, Berlin, 2008.

[16]  H. Faure, Star extreme discrepancy of generalized two-dimensional Hammersley point sets. *Unif. Distrib. Theory* **3**(2), 45–65, 2008.

[17]  H. Faure and H. Chaix, Minoration de discrépance en dimension deux. *Acta Arith.* **76**, 149–164, 1996.

[18]  H. Faure and P. Kritzer, New star discrepancy bounds for $(t, m, s)$-nets and $(t, s)$-sequences. *Monatsh. Math.* **172**, 55–75, 2013.

[19]  H. Faure and C. Lemieux, Generalized Halton sequences in 2008: a comparative study. *ACM Trans. Model. Comp. Sim.* **19**(4), Article 15, 2009.

[20]  H. Faure and C. Lemieux, Improved Halton sequences and discrepancy bounds. *Monte Carlo Methods Appl.* **16**, 231–250, 2010.

[21]  H. Faure and C. Lemieux, Improvements on the star discrepancy of $(t, s)$-sequences. *Acta Arith.* **154**, 61–78, 2012.

[22]  H. Faure and F. Pillichshammer, A generalization of NUT digital $(0, 1)$-sequences and best possible lower bounds for star discrepancy. *Acta Arith.* **158**, 321–340, 2013.

[23]  J. H. Halton, On the efficiency of certain quasi-random sequences of points in evaluating multi-dimensional integrals. *Numer. Math.* **2**, 184–190, 1960.

[24]  J. H. Halton and S. K. Zaremba, The extreme and the $L^2$ discrepancies of some plane sets. *Monatsh. Math.* **73**, 316–328, 1969.

[25] P. Kritzer, A new upper bound on the star discrepancy of (0, 1)-sequences. *Integers* **5**, A11 (electronic), 2005.

[26] P. Kritzer, Improved upper bounds on the star discrepancy of $(t, m, s)$-nets and $(t, s)$-sequences. *J. Complexity* **22**, 336–347, 2006.

[27] P. Kritzer, On some remarkable properties of the two-dimensional Hammersley point set in base 2. *J. Théor. Nombres Bordeaux* **18**, 203–221, 2006.

[28] P. Kritzer, G. Larcher and F. Pillichshammer, A thorough analysis of the discrepancy of shifted Hammersley and van der Corput point sets. *Ann. Math. Pura Appl.* **186**, 229–250, 2007.

[29] L. Kuipers and H. Niederreiter, *Uniform Distribution of Sequences*. John Wiley, New York, 1974. Reprint, Dover Publications, Mineola, NY, 2006.

[30] G. Larcher, On the star-discrepancy of sequences in the unit interval. Submitted, 2013.

[31] G. Larcher and F. Pillichshammer, Sums of distances to the nearest integer and the discrepancy of digital nets. *Acta Arith.* **106**, 379–408, 2003.

[32] C. Lemieux, *Monte Carlo and Quasi-Monte Carlo Sampling*. Springer Series in Statistics. Springer, New York, 2009.

[33] H. Niederreiter, Point sets and sequences with small discrepancy. *Monatsh. Math.* **104**, 273–337, 1987.

[34] H. Niederreiter, *Random Number Generation and Quasi-Monte Carlo Methods*. CBMS-NSF Series in Applied Mathematics, volume 63. SIAM, Philadelphia, PA, 1992.

[35] H. Niederreiter, $(t, m, s)$-nets and $(t, s)$-sequences. In: G. L. Mullen and D. Panario (eds.), *Handbook of Finite Fields*, pp. 619–630. CRC Press, Boca Raton, FL, 2013.

[36] H. Niederreiter and C. P. Xing, Low-discrepancy sequences and global function fields with many rational places. *Finite Fields Appl.* **2**, 241–273, 1996.

[37] H. Niederreiter and C. P. Xing, Quasirandom points and global function fields. In: S. Cohen and H. Niederreiter (eds.), *Finite Fields and Applications*. London Mathematical Society Lecture Note Series, volume 233, pp. 269–296. Cambridge University Press, Cambridge, 1996.

[38] V. Ostromoukhov, Recent progress in improvement of extreme discrepancy and star discrepancy of one-dimensional sequences. In: P. L'Ecuyer and A. Owen (eds.), *Monte Carlo and Quasi-Monte Carlo Methods 2008*, pp. 561–572. Springer, Berlin, 2009.

[39] F. Pillichshammer, Improved upper bounds for the star discrepancy of digital nets in dimension 3. *Acta Arith.* **108**, 167–189, 2003.

[40] F. Pillichshammer, On the discrepancy of (0, 1)-sequences. *J. Number Theory* **104**, 301–314, 2004.

[41] K. F. Roth, On irregularities of distribution. *Mathematika* **1**, 73–79, 1954.

[42] W. M. Schmidt, Irregularities of distribution VII. *Acta Arith.* **21**, 45–50, 1972.

[43] I. H. Sloan and S. Joe, *Lattice Methods for Multiple Integration*. Oxford University Press, New York, 1994.

[44] I. M. Sobol', Distribution of points in a cube and the approximate evaluation of integrals. *USSR Comput. Math. Math. Phys.* 7, 86–112, 1967.

[45] S. Tezuka, Polynomial arithmetic analogue of Halton sequences. *ACM Trans. Model. Comp. Sim.* **3**, 99–107, 1993.

# 5

# On the linear complexity and lattice test of nonlinear pseudorandom number generators

*Domingo Gómez-Pérez and Jaime Gutierrez*
University of Cantabria, Santander

*Dedicated to HN on the occasion of his 70th birthday.*

## Abstract

One of the main contributions which Harald Niederreiter made to mathematics is related to the theory of pseudorandom sequences. In this paper we study several measures for asserting the quality of pseudorandom sequences, involving generalizations of linear complexity and lattice tests and relations between them.

## 5.1 Introduction

Let $\mathbb{F}_q$ be the finite field with $q$ elements, where $q$ is an arbitrary prime power. Throughout the paper, we only consider purely periodic sequences $\mathcal{S} = (s_n) = (s_0, s_1, s_2, \ldots)$ of elements of $\mathbb{F}_q$ and we denote its period by $T$ with $T > 1$.

We recall that the *linear complexity* $\mathcal{L}(\mathcal{S})$ of the sequence $\mathcal{S}$ is the smallest positive integer $L$ for which there exist coefficients $a_0, a_1, \ldots, a_{L-1} \in \mathbb{F}_q$ such that

$$s_{n+L} = a_{L-1}s_{n+L-1} + \cdots + a_1 s_{n-1} + a_0 s_n, \qquad \forall n \geq 0.$$

Note that $\mathcal{L}(\mathcal{S})$ is the length of the shortest linear feedback shift register that can generate $\mathcal{S}$, so the following inequalities $1 \leq \mathcal{L}(\mathcal{S}) \leq T$ hold. The linear complexity of sequences is an important security measure for stream cipher systems (see [2, 11, 16, 24]). A measure closely related to the linear complexity is the lattice test. For a given integer $L \geq 1$, $\mathcal{S}$ passes the $L$-dimensional *lattice test* if the vectors

$$\{\vec{s}_n - \vec{s}_0 \ : \ \vec{s}_n = (s_n, s_{n+1}, \ldots, s_{n+L-1}), \quad \text{for } 0 \le n < T\},$$

span $\mathbb{F}_q^L$. The reason for the existence of this test comes from the use of pseudo-random numbers generated by linear congruences, which were introduced by Lehmer [9]. Although linear generators like the one mentioned are popular, they also have severe deficiencies that make them improper in many applications, such as cryptography. Even more general generators with low linear complexity turned out to be undesirable for more traditional applications in Monte Carlo methods as well, see [13, 14, 18].

One particularly undesirable feature of these pseudorandom number sequences is their coarse lattice structure. This is the reason why Marsaglia [10] proposed a test to measure this special structure. This test was investigated and enhanced by Harald Niederreiter and Arne Winterhof, see [19, 20].

However, this measure is closely related to the linear complexity. Relations between the lattice test and linear complexity for parts of the period are given in [3, 4, 5, 6].

The importance of the relationship comes from the fact that linear complexity is a well-understood concept. For example, the exact value of the number of sequences of a given length and linear complexity on finite fields is known, see [22, Theorem 7.1.6] and [12]. Indeed, the lattice structure has been thoroughly studied in pseudorandom number sequences generated for Monte Carlo methods and stream ciphers (see [7, 14, 15, 17, 21]). The following is a natural generalization of the linear complexity. We define the *quasi-linear complexity* $Q\mathcal{L}(\mathcal{S})$ of the sequence $\mathcal{S}$ as the smallest nonnegative integer $L$ for which there exist coefficients $a_0, a_1, \ldots, a_{L-1} \in \mathbb{F}_q$ and integers $0 < d_1 < \cdots < d_L < T$ such that

$$s_{n+d_L} = a_{L-1}s_{n+d_{L-1}} + \cdots + a_1 s_{n+d_1} + a_0 s_n, \qquad \forall n \ge 0. \qquad (5.1)$$

Obviously, we have $1 \le Q\mathcal{L}(\mathcal{S}) \le \mathcal{L}(\mathcal{S})$ so, in particular, $Q\mathcal{L}(\mathcal{S}) \le T$. We will see that this last concept coincides essentially with the *lattice test* introduced in [21]. For given integers $L \ge 1$, $0 < d_1 < \cdots < d_{L-1} < T$, $\mathcal{S}$ passes the $L$-dimensional *lattice test* with lags $d_1, \ldots, d_{L-1}$ if the vectors

$$\{\vec{s}_n - \vec{s}_0 \ : \ \vec{s}_n = (s_n, s_{n+d_1}, \ldots, s_{n+d_{L-1}}), \quad \text{for } 0 \le n < T\}$$

span $\mathbb{F}_q^L$. The greatest dimension $L$ such that $\mathcal{S}$ satisfies the $L$-dimensional lattice test for all lags $d_1, \ldots, d_{L-1}$ is denoted by $\mathcal{T}(\mathcal{S})$.

The main goal of this paper is to compare the three integers $\mathcal{L}(\mathcal{S})$, $Q\mathcal{L}(\mathcal{S})$ and $\mathcal{T}(\mathcal{S})$. It is divided into five sections. In Section 5.2 we obtain the relation between $Q\mathcal{L}(\mathcal{S})$ and $\mathcal{T}(\mathcal{S})$. The main result is presented in Section 5.3, where we obtain a nontrivial inequality relating the linear complexity and the quasi-linear complexity. Section 5.4 is devoted to applying the results to some pseudorandom number generators, and we show that our result rediscovers a special case of a previous result. Finally, we present an open problem in Section 5.5.

## 5.2 Lattice test and quasi-linear complexity

In this section we compare the two integers $Q\mathcal{L}(\mathcal{S})$ and $\mathcal{T}(\mathcal{S})$ and obtain the main result of the section.

**Theorem 5.1** *With the above notation, we have*

$$\mathcal{T}(\mathcal{S}) \le Q\mathcal{L}(\mathcal{S}) \quad and \quad Q\mathcal{L}(\mathcal{S}) \le 2\mathcal{T}(\mathcal{S}) + 2.$$

*Proof.* We fix the following notation, $Q\mathcal{L}(\mathcal{S}) = M$ and $\mathcal{T}(\mathcal{S}) = L$, so by the definition of $Q\mathcal{L}(\mathcal{S})$, there exist $d_1, \ldots, d_M$ and $a_0, \ldots, a_{M-1} \in \mathbb{F}_q$ such that,

$$s_{n+d_M} = a_{M-1}s_{n+d_{M-1}} + \cdots + a_1 s_{n+d_1} + a_0 s_n, \qquad \forall n \ge 0.$$

Since $(a_0, a_1, \ldots, a_{M-1}, -1)$ is a nonzero vector, then we denote by $H$ the following hyperplane

$$H = \{(x_0, \ldots, x_{M-1}, x_M) \in \mathbb{F}_q^{M+1} : a_0 x_0 + \cdots + a_{M-1}x_{M-1} - x_M = 0\}.$$

Moreover, the vectors $\vec{s}_n = (s_n, s_{n+d_1}, \ldots, s_{n+d_{M-1}}, s_{n+d_M}) \in H$, for $0 \le n < T$. So, $\{\vec{s}_n - \vec{s}_0 : 0 \le n < T\} \subset H$ which implies $L < M + 1$ and thus $\mathcal{T}(\mathcal{S}) \le Q\mathcal{L}(\mathcal{S})$.

On the other hand, by the definition of $\mathcal{T}(\mathcal{S})$, there exist integers $0 < d_1 < \cdots < d_{L-1} < d_L < T$ such that the vector space $V$ generated by $\{\vec{s}_n - \vec{s}_0 : 0 \le n < T\}$ is strictly contained in $\mathbb{F}_q^{L+1}$, where $\vec{s}_n = (s_n, s_{n+d_1}, \ldots, s_{n+d_{L-1}}, s_{n+d_L}) \in \mathbb{F}_q^{L+1}$, for $0 \le n < T$. So, there exits a nonzero vector $\vec{w} = (w_0, \ldots, w_L)$ satisfying $\langle \vec{w}, \vec{s}_n - \vec{s}_0 \rangle = 0$, where $\langle, \rangle$ denotes the usual inner product. We denote by $\delta$ the inner product $\langle \vec{w}, \vec{s}_0 \rangle$, then for $0 \le n$ the following equations hold

$$w_0 s_n + w_1 s_{n+d_1} + \cdots + w_L s_{n+d_L} = \delta,$$
$$w_0 s_{n+1} + w_1 s_{n+1+d_1} + \cdots + w_L s_{n+1+d_L} = \delta.$$
(5.2)

Notice that $w_L \ne 0$ because $\mathcal{T}(\mathcal{S}) = L$, and from Equations (5.2) we have

$$w_L s_{n+d_L+1} = w_0 s_n - w_0 s_{n+1} + \cdots + w_{L-1} s_{n+d_{L-1}} - w_{L-1} s_{n+1+d_{L-1}} + w_L s_{n+d_L}.$$

Now, we distinguish two cases: the previous equation is trivial or not. Notice that if the previous equation is not trivial, then it is of the form of (5.1). In the other case, the lags satisfy the following relation,

$$d_i = i \bmod T, \quad i = 1, \dots, L.$$

This implies that $L = T - 1$ so the inequality is satisfied trivially.    $\square$

The following example shows that we cannot relax the first of the inequalities.

**Example 5.2** *The following sequence is a defined in any field of odd characteristic, so these elements $\{0, 1, -1\}$ are different. Take $\mathcal{S} = (s_0, s_1, \dots)$ to be the sequence with even period defined by the following function,*

$$s_i = \begin{cases} -1 & \text{if } i = T/2 - 1 \\ 1 & \text{if } i = T - 1 \\ 0 & \text{otherwise.} \end{cases}$$

*It is clear that $Q\mathcal{L}(\mathcal{S}) = 1 = \mathcal{T}(\mathcal{S})$, the reason is that the sequence satisfies the following recurrence:*

$$s_{n+T/2} = -s_n, \quad \forall n \geq 0.$$

*It is even easy to see that $\mathcal{L}(\mathcal{S}) = T/2$.*

## 5.3  Quasi-linear and linear complexity

In this section we give a relationship between two measures, the linear complexity and the quasi-linear complexity, under the extra condition that the period is a power of a prime number, without any extra conditions on the field $\mathbb{F}_q$.

**Theorem 5.3** *If $T$ is a power of prime number $P$, $T = P^t$, then the following inequality holds*

$$Q\mathcal{L}(\mathcal{S}) + 1 \geq \frac{\log T}{t\,(\log T - \log \mathcal{L}(\mathcal{S}) + 2)},$$

*where* $\log$ *denotes the binary logarithm.*

We need the following result, which is proved in [25] for the linear complexity.

**Lemma 5.4** *Let a be a positive integer, we denote by $\mathcal{S}_a$ the sequence $(s_{na}) = (s_0, s_a, s_{2a}, \ldots)$. If $\gcd(a, T) = 1$ then $\mathcal{S}_a$ has period $T$, $\mathcal{L}(\mathcal{S}) = \mathcal{L}(\mathcal{S}_a)$, $Q\mathcal{L}(\mathcal{S}) = Q\mathcal{L}(\mathcal{S}_a)$ and $T(\mathcal{S}) = T(\mathcal{S}_a)$.*

*Proof.* We prove only $Q\mathcal{L}(\mathcal{S}) = Q\mathcal{L}(\mathcal{S}_a)$ and the proof of the other properties is done similarly. We write $L = Q\mathcal{L}(\mathcal{S})$, then there exist coefficients $a_0, a_1, \ldots, a_{L-1} \in \mathbb{F}_q$ and integers $0 < d_1 < \cdots < d_L < T$ satisfying Equation (5.1), for all $0 \leq n$. Evaluating Equation (5.1) in the integers of the form $na$, we have

$$s_{an+d_L} = a_{L-1}s_{an+d_{L-1}} + \cdots + a_1 s_{an+d_1} + a_0 s_{an}, \quad \forall n \geq 0.$$

We take the positive integer $1 \leq r < T$ such that $ar \equiv 1 \bmod T$, then there exists a positive integer $\lambda_i$ such that $d_i + \lambda_i T = ard_i$, for $i = 0, \ldots, L$. Since $T$ is the period, we obtain $s_{a(n+rd_i)} = s_{an+ard_i} = s_{an+d_i+\lambda_i T} = s_{an+d_i}$. So, we obtain:

$$s_{a(n+rd_L)} = a_{L-1}s_{a(n+rd_{L-1})} + \cdots + a_1 s_{a(n+rd_1)} + a_0 s_{an}.$$

This implies $Q\mathcal{L}(\mathcal{S}_a) \leq Q\mathcal{L}(\mathcal{S})$. To conclude the proof, we consider the sequence $\mathcal{S}_{ar}$, i.e., $(s_{nar}) = (s_0, s_{ar}, s_{2ar}, \ldots)$. By the same argument used in the above part of the proof, we have $Q\mathcal{L}(\mathcal{S}_{ar}) \leq Q\mathcal{L}(\mathcal{S}_a)$. Clearly, $(s_{nar}) = (s_n)$, because $ar \equiv 1 \bmod T$. $\qquad \square$

Another trivial remark, which we will use in the proof, is the following connecting the values of the lags and the linear complexity.

**Remark 5.5** Let $\Delta$ be a positive integer, $a_0, a_1, \ldots, a_{L-1}, a_L$ nonzero elements of $\mathbb{F}_q$ and integers satisfying $(-\Delta) \leq d_1 < d_2 < \cdots < d_L \leq \Delta$ and suppose that the sequence $\mathcal{S}$ satisfies the quasi-linear recurrence $a_L s_{n+d_L} = a_{L-1}s_{n+d_{L-1}} + \cdots + a_1 s_{n+d_1} + a_0 s_n$, for all $\Delta \leq n$, then it is trivial that $\mathcal{L}(\mathcal{S}) \leq 2\Delta$.

Lattice theory will play an important role in the proof of Theorem 5.3, especially the well-known Minkowski theorem.

Let $\{\vec{b}_1, \ldots, \vec{b}_s\} = B$ be a set of linearly independent vectors in $\mathbb{R}^r$. The set

$$\Lambda = \{\vec{z} : \vec{z} = c_1\vec{b}_1 + \cdots + c_s\vec{b}_s, \quad c_1, \ldots, c_s \in \mathbb{Z}\}$$

is called an *s-dimensional lattice* with *basis* $\{\vec{b}_1, \ldots, \vec{b}_s\}$.

To each lattice $\Lambda$ one can naturally associate its *volume*

$$\text{vol}\,(\Lambda) = \left( \det \left( \langle \vec{b}_i, \vec{b}_j \rangle_{i,j=1}^s \right) \right)^{1/2},$$

which does not depend on the choice of the basis $\{\vec{b}_1, \ldots, \vec{b}_s\}$.

For a vector $\vec{u}$, let $\|\vec{u}\|$ denote its *infinity norm*. The famous Minkowski theorem, see Theorem 5.3.6 in Section 5.3 of [8], gives the upper bound

$$\min \left\{ \|\vec{z}\| : \ \vec{z} \in \Lambda \setminus \{\vec{0}\} \right\} \le \text{vol}\,(\Lambda)^{1/s} \qquad (5.3)$$

on the shortest nonzero vector in any $s$-dimensional lattice $\Lambda$ in terms of its volume. Now, we have all the ingredients to prove Theorem 5.3.

*Proof of Theorem 5.3.* We write $M = Q\mathcal{L}(\mathcal{S})$, then there exits an equation of the form (5.1). By definition of $Q\mathcal{L}(\mathcal{S})$, there exist coefficients $b_0, b_1, \ldots, b_{M-1} \in \mathbb{F}_q$ and integers $0 < d_1 < \cdots < d_M < T$ such that

$$s_{n+d_M} = b_{M-1} s_{n+d_{M-1}} + \cdots + b_1 s_{n+d_1} + b_0 s_n, \qquad \forall n \ge 0. \qquad (5.4)$$

We introduce the following notation,

$$d_i = \overline{d_i} P^{t-1} + r_i, \quad 0 \le r_i < P^{t-1}, \quad 0 \le \overline{d_i} < P, \quad i = 1, \ldots, M, \quad (5.5)$$

and consider the lattice $\Lambda$ generated by the columns of the following matrix $B \in \mathbb{Z}^{M+1 \times M+1}$

$$\begin{pmatrix} \overline{d_M} & 0 & \ldots & 0 & P \\ \overline{d_{M-1}} & 0 & \ldots & P & 0 \\ \vdots & \vdots & \vdots & \vdots & \vdots \\ \overline{d_1} & P & \ldots & 0 & 0 \\ 1 & 0 & \ldots & 0 & 0 \end{pmatrix}.$$

Clearly, the volume of the lattice is $\text{vol}\,(\Lambda) = P^M$, so, by the Minkowski theorem (5.3), if $\vec{v}$ is the shortest vector in the lattice with infinity norm, then

$$\|\vec{v}\| \le P^{1 - \frac{1}{M+1}}. \qquad (5.6)$$

Since $\vec{v} \in \Lambda$ there exist integers $\lambda_i, i = 1, \ldots, M$ and a positive integer $R$ such that

$$\vec{v} = (R\overline{d_M} + \lambda_M P, \ldots, R\overline{d_1} + \lambda_1 P, R).$$

Notice that $\gcd(R, P) = \gcd(R, T) = 1$, because $R < P$ and $P$ is a prime. Now, we consider $a$ satisfying $aR \equiv 1 \bmod T$ and the sequence

$\mathcal{S}_a = (s_0, s_a, s_{2a}, \ldots,) = (s_0', s_1', s_2' \ldots)$. From Equation (5.4), $\mathcal{S}_a$ satisfies the following quasi-linear recurrence:

$$s_{n+Rd_M}' = b_{M-1}s_{n+Rd_{M-1}}' + \cdots + b_1's_{n+Rd_1}' + b_0s_n', \qquad \forall n \geq 0. \quad (5.7)$$

On the other hand, since $T$ is the period of the sequence $\mathcal{S}_a$ and recalling Equation (5.5), we have,

$$s_{n+Rd_i}' = s_{n+\overline{Rd_i}P^{t-1}+Rr_i}' = s_{n+\overline{Rd_i}P^{t-1}+Rr_i+\lambda_i P^t}' = s_{n+(\overline{Rd_i}+\lambda_i P)P^{t-1}+Rr_i}'$$

for all $i = 1, \ldots, M$. Then from (5.7), we get

$$s_{n+Rd_M+\lambda_M T}' = b_{M-1}s_{n+Rd_{M-1}+\lambda_{M-1}T}' + \cdots + b_1 s_{n+Rd_1+\lambda_1 T}' + b_0 s_n', \qquad \forall n \geq 0.$$

By the bound (5.6), we have for $i = 1, \ldots, M$,

$$|Rd_i + \lambda_i T| = |(\overline{Rd_i} + \lambda_i P)P^{t-1} + Rr_i| \leq 2P^{t-1}P^{1-\frac{1}{M+1}} = 2T^{1-\frac{1}{t(M+1)}}.$$

By Remark 5.5, $\mathcal{L}(\mathcal{S}_a) \leq 4T^{1-\frac{1}{t(M+1)}}$. Now, using Theorem 5.4

$$\mathcal{L}(\mathcal{S}) = \mathcal{L}(\mathcal{S}_a) \leq 4T^{1-\frac{1}{t(M+1)}},$$

and operating we obtain the result.                                    □

This result generalizes the one presented in [1]. Indeed, in the particular case that $T$ is a prime number, we can give this improved result.

**Corollary 5.6** *If $T$ is a prime number, then the following inequality*

$$Q\mathcal{L}(\mathcal{S}) \geq \frac{\log T}{\log T - \log \mathcal{L}(\mathcal{S}) + 1},$$

*holds, where* log *denotes the binary logarithm.*

## 5.4 Applications of our results

This bound is very general and is applicable to several pseudorandom number generators. Apart from the fact that the period must be a power of a prime number, the other condition required to obtain a nontrivial result for the lattice test or the quasi-linear complexity is that the linear complexity of the pseudorandom number generator is known to be large.

There are several sequences which have large linear complexity, nearly as big as the period, see surveys [26, 27]. For example, if $\mathcal{S}$ is the inverse recursive generator and has prime period, then $S$ has linear complexity greater than $(T-1)/2$. This implies that,

$$QL(S) \geq \frac{\log T}{3}.$$

Similar bounds can be used to find bounds for the Legendre sequence and the Sidelnikov sequence. Here, we want to comment an another application of our result where the best results known are of similar strength.

Pirsic and Winterhof [23] studied lattice tests for digital explicit inversive generators and they obtained bounds, even in parts of the sequence. We cite their result only in the case of full period.

**Theorem 5.7** *Let $S$ be a sequence arising from a digital explicit inverse generator defined over $\mathbb{F}_q$ with $q = p^t$, then we have that,*

$$T(S) \geq \frac{\log T - \log \log T - 1}{t-1} - 1,$$

*if $t > 1$. For $t = 1$ the inequality*

$$T(S) \geq \frac{T}{2} - 1,$$

*holds.*

To apply Theorem 5.3, we need the following bound from [28]. We cite it restricted to the special case of a sequence arising from a digital explicit inverse generator.

**Lemma 5.8** *Let $S$ be a sequence arising from a digital explicit inversive generator defined over $\mathbb{F}_q$. Then we have*

$$L(S) \geq \frac{q(p-1)}{p} \geq \frac{q}{2}.$$

This result and a direct application of Theorem 5.3 gives

$$QL(S) \geq \frac{\log T}{3t} - 1.$$

Using that $QL(S) \leq T(S)$, we obtain a lower bound which is of the same order as the result obtained in Theorem 5.7.

Although our bound seems to be weak, it is also quite general. Indeed, for sequences defined by Fermat quotients, we know the exact value of the quasi-linear complexity, which is two, and our bound only gives that the quasi-linear complexity is greater than one.

## 5.5  An open problem

We think that Theorem 5.3 can be formulated for sequences of period $T$ under some restrictions, but not necessarily power of a prime number. However, Example 5.2 shows that it is not true for arbitrary $T$, but software computations show that our bounds hold in many cases and we think that, under some mild restrictions, it should be possible to prove a lower bound in the quasi-linear complexity depending only on the linear complexity and the period. Also, we would like to determine a framework in which to study the real value of the quasi-linear complexity, as in the linear complexity case.

## Acknowledgements

The authors want to thank Arne Winterhof for very valuable discussions during our visit to Banff International Research Station and for the preparation of the paper. We also want to thank the anonymous referee for comments which improved the paper. This work is supported in part by the Spanish Ministry of Science, project MTM2011-24678.

## References

[1] Z. Chen, D. Gomez and G. Pirsic, On lattice profile of the elliptic curve linear congruential generators. *Period. Math. Hung.* **68**, 1–12, 2014.

[2] T. W. Cusick, C. Ding and A. Renvall, *Stream Ciphers and Number Theory*. North-Holland Mathematical Library, volume 55. North-Holland, Amsterdam, 1998.

[3] G. Dorfer, Lattice profile and linear complexity profile of pseudorandom number sequences. In: G. L. Mullen, A. Poli and H. Stichtenoth (eds.), *International Conference on Finite Fields and Applications*. Lecture Notes in Computer Science, volume 2948, pp. 69–78. Springer, 2003.

[4] G. Dorfer and A. Winterhof, Lattice structure and linear complexity profile of nonlinear pseudorandom number generators. *Appl. Algebra Eng. Commun. Comput.* **13**(6), 499–508, 2003.

[5] G. Dorfer and A. Winterhof, Lattice structure of nonlinear pseudorandom number generators in parts of the period. *Monte Carlo and quasi-Monte Carlo Methods 2002*, pp. 199–211. Springer, New York, 2004.

[6] G. Dorfer, W. Meidl and A. Winterhof, Counting functions and expected values for the lattice profile at $n$. *Finite Fields Appl.* **10**(4), 636–652, 2004.

[7] J. Eichenauer-Herrmann, E. Herrmann and S. Wegenkittl, A survey of quadratic and inversive congruential pseudorandom numbers. *Monte Carlo and quasi-Monte Carlo Methods*. Lecture Notes in Statistics, volume 127, pp. 66–97. Springer, Salzburg, 1996.

[8]   M. Grötschel, L. Lovász and A. Schrijver, *Geometric Algorithms and Combinatorial Optimization*. Springer, Berlin, 1993.

[9]   D. H. Lehmer, Mathematical methods in large-scale computing units. *Proceedings of a Second Symposium on Large-Scale Digital Calculating Machinery, 1949*, pp. 141–146. Harvard University Press, Cambridge, MA, 1951.

[10]  G. Marsaglia, The structure of linear congruential sequences. *Applications of Number Theory to Numerical Analysis (Proc. Symp., Univ. Montreal, Montreal, 1971)*, pp. 249–285. Academic Press, New York, 1972.

[11]  A. J. Menezes, P. C. van Oorschot and S. A. Vanstone, *Handbook of Applied Cryptography*. CRC Press, Boca Raton, FL, 1997.

[12]  H. Niederreiter, The linear complexity profile and the jump complexity of keystream sequences. In: I. Damgård (ed.), *EUROCRYPT*. Lecture Notes in Computer Science, volume 473, pp. 174–188. Springer, 1990.

[13]  H. Niederreiter, New methods for pseudorandom numbers and pseudorandom vector generation. *Winter Simulation Conference*, pp. 264–269. ACM Press, 1992.

[14]  H. Niederreiter, *Random Number Generation and quasi-Monte Carlo Methods*. CBMS-NSF Regional Conference Series in Applied Mathematics, volume 63. Society for Industrial and Applied Mathematics (SIAM), Philadelphia, PA, 1992.

[15]  H. Niederreiter. New developments in uniform pseudorandom number and vector generation. *Monte Carlo and quasi-Monte Carlo Methods*. Lecture Notes in Statistics, volume 107, pp. 87–120. Springer, Las Vegas, NV, 1994.

[16]  H. Niederreiter, Some computable complexity measures for binary sequences. *Proc. Int. Conf. on Sequences and their Applications (SETA'98)*, pp. 67–78. Springer, Singapore, 1999.

[17]  H. Niederreiter, Linear complexity and related complexity measures for sequences. In: T. Johansson and S. Maitra (eds.), *INDOCRYPT*. Lecture Notes in Computer Science, volume 2904, pp. 1–17. Springer, 2003.

[18]  H. Niederreiter and I. Shparlinski, Recent advances in the theory of nonlinear pseudorandom number generators. In: K.-T. Fang, Harald Niederreiter and F.J. Hickernell (eds.), *Monte Carlo and Quasi-Monte Carlo Methods 2000*, pp. 86–102. Springer, Berlin, 2002.

[19]  H. Niederreiter and A. Winterhof, On the lattice structure of pseudorandom numbers generated over arbitrary finite fields. *Appl. Algebra Eng. Commun. Comput.* **12**(3), 265–272, 2001.

[20]  H. Niederreiter and A. Winterhof, Lattice structure and linear complexity of nonlinear pseudorandom numbers. *Appl. Algebra Eng. Commun. Comput.* **13**(4), 319–326, 2002.

[21]  H. Niederreiter and A. Winterhof, On the structure of inversive pseudorandom number generators. *Applied Algebra, Algebraic Algorithms and Error-Correcting Codes*. Lecture Notes in Computer Science, volume 4851, pp. 207–216. Springer, 2007.

[22]  H. Niederreiter and C. Xing, *Rational Points on Curves over Finite Fields: Theory and Applications*. London Mathematical Society Lecture Note Series, volume 285. Cambridge University Press, Cambridge, 2001.

[23]  G. Pirsic and A. Winterhof, On the structure of digital explicit nonlinear and inversive pseudorandom number generators. *J. Complexity* **26**(1), 43–50, 2010.

[24]   R. Rueppel, Stream ciphers. *Contemporary Cryptology: The Science of Information Integrity*, pp. 65–134. IEEE Press, New York, 1992.

[25]   I. Shparlinski, On the uniformity of distribution of the Naor Reingold pseudorandom function. *Finite Field Appl.* **7**(2), 318–326, 2001.

[26]   A. Topuzoğlu and A. Winterhof, Pseudorandom sequences. *Topics in Geometry, Coding Theory and Cryptography*. Algebra and Applications, volume 6, pp. 135–166. Springer, Dordrecht, 2007.

[27]   A. Winterhof, Recent results on recursive nonlinear pseudorandom number generators (invited paper). In: C. Carlet and A. Pott (eds.), *Sequences and their Applications, SETA 2010*. Lecture Notes in Computer Science, volume 6338, pp. 113–124. Springer, Berlin, 2010.

[28]   A. Winterhof and W. Meidl, On the linear complexity profile of explicit nonlinear pseudorandom numbers. *Inf. Process. Lett.* **85**, 13–18, 2003.

# 6

# A heuristic formula estimating the keystream length for the general combination generator with respect to a correlation attack

*Rainer Göttfert*
Infineon Technologies AG, Munich

*Dedicated to Harald Niederreiter on the occasion of his 70th birthday.*

## Abstract

A formula is derived estimating the keystream length for the general combination generator with regard to a correlation attack that combines Siegenthaler's correlation attack with Matsui's linear cryptanalysis.

## 6.1 The combination generator

The general combination generator consists of $n$ binary nonsingular feedback shift registers (FSRs) and a Boolean combining function $F : \mathbb{F}_2^n \to \mathbb{F}_2$. The shift registers are initialized with the key $K$ and subsequently produce individual output sequences $\sigma_1, \ldots, \sigma_n$ of respective periods $p_1, \ldots, p_n$. The $n$ sequences produced by the $n$ shift registers are compressed into one single sequence $\zeta = (z_i)_{i=0}^{\infty}$, called the keystream, by the combining function $F$. That is,

$$\zeta = F(\sigma_1, \ldots, \sigma_n).$$

## 6.2 The model

To analyze the behavior of the combination generator in conjunction with correlation attacks, the combination generator is emulated by the following model. For $j = 1, \ldots, n$, the FSR sequence $\sigma_j$ is modeled by the sequence

$$\mathbf{X}_j = (X_{j,0}, X_{j,1}, \ldots, X_{j,p_j-1})^{\infty}$$

of statistically independent and symmetrically distributed binary valued random variables $X_{j,k}$, $k = 0, 1, \ldots, p_j - 1$, that are repeated periodically. Random variables belonging to different sequences are assumed to be statistically independent as well.

The keystream $\zeta = (z_i)_{i=0}^{\infty}$ is modeled by the sequence $Z = (Z_i)_{i=0}^{\infty}$ of random variables $Z_i$ defined by

$$Z_i = F(X_{1,i}, \ldots, X_{n,i}) \quad \text{for } i = 0, 1, \ldots.$$

Using a more compact notation, we write

$$\mathbf{Z} = F(\mathbf{X}_1, \ldots, \mathbf{X}_n).$$

## 6.3 Preliminaries

We briefly recall some notions that will appear in later parts of the chapter. A *Boolean function* in $n$ variables is a mapping from $\mathbb{F}_2^n$ into $\mathbb{F}_2$. The Boolean function $f$ is called balanced if $f(\mathbf{a}) = 1$ for exactly $2^{n-1}$ arguments $\mathbf{a} \in \mathbb{F}_2^n$. A Boolean function $f$ in $n$ variables is *correlation immune* of order $c$ if knowing the value of any $c$ of the input variables of $f$ gives no additional information on the value of the output of $f$. The *Hamming distance* $d(f, g)$ between two Boolean functions $f$ and $g$ in $n$ variables is defined by $d(f, g) = |\{\mathbf{a} \in \mathbb{F}_2^n : f(\mathbf{a}) \neq g(\mathbf{a})\}|$. The *nonlinearity* of $f$ is defined by

$$\mathrm{NL}(f) = \min_{l} d(f, l),$$

where $l$ runs through all the $2^{n+1}$ affine functions in $n$ variables.

An $n$-stage binary feedback shift register (FSR) is called *nonsingular* if its feedback function $F$ has the form

$$F(x_0, x_1, \ldots, x_{n-1}) = x_0 + f(x_1, \ldots, x_{n-1}),$$

where $f$ is an arbitrary Boolean function of the $n - 1$ variables $x_1, \ldots, x_{n-1}$. The next state function of a nonsingular FSR is bijective. Any output sequence of a nonsingular FSR is purely periodic.

## 6.4 The correlation attack

The following correlation attack was proposed in [2], extended in [3] and [1], and supported with proofs in [1].

Let $m$ be an integer with $1 \leq m \leq n$, and let $M = \{j_1, \ldots, j_m\}$ be a subset of $\{1, \ldots, n\}$ of cardinality $|M| = m$. Decompose $M$ into $k$ pairwise disjoint subsets $M_1, \ldots, M_k$ and compute the least common multiples

$$q_j = \text{lcm}(p_t : t \in M_j), \qquad 1 \le j \le k.$$

Let $i_1, \ldots, i_k$ be arbitrary positive integers, and define $r_j = i_j q_j$ for $1 \le j \le k$. The binary polynomial

$$g(x) = \prod_{j=1}^{k} (x^{r_j} - 1) = \sum_{d \in D} x^d \tag{6.1}$$

is a characteristic polynomial of the sequence

$$\mathbf{X} = \mathbf{X}_{j_1} + \cdots + \mathbf{X}_{j_m},$$

which means that the operator $g(T)$ annihilates the sequence $\mathbf{X}$, where $T : (s_i)_{i=0}^{\infty} \mapsto (s_{i+1})_{i=0}^{\infty}$ is the shift operator.

The linear operator $g(T)$ applied to $\mathbf{Z} = (Z_i)_{i=0}^{\infty}$ yields the sequence $g(T)\mathbf{Z} = \mathbf{Y} = (Y_i)_{i=0}^{\infty}$ of random variables $Y_i = \sum_{d \in D} Z_{i+d}, i = 0, 1, \ldots$.

If $e$ is an integer and $b$ is a positive integer, then $e \bmod b$ denotes the uniquely determined integer $a$ with $0 \le a \le b - 1$ and $a \equiv e \bmod b$. For a set $E$ of nonnegative integers and $b \ge 1$, we define $E_{\bmod b} = \{e \bmod b : e \in E\}$.

The following theorem was proved in [1].

**Theorem 6.1** *Let $F : \mathbb{F}_2^n \to \mathbb{F}_2$ be balanced and correlation immune of order $c \ge 0$. If $m \le c + 1$ and $|D_{\bmod p_j}| = 2^k$ for all $1 \le j \le n$ with $j \notin M$, then*

$$\Pr(Y_i = 0) = \frac{1}{2}\left(1 + \varepsilon^{2^k}\right) \qquad \text{for } i = 0, 1, \ldots,$$

*where $\varepsilon = 2\Pr(F = L) - 1$ is the correlation coefficient between the combining function $F(x_1, \ldots, x_n)$ and the linear function $L(x_1, \ldots, x_n) = x_{j_1} + \cdots + x_{j_m}$.*

Consider the polynomial $g(x)$ in (6.1). Two properties of $g(x)$ are significant for the correlation attack: (i) the degree $\deg(g)$ of $g$; (ii) the weight $w(g)$ of $g$.

We call the polynomial $g(x)$ nondegenerate if $|D_{\bmod p_j}| = 2^k$ for all $j \in \{1, \ldots, n\}$ with $j \notin M$. If $g(x)$ is nondegenerate, then $w(g) = 2^k$ (while the converse is not true).

To simplify matters we will assume in the following that all deployed characteristic polynomials $g$ of the type in (6.1) are nondegenerate.

In the correlation attack, the linear operator $g(T)$ is applied to a keystream segment $(z_0, z_1, \ldots, z_{A-1})$ of length $A$. This yields a sequence $(y_0, y_1, \ldots, y_{B-1})$ of length

$$B = A - \deg(g). \tag{6.2}$$

In the model this corresponds to applying the operator $g(T)$ to $(Z_0, Z_1, \ldots, Z_{A-1})$ producing a sequence of $B$ dependent random variables

$(Y_0, Y_1, \ldots, Y_{B-1})$. The element $y_i$ can be regarded as the outcome of the random variable $Y_i$. We call the element $y_i$ a *decision bit*. The larger the degree of $g(x)$, the fewer decision bits we get according to (6.2).

All the random variables $Y_0, Y_1, \ldots$ are equally strong biased towards zero. In fact, by Theorem 6.1,

$$\Pr(Y_i = 0) = \frac{1}{2}\left(1 + \varepsilon^{w(g)}\right) \qquad \text{for } 0 \le i \le B - 1. \qquad (6.3)$$

Thus, the smaller the weight of $g(x)$, the greater the bias, and the more useful is $Y_i$ for the correlation attack. In setting up the most powerful characteristic polynomial $g(x)$, the effect of both $\deg(g)$ and $w(g)$, must be taken into account.

The correlation attack requires a sufficiently large number of decision bits (among which the ratio of zeros versus ones is pivotal). How many decision bits are needed for the attack? For independent decision bits (i.e., decision bits that are realizations of statistically independent random variables) the answer is well known (e.g., see Theorem 2 in [1]). For a distinguishing attack, the least number $n_0$ of independent decision bits required in the attack is given by

$$n_0 = \varepsilon^{-2^{k+1}}. \qquad (6.4)$$

For a guess and determine attack in which the states of $r$ shift registers of lengths $N_1, \ldots, N_r$ are guessed, the least number $n_0$ of required decision bits is given by

$$n_0 = 2(N_1 + \cdots + N_r) \ln 2\varepsilon^{-2^{k+1}}. \qquad (6.5)$$

If that many independent decision bits are known, the states of the $r$ target shift registers can be determined with probability greater than $1/2$.[1]

Notice, however, that the random variables $Y_i$ in $(Y_i)_{i=0}^{\infty} = g(T)(Z_i)_{i=0}^{\infty}$ are not statistically independent. Therefore, even the value for $n_0$ given in (6.5) may not necessarily be sufficient.[2]

## 6.5 The formula

An idea already applied in [1] is to use not just one but several characteristic polynomials on the known keystream segment. Let

$$g(x) = (x^{q_1} - 1) \cdots (x^{q_k} - 1)$$

---

[1] Notice that in the guess and determine attacks, in [2] and [3] the incorrect value (6.4) was used, leading to incorrect estimations for the data complexity of the attacks.
[2] Although computer simulations seem to support its sufficiency. See [1].

be a well-chosen characteristic polynomial. Let $n_0$ be taken from (6.4) or (6.5), depending on whether a distinguishing or a guess and determine attack is planned. Divide the periods $q_1, \ldots, q_k$ into two disjoint subsets. The first subset contains all periods that are smaller than $n_0$. The second subset contains all periods that are greater than or equal to $n_0$. For ease of notation let us assume that the first subset contains the periods $q_1, \ldots, q_h$, and the second subset contains the periods $q_{h+1}, \ldots, q_k$. Consider the collection of characteristic polynomials

$$g_{i_1 \ldots i_h} = (x^{i_1 q_1} - 1) \cdots (x^{i_h q_h} - 1)(x^{q_{h+1}} - 1) \cdots (x^{q_k} - 1) \qquad (6.6)$$

with $1 \leq i_1 \leq u_1, \ldots, 1 \leq i_h \leq u_h$. The objective is to determine the integer values $u_1, \ldots, u_h$ and the amount of keystream needed to recruit $n_0$ decision bits.

**Theorem 6.2** *From the keystream segment* $(z_0, z_1, \ldots, z_{A-1})$ *of length*

$$A = \frac{h+1}{2} \sqrt[h+1]{2n_0 q_1 \cdots q_h} + \frac{1}{2}(q_1 + \cdots + q_h) + q_{h+1} + \cdots + q_k, \qquad (6.7)$$

$n_0$ *decision bits can be derived.*

*Proof.* Consider the keystream segment $(z_0, z_1, \ldots, z_{A-1})$ of length $A$. By applying the linear operators $g_{i_1 \ldots i_h}(T)$ to the given keystream segment of length $A$, we obtain $A - \deg(g_{i_1 \ldots i_h}(x))$ decision bits. By applying all $u = u_1 \cdots u_h$ linear operators $g_{i_1 \ldots i_h}(T)$, $1 \leq i_1 \leq u_1, \ldots, 1 \leq i_h \leq u_h$, we get

$$\sum_{i_1=1}^{u_1} \cdots \sum_{i_h=1}^{u_h} [A - \deg(g_{i_1 \ldots i_h}(x))]$$

decision bits. We need a total of $n_0$ decision bits. Therefore, we set

$$n_0 = \sum_{i_1=1}^{u_1} \cdots \sum_{i_h=1}^{u_h} [A - \deg(g_{i_1 \ldots i_h}(x))]$$

$$= \sum_{i_1=1}^{u_1} \cdots \sum_{i_h=1}^{u_h} [A - (i_1 q_1 + \cdots + i_h q_h + q_{h+1} + \cdots + q_k)]$$

$$= u_1 \cdots u_h A - \sum_{i_1=1}^{u_1} \cdots \sum_{i_h=1}^{u_h} (i_1 q_1 + \cdots + i_h q_h + q_{h+1} + \cdots + q_k)$$

$$= uA - \left[ q_1 u \frac{u_1 + 1}{2} + \cdots + q_h u \frac{u_h + 1}{2} + q_{h+1} u + \cdots + q_k u \right].$$

It follows that

$$uA = n_0 + u \sum_{j=1}^{h} q_j \frac{u_j + 1}{2} + u \sum_{j=h+1}^{k} q_j.$$

Therefore,

$$A = \frac{n_0}{u} + \sum_{j=1}^{h} q_j \frac{u_j + 1}{2} + \sum_{j=h+1}^{k} q_j. \tag{6.8}$$

We regard $A$ as a function of the variables $u_1, \ldots, u_h$. Although the variables $u_1, \ldots, u_h$ represent integers, we treat them for the moment as real variables. We want to determine the minimum of the real valued function $A = A(u_1, \ldots, u_h)$ (defined on some domain of $\mathbb{R}^h$). Therefore we compute the partial derivatives of the first order and set them to zero:

$$\frac{\partial A}{\partial u_j} = \frac{-n_0}{u u_j} + \frac{q_j}{2} = 0 \quad \text{for } 1 \le j \le h.$$

We get

$$\frac{q_j u_j}{2} = \frac{n_0}{u} \quad \text{for } 1 \le j \le h. \tag{6.9}$$

This implies

$$u_j = \frac{q_1}{q_j} u_1 \quad \text{for } 1 \le j \le h. \tag{6.10}$$

Next we want to determine $u_1$. From (6.9), we get

$$u_1^2 u_2 \cdots u_h = \frac{2n_0}{q_1}.$$

Using the relations in (6.10), we obtain

$$u_1^{h+1} = \frac{2n_0 q_2 \cdots q_h}{q_1^h}.$$

Hence,

$$u_1 = \sqrt[h+1]{\frac{2n_0 q_2 \cdots q_h}{q_1^h}} = \frac{1}{q_1} \sqrt[h+1]{2n_0 q_1 \cdots q_h}. \tag{6.11}$$

According to (6.10), we have

$$q_j u_j = q_1 u_1 \quad \text{for } 1 \le j \le h.$$

In conjunction with (6.11) this yields

$$u_j = \frac{1}{q_j} \sqrt[h+1]{2n_0 q_1 \cdots q_h} \quad \text{for } 1 \le j \le h. \tag{6.12}$$

We are now able to determine $A$. We write (6.8) in the form

$$A = \frac{n_0}{u} + \sum_{j=1}^{h} \frac{q_j u_j}{2} + \frac{1}{2} \sum_{j=1}^{h} q_j + \sum_{j=h+1}^{k} q_j.$$

According to (6.9), the first $h + 1$ summands

$$\frac{n_0}{u}, \; \frac{q_1 u_1}{2}, \; \ldots, \; \frac{q_h u_h}{2}$$

are identical with each other. According to (6.12), they are all equal to

$$\frac{1}{2} \sqrt[h+1]{2 n_0 q_1 \cdots q_h}$$

and the formula follows.                                          □

Since the produced decision bits are not statistically independent, it is not guaranteed that the correlation attack can be mounted on a keystream segment of the length described by the formula in the theorem. However, if the periods $q_1 \cdots q_h$ are not much smaller than $n_0$, computer simulations suggest that this is the case. If the numbers $q_1 \cdots q_h$ are much smaller than $n_0$ the dependencies among decision bits increase and the formula gains the flavor of a lower bound.

## Acknowledgement

The author is very grateful to Arne Winterhof and Gerhard Larcher for their generous handling of deadline matters, and to the anonymous referee for valuable comments.

## References

[1]  R. Göttfert and B. Gammel, On the frame length of Achterbahn-128/80. *Proceedings of the 2007 IEEE Information Theory Workshop on Information Theory for Wireless Networks*, pp. 1–5. IEEE, New York, 2007.

[2]  T. Johansson, W. Meier and F. Muller, Cryptanalysis of Achterbahn. *FSE 2006*. Lecture Notes in Computer Science, volume 4047, pp. 1–14. Springer-Verlag, Berlin, 2006.

[3]  M. Naya-Plasencia, Cryptanalysis of Achterbahn-128/80. *SASC 2007, The State of the Art of Stream Ciphers (Bochum, 2007)*, Workshop Record, pp. 139–151.

# 7

# Point sets of minimal energy

*Peter J. Grabner*
Technische Universität, Graz

*Dedicated to Harald Niederreiter on the occasion of his 70th birthday.*

## Abstract

For a given compact manifold $M \subset \mathbb{R}^{d+1}$ and a set of $N$ distinct points $X_N = \{x_1, \ldots, x_N\} \subset M$, the Riesz $s$-energy is defined as $E_s(X_N) = \sum_{i \neq j} \|x_i - x_j\|^{-s}$. A configuration $X_N^*$, which minimizes $E_s$ amongst all $N$-point configurations, is called a minimal energy configuration. The motivation for studying such configurations comes from chemistry and physics, where self-organization of mutually repelling particles under inverse power laws occurs. For fixed $s$ and $N \to \infty$ the distribution $\frac{1}{N} \sum_{j=1}^{N} \delta_{x_j^*}$ approaches a continuous limiting measure, which can be described by classical potential theory, if $s < \dim(M)$, and which is a normalized Hausdorff measure, if $s \geq \dim(M)$ by a recent result of D. P. Hardin and E. B. Saff.

We collect recent results on the discrepancy of minimal energy point sets on manifolds, especially the sphere.

## 7.1 Introduction

Different types of constructions have been used to find "good" configurations of $N$ points $X_N = \{\mathbf{x}_1, \ldots, \mathbf{x}_N\}$ on a manifold $M$, especially the unit sphere $\mathbb{S}^d$ in $\mathbb{R}^{d+1}$. Of course the construction depends on what quantitative measure is used for the configuration. In this survey we will discuss two such measures and their interrelation. We will mostly emphasize the latest results and developments in the context of minimal energy point configurations.

The **discrepancy** of a point set $X_N$ is given by

$$D(X_N) = \sup_{C} \left| \frac{1}{N} \sum_{n=1}^{N} \chi_C(\mathbf{x}_n) - \sigma(C) \right|, \tag{7.1}$$

where the supremum is extended over a system of Riemann-measurable sub-sets of $M$ (for instance all spherical caps in the case of the sphere), $\chi_C$ denotes the indicator function of the set $C$, and $\sigma$ is a normalized measure on $M$ (the normalized surface area measure in the case of the sphere). This is a classical measure for the quality of a finite point distribution approximating a measure, which has been studied intensively in the theory of uniform distribution (cf. [21, 31]) as well as in the theory of irregularities of distribution (cf. [2]).

The **energy** of a point set $X_N$ is defined as

$$E_s(X_N) = \sum_{\substack{i,j=1 \\ i \neq j}}^{N} \|\mathbf{x}_i - \mathbf{x}_j\|^{-s} \tag{7.2}$$

for a positive real parameter $s$. This is a discrete version of the energy integral

$$I_s(\mu) = \iint_{M \times M} \|\mathbf{x} - \mathbf{y}\|^{-s} \, d\mu(\mathbf{x}) \, d\mu(\mathbf{y}). \tag{7.3}$$

The discrete distributions of point sets $X_N^*$ minimizing $E_s$,

$$\nu_N = \frac{1}{N} \sum_{n=1}^{N} \delta_{\mathbf{x}_n^*},$$

are studied then for $N \to \infty$. The minimization of such discrete energy expressions is a problem attributed to Fekete. Minimal energy point sets are thus called Fekete-points. The case $s < \dim(M)$ can be investigated by methods from classical potential theory (cf. [32]). In this case the unique minimizer $\mu_M^{(s)}$ of $I_s(\mu)$ is the weak limit of the measures $\nu_N$ (cf. [32]). For $s \geq \dim(M)$, the situation changes completely. The corresponding energy integral diverges for all probability measures. Techniques from geometric measure theory were applied in [6, 7, 27, 28] to show that the limiting distribution $\mu_M^{(s)}$ of the minimal energy distributions is the normalized $\dim(M)$-dimensional Hausdorff measure on $M$, if $M$ is rectifiable.

## 7.2 Generalized energy and uniform distribution on the sphere

Using mutually repelling forces on $N$ particles to distribute them on a surface $M$ is a rather compelling idea. The motivation for this could be taken from physical experiments, where electric charges distribute themselves in a way that minimizes the sum of the mutual energies (7.2) for $s = \dim(M) - 1$ (cf. [22, 39]). The study of the precise distribution of the charges is the subject of classical potential theory (cf. [32]), which shows that the energy integral (7.3) has a unique minimizer amongst all Borel probability measures supported on $M$; in the case $s = \dim(M) - 1$ this is the harmonic measure on $M$. The minimizing measure depends highly on the curvature of the surface and the value of the parameter $s$, and thus differs from the surface measure, except for surfaces with high symmetry, like the sphere. For values of $s \neq \dim(M) - 1$ (and $\dim(M) \neq 2, 3$), there is no physical experiment which can be used to describe the charge distribution, nevertheless, the intuition and the result remain the same – there exists a unique equilibrium measure depending on $s$ on $M$ – if $s < \dim(M)$.

In the following we will study the general potential theoretic situation of a strictly positive definite continuous kernel $g : [-1, 1) \to \mathbb{R}$ (cf. [32]). Here a function $g$ is called positive definite, if the *energy integral*

$$I_g(\mu) = \iint_{\mathbb{S}^d \times \mathbb{S}^d} g(\langle \mathbf{x}, \mathbf{y} \rangle) \, d\mu(\mathbf{x}) \, d\mu(\mathbf{y}) \geq 0, \qquad (7.4)$$

for all signed Borel measures $\mu$ on $\mathbb{S}^d$, and $I_g(\mu)$ is finite for at least one Borel measure $\mu$. It is called *strictly* positive definite, if equality in (7.4) only occurs for the zero measure.

Let $g$ be given by its Laplace expansion (cf. [35])

$$g(t) = \sum_{n=0}^{\infty} a_n Z(d, n) P_n^{(d)}(t) \qquad (7.5)$$

in terms of the Legendre–Gegenbauer polynomials $P_n^{(d)}$. These are the orthogonal polynomials with respect to the weight function $(1-t^2)^{\frac{d-2}{2}}$ normalized so that $P_n^{(d)}(1) = 1$. Then the requirement of strict positive definiteness is expressed by the strict positivity of all coefficients $a_n$ (cf. [42]). Furthermore, $Z(d, n)$ denotes the dimension of the space of spherical harmonics of degree $n$ on $\mathbb{S}^d$,

$$Z(d, n) = \frac{2n + d - 1}{d - 1} \binom{n + d - 2}{d - 2}.$$

Assume further that $g$ is integrable,

$$\forall \mathbf{y} \in \mathbb{S}^d : \int_{\mathbb{S}^d} g(\langle \mathbf{x}, \mathbf{y} \rangle)\,\mathrm{d}\sigma_d(\mathbf{x}) = a_0 \omega_d,$$

where $\sigma_d$ denotes the surface measure on $\mathbb{S}^d$ and $\omega_d = \sigma_d(\mathbb{S}^d)$.

Under the assumptions of continuity on $[-1, 1)$ and integrability, the function $g$ is represented by (7.5) in the sense of Abel summability, namely

$$\lim_{r \to 1-} \sum_{n=0}^{\infty} r^n a_n Z(d, n) P_n^{(d)}(t) = g(t); \qquad (7.6)$$

this relation holds uniformly on any interval $[-1, 1-\varepsilon]$ for $\varepsilon > 0$ by positivity of the Poisson kernel

$$\sum_{n=0}^{\infty} r^n Z(d, n) P_n^{(d)}(\langle \mathbf{x}, \mathbf{y} \rangle) = \frac{1 - r^2}{\|\mathbf{x} - r\mathbf{y}\|^{d+1}}.$$

**Remark 7.1** Notice that the series (7.5) can diverge for certain kernel functions; for instance, the series diverges for $g(t) = (1 - t)^{-s/2}$ for $\frac{d+1}{2} < s < d$, which corresponds to the classical Riesz kernels. Thus we had to use a summation method in order to ensure convergence. We chose Abel summation for simplicity. For a comprehensive discussion of applications of summation methods to Laplace series, we refer to [3].

In general $g$ will have a singularity at $t = 1$, namely

$$\lim_{t \to 1-} g(t) = +\infty,$$

which also means that the series

$$\sum_{n=0}^{\infty} Z(d, n) a_n$$

diverges. If $g$ is continuous on $[-1, 1]$, we call $g$ a *regular* kernel, whereas if it has a singularity at $t = 1$ but is still integrable, we call $g$ *singular*.

For a regular or singular kernel $g$, the energy integral (7.4) is uniquely minimized by $\frac{1}{\omega_d}\sigma_d$ amongst all Borel probability measures on $\mathbb{S}^d$. By our assumptions on $g$ we have for every probability measure $\nu$

$$a_0 \leq \lim_{r \to 1-} \iint_{\mathbb{S}^d \times \mathbb{S}^d} \sum_{n=0}^{\infty} r^n a_n Z(d, n) P_n^{(d)}(\langle \mathbf{x}, \mathbf{y} \rangle)\,\mathrm{d}\nu(\mathbf{x})\,\mathrm{d}\nu(\mathbf{y})$$

$$= \iint_{\mathbb{S}^d \times \mathbb{S}^d} g(\langle \mathbf{x}, \mathbf{y} \rangle)\,\mathrm{d}\nu(\mathbf{x})\,\mathrm{d}\nu(\mathbf{y}),$$

where the second equality follows from the uniform convergence in (7.6). Equality holds, if and only if

$$\iint_{\mathbb{S}^d \times \mathbb{S}^d} P_n^{(d)}(\langle \mathbf{x}, \mathbf{y} \rangle) \, d\nu(\mathbf{x}) \, d\nu(\mathbf{y}) = 0$$

for $n \geq 1$, which is equivalent to $\nu = \frac{1}{\omega_d} \sigma_d$.

For a finite set $X_N = \{\mathbf{x}_1, \ldots, \mathbf{x}_N\} \subset \mathbb{S}^d$ of $N$ (pairwise distinct!) points we define the $g$-energy as

$$E_g(X_N) = \sum_{i \neq j} g(\langle \mathbf{x}_i, \mathbf{x}_j \rangle). \tag{7.7}$$

Furthermore, we denote by

$$\mathcal{E}_g(\mathbb{S}^d, N) = \min_{X_N} E_g(X_N)$$

the minimal $g$-energy of an $N$-point set on the sphere $\mathbb{S}^d$. We denote point sets minimizing the energy by $X_N^*$ (suppressing the dependence on $g$ in this notation). To any point set $X_N$ we associate the measure

$$\nu_N = \frac{1}{N} \sum_{i=1}^{N} \delta_{\mathbf{x}_i}.$$

**Theorem 7.2** *Let $g$ be a strictly positive definite regular or singular integrable kernel function, which is continuous on $[-1, 1)$. Let $(X_N)_N$ be a sequence of point sets on the unit sphere $\mathbb{S}^d$ such that*

$$\lim_{N \to \infty} \frac{1}{N^2} E_g(X_N) = \frac{1}{\omega_d} \int_{\mathbb{S}^d} g(\langle \mathbf{x}, \mathbf{y} \rangle) \, d\sigma_d(\mathbf{x}) = a_0. \tag{7.8}$$

*Then the associated measures $\nu_N$ tend weakly to the normalized surface measure $\frac{1}{\omega_d} \sigma_d$. In particular the measures associated to a sequence of energy minimizing configurations $(X_N^*)_N$ tend to $\frac{1}{\omega_d} \sigma_d$.*

*Proof.* We first notice that $I_g(\mu)$ is uniquely minimized amongst all Borel probability measures by the normalized surface measure $\frac{1}{\omega_d} \sigma_d$. This is an immediate consequence of the addition theorem for spherical harmonics (cf. [35]) and the fact that every measure on $\mathbb{S}^d$ is characterized by its Fourier coefficients.

We will only give a proof of the theorem for singular $g$; the case of regular kernels can be treated in a much simpler way by harmonic analysis. Assume that we have a sequence of point sets $(X_N)_N$ such that the associated measures $\nu_N$ weakly tend to a limiting measure $\nu$. Then we also have

$$\frac{1}{N^2}\sum_{i\neq j}\delta_{(\mathbf{x}_i,\mathbf{x}_j)} \rightharpoonup \nu\otimes\nu.$$

For $M > 0$, let $g_M(t) = \min(g(t), M)$. Then $(g_M)_M$ is a pointwise monotonically increasing family of continuous functions; we have

$$\lim_{M\to\infty}\lim_{N\to\infty}\frac{1}{N^2}\sum_{i\neq j}g_M(\langle\mathbf{x}_i,\mathbf{x}_j\rangle) = \lim_{M\to\infty}\iint_{\mathbb{S}^d\times\mathbb{S}^d} g_M(\langle\mathbf{x},\mathbf{y}\rangle)\,d\nu(\mathbf{x})\,d\nu(\mathbf{y})$$

$$= \iint_{\mathbb{S}^d\times\mathbb{S}^d} g(\langle\mathbf{x},\mathbf{y}\rangle)\,d\nu(\mathbf{x})\,d\nu(\mathbf{y}).$$

On the other hand, by the monotonicity of $(g_M)_M$, we have

$$\lim_{N\to\infty}\frac{1}{N^2}\sum_{i\neq j}g_M(\langle\mathbf{x}_i,\mathbf{x}_j\rangle) \leq \liminf_{N\to\infty}\frac{1}{N^2}\sum_{i\neq j}g(\langle\mathbf{x}_i,\mathbf{x}_j\rangle),$$

from which we derive

$$\iint_{\mathbb{S}^d\times\mathbb{S}^d} g(\langle\mathbf{x},\mathbf{y}\rangle)\,d\nu(\mathbf{x})\,d\nu(\mathbf{y}) \leq \liminf_{N\to\infty}\frac{1}{N^2}\sum_{i\neq j}g(\langle\mathbf{x}_i,\mathbf{x}_j\rangle).$$

Now take a sequence $(X_N)_N$ satisfying the assumptions of the theorem and assume that $\nu$ is cluster point of the sequence of measures $(\nu_N)_N$; such a cluster point exists by the Banach–Alaoglu theorem. Since $I_g(\nu) = a_0 = I_g(\frac{1}{\omega_d}\sigma_d)$ and the fact that the energy integral (7.4) is uniquely minimized by $\frac{1}{\omega_d}\sigma_d$, we obtain $\nu = \frac{1}{\omega_d}\sigma_d$. Since this is the only possible cluster point and under our assumptions we have $\nu_N \rightharpoonup \frac{1}{\omega_d}\sigma_d$. Applying this argument to a sequence $(X_N^*)_N$ of energy minimizing point sets gives the second assertion. $\qquad\square$

We recall an averaging argument from [38], which shows the existence of point sets $X_N$ with $E_g(X_N) \leq a_0 N^2$. Let $(D_i)_{i=1}^N$ be an area regular partition of $\mathbb{S}^d$, namely a collection of closed subsets of $\mathbb{S}^d$ satisfying

(i) $\displaystyle\bigcup_{i=1}^N D_i = \mathbb{S}^d$,

(ii) $D_i^\circ \cap D_j^\circ = \emptyset$ for $1 \leq i < j \leq N$,

(iii) $\sigma_d(D_i) = \dfrac{\omega_d}{N}$ for $1 \leq i \leq N$,

(iv) $\sigma_d(\partial D_i) = 0$ for $1 \leq i \leq N$.

Then integrating

$$\sum_{i\neq j} g(\langle \mathbf{x}_i, \mathbf{x}_j\rangle)$$

with respect to the product of the measures $\sigma_i^* = \frac{N}{\omega_d}\sigma_d|_{D_i}$ (restriction of $\sigma_d$ to $D_i$) gives

$$\int_{D_1}\cdots\int_{D_N}\sum_{i\neq j} g(\langle \mathbf{x}_i, \mathbf{x}_j\rangle)\, d\sigma_1^*(\mathbf{x}_1)\cdots d\sigma_N^*(\mathbf{x}_N)$$

$$= N^2 I_g\left(\frac{1}{\omega_d}\sigma_d\right) - \frac{N^2}{\omega_d^2}\sum_{i=1}^{N}\iint_{D_i\times D_i} g(\langle \mathbf{x}, \mathbf{y}\rangle)\, d\sigma_d(\mathbf{x})\, d\sigma_d(\mathbf{y}).$$

Since the value on the right hand side is smaller than $a_0 N^2$, this shows the existence of point sets with small energy.

**Remark 7.3** Notice that Bondarenko *et al.* [4], in the course of proving the existence of well-separated spherical designs of optimal asymptotic growth order [5], showed the existence of area regular partitions with geodesically convex sets $D_i$ and diameter $\operatorname{diam}(D_i) = \mathcal{O}(N^{-1/d})$ (the optimal order).

**Remark 7.4** Notice that the classical Riesz kernels $g_s(\langle \mathbf{x}, \mathbf{y}\rangle) = \|\mathbf{x}-\mathbf{y}\|^{-s}$ for $0 < s < d$ satisfy the hypotheses of Theorem 7.2. Thus in the classical potential theoretic situation the discrete minimizing configurations are asymptotically uniformly distributed (see also [32]). Furthermore, the cases $s = 0$ with the modified kernel function $g_0(\langle \mathbf{x}, \mathbf{y}\rangle) = \log\frac{1}{\|\mathbf{x}-\mathbf{y}\|}$ and $-2 < s < 0$ with $g_s(\langle \mathbf{x}, \mathbf{y}\rangle) = 2^{-s} - \|\mathbf{x}-\mathbf{y}\|^{-s}$ are covered by this theorem.

**Remark 7.5** Discrete energies on the sphere $\mathbb{S}^d$ have also been studied for negative values of $s > -2$, i.e., positive exponents of the distance (cf. [36, 44, 45, 47]). For the case $s \leq -2$, the kernel has to be modified further, to ensure the positivity of all coefficients in its Laplace expansion; furthermore, in the case of $s = -2k$ being a negative even integer, the kernel $g_{-2k}(\langle \mathbf{x}, \mathbf{y}\rangle) = \|\mathbf{x}-\mathbf{y}\|^{2k}\log\frac{1}{\|\mathbf{x}-\mathbf{y}\|} + p_k(\langle \mathbf{x}, \mathbf{y}\rangle)$ is used instead, where the function $p_k$ has to be added to ensure positive definiteness; its Laplace series is given in [12]. Without the logarithmic factor the kernel would be a polynomial in this case.

**Remark 7.6** Discrete energies for negative values of $s$ play an important role in the study of numerical integration errors for certain Sobolev spaces $H$. In this context the energy can be interpreted as the square of the worst case error of integration on the underlying function space $H$

$$\mathrm{wce}_H(X_N) = \sup_{\substack{f \in H \\ \|f\| \leq 1}} \left| \frac{1}{N} \sum_{n=1}^{N} f(\mathbf{x}_n) - \frac{1}{\omega_d} \int_{\mathbb{S}^d} f(\mathbf{x}) \, d\sigma_d(\mathbf{x}) \right|.$$

It is a general feature of reproducing kernel Hilbert spaces that the square of the worst case error can be expressed in terms of a generalized discrete energy. This can be seen as a generalization of Stolarsky's invariance principle (cf. [9, 13, 44, 45]), which relates the sum of distances of the point set $X_N$ (this is the case $s = -1$) to the $L^2$ discrepancy defined in (7.15) below. In turn the $L^2$ discrepancy and similar energy functionals with regular kernels can be interpreted as the mean square integration error, if the function space is equipped with an appropriate Wiener measure (cf. [25]). Furthermore, the relation between the worst case integration error in Sobolev spaces and general energy functionals is used in [16] to define sequences of QMC designs $(X_N)_N$ as sequences of point sets achieving optimal order of magnitude for the integration error.

## 7.3 Hyper-singular energies and uniform distribution

For $s \geq \dim(M)$, the situation changes completely. The energy integral (7.3) diverges for all measures $\mu$. But minimizing point configurations of the energy sum (7.2) can still be studied. By a result of Hardin and Saff [27, 28] from 2004, the energy mimimizing points distribute asymptotically according to the normalized surface measure for a rectifiable manifold. In the following, $\mathcal{H}_d$ will denote the $d$-dimensional Hausdorff measure, normalized so that $\mathcal{H}_d([0, 1]^d) = 1$.

**Theorem 7.7** (Theorems 2.1 and 2.2 in [28]) *Let $A \subset \mathbb{R}^d$ be a compact set and $s > d$. Let*

$$\mathcal{E}_s(A, N) = \min_{X_N \subset A} E_s(X_N)$$

*be the minimal $s$-energy of a point set $X_N \subset A$. Then the limit $\lim_{N \to \infty} \mathcal{E}_s(A, N) N^{-1-s/d}$ exists and is given by*

$$\lim_{N \to \infty} \frac{1}{N^{1+s/d}} \mathcal{E}_s(A, N) = \frac{C_d(s)}{\mathcal{H}_d(A)^{s/d}}, \tag{7.9}$$

*where $C_d(s)$ is a positive constant depending only on $s$ and $d$; the constant occurs as the limit for the case $A = U_d = [0, 1]^d$,*

$$C_d(s) = \lim_{N \to \infty} \frac{\mathcal{E}(U_d, N)}{N^{1+s/d}}.$$

*Furthermore, if A has positive d-dimensional Hausdorff-measure $\mathcal{H}_d(A) > 0$ and $(X_N)_N$ $(X_N = \{\mathbf{x}_1, \ldots, \mathbf{x}_N\} \subset A)$ is a sequence of point sets with*

$$\lim_{N \to \infty} \frac{1}{N^{1+s/d}} E_s(X_N) = \frac{C_d(s)}{\mathcal{H}_d(A)^{s/d}},$$

*then the corresponding measures $\nu_N$ tend weakly to the normalized Hausdorff measure on A,*

$$\frac{1}{N} \sum_{i=1}^{N} \delta_{\mathbf{x}_i} \to \frac{\mathcal{H}_d|_A}{\mathcal{H}_d(A)}. \tag{7.10}$$

**Remark 7.8** In the case $s = d$ a similar result holds with $N^{1+s/d}$ replaced by $N^2 \log N$. In this case also the constant $C'_d(d)$ is explicitly known, namely

$$C'_d(d) = \mathcal{H}_d(B_d) = \frac{\pi^{d/2}}{\Gamma\left(\frac{d}{2} + 1\right)},$$

where $B_d$ is the $d$-dimensional unit ball in $\mathbb{R}^d$.

This result has several interesting features that we shall discuss briefly:

- the limiting measure is independent of $s$, which is in obvious contrast to the case $s < d$, where the measure depends highly on $s$ (except for manifolds of high symmetry such as the sphere; this is the reason why we restricted our discussion to the sphere in Section 7.2);
- the limiting measure has a geometrical interpretation ("surface");
- the proof shows that in contrast to the situation for $s < d$ only local (short range) interactions contribute to the asymptotic behavior of $E_s(X_N)$.

The proof of the existence of the limit

$$\lim_{N \to \infty} \frac{1}{N^{1+s/d}} \mathcal{E}_s(N)$$

for $s > d$ and the fact that the corresponding distribution measures tend to the normalized surface area measure, the normalized Hausdorff measure $\mathcal{H}_d$ on $A$, is rather intricate and technical; we refer the reader to [28]. Furthermore, there is an excellent survey in [27] on the result, which recalls the proof of the result for the unit cube.

### 7.3.1 Facts and conjectures about energy

In [10, 30, 38, 40, 48, 50] the asymptotic behavior of the minimal energy of $N$-point configurations on the sphere $\mathbb{S}^d$ for all positive values of the parameter

$s$ was studied. For the case $0 < s < d$, it was shown that there exist positive constants $C_1, C_2$ such that

$$I_s \left( \frac{1}{\omega_d} \sigma_d \right) N^2 - C_1 N^{1+s/d} \leq \mathcal{E}_s(\mathbb{S}^d, N) \leq I_s \left( \frac{1}{\omega_d} \sigma_d \right) N^2 - C_2 N^{1+s/d}.$$

$$(7.11)$$

For $s = d$, a phase change occurs, namely

$$\mathcal{E}_d(\mathbb{S}^d, N) \sim C N^2 \log N,$$

this could be explained by the "collapse" of the two terms in the lower and upper bounds in (7.11) ($1 + s/d = 2$ in this case). The coincidence of two asymptotic terms often produces phase change phenomena. The behavior of the minimal energy for $s \geq d$, the hyper-singular case, was only understood after the work of Hardin and Saff [28] which we stated as Theorem 7.7. The proof shows the existence of the constant $C_d(s)$, the exact value of this constant is still conjectural.

For $d = 2$, the value $C_2(s)$ ($s > 2$) is conjectured to be

$$C_2(s) = \left( \frac{\sqrt{3}}{2} \right)^{s/2} \zeta_{\mathsf{A}_2}(s),$$

where $\zeta_{\mathsf{A}_2}$ denotes the Epstein zeta function,

$$\zeta_{\mathsf{A}_2}(s) = \sum_{\substack{\mathbf{z} \in \mathsf{A}_2 \\ \mathbf{z} \neq \mathbf{0}}} \|\mathbf{z}\|^{-s},$$

of the hexagonal lattice $\mathsf{A}_2$ spanned by the vectors $(1, 0)$ and $(\frac{1}{2}, \frac{1}{2}\sqrt{3})$. The conjecture is supported by the fact that the hexagonal lattice is the solution of the two-dimensional best packing problem, as well as the fact that this lattice is universally optimal (cf. Section 7.5) among two-dimensional lattices [34]. Furthermore, it was conjectured [15] that the inequality (7.11) can be sharpened to an asymptotic relation

$$\mathcal{E}_s(\mathbb{S}^d, N) = I_s \left( \frac{1}{\omega_d} \sigma_d \right) N^2 - \frac{C_d(s)}{\mathcal{H}_d(\mathbb{S}^d)^{s/d}} N^{1+s/d} + o \left( N^{\min(2, 1+s/d)} \right),$$

valid for $s \in (-2, 0) \cup (0, d) \cup (d, d + 1)$; the constant $C_d(s)$ for $s$ in this range is conjectured to be the analytic continuation of $C_d(s)$ for $s > d$ to the complex plane; the value $I_s(\frac{1}{\omega_d} \sigma_d)$ has to be interpreted as the analytic continuation of the expression $I_s(\frac{1}{\omega_d} \sigma_d)$ for $s < d$ to complex values of $s$. This was coined "the principle of analytic continuation" in [15]. Notice that the first two asymptotic terms change their role at $s = d$; for this value of $s$, $I_s(\frac{1}{\omega_d} \sigma_d)$ and $C_d(s)$ have a singularity, which is mirrored by a $N^2 \log N$ term in the asymptotic expression. The conjecture includes the value of the constant $C_d(s)$ in terms of the Epstein zeta function of a lattice minimizing the energy (for $s \geq 0$, $s \neq d$). It is supported by the fact that a corresponding principle holds for $d = 1$ (cf. [14]). Moreover, for $d = 2, 4, 8, 24$, it is conjectured that $C_d(s) = |\Lambda_d|^{s/d} \zeta_{\Lambda_d}(s)$, where $\Lambda_d$ denotes, respectively, the hexagonal lattice $A_2$, $D_4$, $E_8$, and the Leech lattice $\Lambda_{24}$.

For dimensions $d \geq 3$, the situation seems to be much more complicated. In general it is known that the limit

$$\lim_{s \to \infty} C_d(s)^{1/s}$$

exists and is related to the best packing constant in dimension $d$.

Of course, not all Voronoï cells of a minimal energy configuration on the sphere $\mathbb{S}^2$ can be hexagonal; there have to be at least 12 pentagons by Euler's polyhedral formula. Numerical experiments with large numbers of points show that not only pentagonal cells, but also structures of pentagons and heptagons occur, which seem to organize themselves along curves, called "scars" (cf. [1, 8, 27]).

## 7.4  Discrepancy estimates

The discrepancy given by (7.1) is an easy to understand concept; $D(X_N)$ just measures the maximal deviation of the discrete distribution from the limiting distribution $\sigma$ (in statistics this is called the Kolmogorov–Smirnov statistics). On the other hand, the precise value of the discrepancy of a point set is rather difficult to compute. Thus discrepancy is usually estimated rather than computed directly. In the simplest one-dimensional case there are two classical estimates for discrepancy, namely the Erdős–Turán inequality and LeVeque's inequality (cf. [31]). Both inequalities have been generalized to the spherical case and used for estimating the discrepancy of point sets constructed by various methods.

The $\mathbb{S}^d$ version of the Erdős–Turán inequality has been given independently by Grabner [24] and by Li and Vaaler [33] and reads as

$$D(X_N) \leq \frac{C_1(d)}{M} + \sum_{\ell=1}^{M} \frac{C_2(d)}{\ell} \sum_{m=1}^{Z(d,\ell)} \frac{1}{N} \left| \sum_{n=1}^{N} Y_{\ell,m}(\mathbf{x}_n) \right|, \qquad (7.12)$$

valid for all positive integer values of $M$. Here $C_1(d)$ and $C_2(d)$ denote (explicitly known) constants, $Y_{\ell,m}$ ($m = 1, \ldots, Z(d, \ell)$) denote an orthonormal system of real spherical harmonics of order $\ell$, and $Z(d, \ell)$ denotes the dimension of the space of these spherical harmonics.

Only recently, a spherical version of the LeVeque inequality was found (cf. [36]):

$$D(X_N) \leq A(d) \left( \sum_{\ell=0}^{\infty} \ell^{-(d+1)} \sum_{m=1}^{Z(d,\ell)} \left( \frac{1}{N} \sum_{n=1}^{N} Y_{\ell,m}(\mathbf{x}_n) \right)^2 \right)^{\frac{1}{d+2}} \qquad (7.13)$$

with an explicit constant $A(d)$. Both inequalities (7.12) and (7.13) specialize to their classical versions for $d = 1$. It is interesting to mention that the LeVeque inequality also has an opposite version providing a lower bound (cf. [36]):

$$D(X_N) \geq B(d) \left( \sum_{\ell=0}^{\infty} \ell^{-(d+1)} \sum_{m=1}^{Z(d,\ell)} \left( \frac{1}{N} \sum_{n=1}^{N} Y_{\ell,m}(\mathbf{x}_n) \right)^2 \right)^{1/2} \qquad (7.14)$$

with some explicit positive constant $B(d)$. It should also be mentioned that the expression raised to the $d + 2$ power in (7.13) is equivalent to the $L^2$ discrepancy

$$\int_0^\pi \int_{\mathbb{S}^d} \left( \frac{1}{N} \sum_{n=1}^N 1_{C(\mathbf{x}_n,\varphi)}(\mathbf{x}) - \frac{1}{\omega_d} \sigma_d(C(\cdot,\varphi)) \right)^2 \sin(\varphi)^{d-1} \, d\sigma_d(\mathbf{x}) \, d\varphi, \qquad (7.15)$$

where $C(\mathbf{x}, \varphi) = \{\mathbf{y} \in \mathbb{S}^d \mid \langle \mathbf{x}, \mathbf{y} \rangle \geq \cos(\varphi)\}$ denotes the spherical cap centered at $\mathbf{x}$ with angle $\varphi$. A similar inequality relating the discrepancy $D(X_N)$ to the $L^2$ discrepancy has been given in the Euclidean case [37] and in the general case of a metric space [46], which specializes to an inequality similar to (7.13) in the case of the sphere.

Although the limiting distribution of minimal energy point sets $X_N^*$ for $s \geq d$ on the sphere has been determined [28], almost nothing is known about quantitative results. The only, and very weak, estimate for the discrepancy of minimal energy point sets in the singular case is due to Damelin and Grabner [20] and gives

$$D(X_N^*) = \mathcal{O}\left( \sqrt{\frac{\log \log N}{\log N}} \right) \qquad (7.16)$$

for $s = d$. This estimate is proved by approximating the $d$-energy by a limiting process $s \to d-$.

In the harmonic case $s = d - 1$, Götz [23] proved that minimal energy configurations $X_N^*$ satisfy

$$D(X_N^*) \ll N^{-1/d} \log N,$$

solving a conjecture of Korevaar (cf. [29]), up to the logarithmic factor. This improves the exponent $-\frac{1}{2d}$ given by Sjögren [43].

In [36] the LeVeque type inequality (7.13) for the spherical cap discrepancy was proved and applied to minimal energy point sets $X_N^*$ for $g_s$ with $-1 < s < 0$. This gives bounds for the discrepancy

$$D(X_N^*) \ll N^{-\frac{d-s}{d(d+2)}}. \tag{7.17}$$

For $s = 0$, the bound

$$D(X_N^*) \ll N^{-\frac{1}{d+2}} \tag{7.18}$$

had been obtained [11].

It should be mentioned that from the theory of irregularities of distribution [2] it is known that for all sets $X_N \subset \mathbb{S}^d$ the inequality

$$D(X_N) \gg N^{-1/2-1/2d} \tag{7.19}$$

holds for the spherical cap discrepancy. Inequality (7.14) together with lower bounds for the energy from [48] was used to reprove this result in [36]. This indicates the sharpness of the inequality (7.14) as a lower bound for the discrepancy in terms of sums over spherical harmonics.

It is also known that inequality (7.19) is best possible up to a factor $\sqrt{\log N}$. The existence of point sets $X_N$ with

$$D(X_N) \ll N^{-1/2-1/2d}\sqrt{\log N}$$

uses a probabilistic argument resembling the averaging argument that we used to prove the existence of point sets of small energy; up to now, no explicit construction of such a point set is known. All the known estimates for the discrepancy of point sets differ from the lower bound (7.19) by a power of $N$. The bounds (7.16), (7.17), and (7.18) should be compared to (7.19).

**Remark 7.9** Discrepancy estimates for point sets of minimal energy are known in the case of $-2 < s \le 0$ (giving sums of positive powers of the distance) from the work of Wagner [49]. These results have been partly rediscovered and refined [36]. Estimates for the discrepancy in terms of $g$-energy of the point set for singular $g$ satisfying an additional technical hypothesis have also been given [20]. This gives estimates for the discrepancy of point sets minimizing the Riesz $s$-energy for $-2 < s \le d$. All these estimates have the disadvantage that they have been derived via harmonic analysis, and this method has to use estimates for the Fourier coefficients of certain functions in an unfavorable way; thus it can be expected that these estimates for the discrepancy of minimal energy point sets are far from the correct order of magnitude.

Furthermore, nothing is known about the discrepancy of energy minimizing point sets for $s > d$. This is because harmonic analysis is not applicable in this case.

## 7.5  Some remarks on lattices

As was pointed out in the discussion of the local structure of minimal energy point configurations at the end of Section 7.3, there seems to be an intricate connection to lattices which minimize a corresponding energy functional. Therefore, we add a short description of the appropriate notions for lattices and the state of knowledge about them.

The optimal density of sphere packings in $\mathbb{R}^d$ is a classical question that found new interest with Hales's proof of the Kepler conjecture [26]. New upper bounds for the density of sphere packings in dimensions $3 < d \leq 36$ could be derived from linear programming bounds based on Fourier transforms [17]. This led to the definition of *universally optimal lattices* as those lattices $\Lambda$ which minimize

$$\sum_{\lambda \in \Lambda} e^{-t\|\lambda\|^2}$$

for all real parameters $t > 0$ amongst all lattices of covolume 1.

As was pointed out in Section 7.3, the conjectured local structure of minimal energy point sets in the case $s > \dim(M)$ is related to lattices $\Lambda$ of covolume 1 which minimize the Riesz energy

$$\sum_{\lambda \in \Lambda \setminus \{0\}} \|\lambda\|^{-s} \tag{7.20}$$

for $s > d$; this equals the classical Epstein zeta function of the lattice $\Lambda$. For recent progress on minimization of values of the Epstein zeta function we refer to [41]. By Mellin transform, universally optimal lattices also minimize the Riesz energy (7.20).

The sphere packing problem is naturally related to the study of periodic point sets with minimal energy. As in the case of spherical codes, one may ask whether there exist *universally optimal periodic sets*, that is, periodic sets that minimize the energy

$$\sum_{\lambda \in \Lambda} f(\|\lambda\|) \tag{7.21}$$

for all completely monotonic functions $f$. Up to now, no such universally optimal periodic set is known. However, exceptional lattices such as the hexagonal lattice $A_2$, the root lattice $E_8$, and the 24-dimensional Leech lattice $\Lambda_{24}$ are conjectured to be examples (see [18]). Some candidates for universally optimal lattices have been found numerically (cf. [19]), proofs are still missing.

## Acknowledgement

The author is grateful to an anonymous referee for many valuable remarks, which greatly improved the presentation of the material.

The author is supported by the Austrian Science Fund FWF projects F5503 (part of the Special Research Program (SFB) "Quasi-Monte Carlo Methods: Theory and Applications") and W1230 (Doctoral Program "Discrete Mathematics").

## References

[1]  A. R. Bausch, M. J. Bowick, A. Cacciuto, A. D. Dinsmore, M. F. Hsu, D. R. Nelson, M. G. Nikolaides, A. Travesset and D. A. Weitz, Grain boundary scars and spherical crystallography. *Science* **299**, 1716–1718, 2003.

[2]  J. Beck and W. W. L. Chen, *Irregularities of Distribution*. Cambridge Tracts in Mathematics, volume 89. Cambridge University Press, Cambridge, 1987.

[3]  H. Berens, P. L. Butzer and S. Pawelke, Limitierungsverfahren von Reihen mehrdimensionaler Kugelfunktionen und deren Saturationsverhalten. *Publ. Res. Inst. Math. Sci. Ser. A* **4**, 201–268, 1968–1969.

[4]  A. V. Bondarenko, D. Radchenko and M. S. Viazovska, Optimal asymptotic bounds for spherical designs, *Ann. Math. (2)* **178**(2), 443–452, 2013.

[5]  A. V. Bondarenko, D. Radchenko and M. S. Viazovska, *Well separated spherical designs*, arXiv:1303.5991, March 2013.

[6]  S. V. Borodachov, D. P. Hardin and E. B. Saff, Asymptotics of best-packing on rectifiable sets. *Proc. Am. Math. Soc.* **135**(8), 2369–2380, 2007 (electronic).

[7]  S. V. Borodachov, D. P. Hardin and E. B. Saff, Asymptotics for discrete weighted minimal Riesz energy problems on rectifiable sets. *Trans. Am. Math. Soc.* **360**(3), 1559–1580, 2008 (electronic).

[8]  M. Bowick, A. Cacciuto, D. R. Nelson and A. Travesset, Crystalline order on a sphere and the generalized Thomson problem. *Phys. Rev. Lett.* **89**, 185502, 2002.

[9]  J. S. Brauchart, Invariance principles for energy functionals on spheres. *Monatsh. Math.* **141**(2), 101–117, 2004.

[10]  J. S. Brauchart, About the second term of the asymptotics for optimal Riesz energy on the sphere in the potential-theoretical case. *Integral Transforms Spec. Funct.* **17**(5), 321–328, 2006.

[11]  J. S. Brauchart, Optimal logarithmic energy points on the unit sphere. *Math. Comp.* **77**(263), 1599–1613, 2008.

[12] J. S. Brauchart and J. Dick, A characterization of Sobolev spaces on the sphere and an extension of Stolarsky's invariance principle to arbitrary smoothness. *Constr. Approx.* **38**(3), 397–445, 2013.

[13] J. S. Brauchart and J. Dick, A simple proof of Stolarsky's invariance principle. *Proc. Am. Math. Soc.* **141**(6), 2085–2096, 2013.

[14] J. S. Brauchart, D. P. Hardin and E. B. Saff, The Riesz energy of the $N$th roots of unity: an asymptotic expansion for large $N$. *Bull. London Math. Soc.* **41**(4), 621–633, 2009.

[15] J. S. Brauchart, D. P. Hardin and E. B. Saff, The next-order term for optimal Riesz and logarithmic energy asymptotics on the sphere. *Recent Advances in Orthogonal Polynomials, Special Functions, and their Applications.* Contemporary Mathematics, volume 578, pp. 31–61. American Mathematical Society. Providence, RI, 2012.

[16] J. S. Brauchart, E. B. Saff, I. H. Sloan and R. S. Womersley, QMC designs: optimal order quasi Monte Carlo integration schemes on the sphere. *Math. Comp.*, 2014, to appear.

[17] H. Cohn and N. Elkies, New upper bounds on sphere packings. I. *Ann. Math. (2)* **157**(2), 689–714, 2003.

[18] H. Cohn and A. Kumar, Universally optimal distribution of points on spheres. *J. Am. Math. Soc.* **20**(1), 99–148, 2007.

[19] H. Cohn, A. Kumar and A. Schürmann, Ground states and formal duality relations in the Gaussian core model. *Phys. Rev. E* **80**, 061116, 2009.

[20] S. B. Damelin and P. J. Grabner, Energy functionals, numerical integration and asymptotic equidistribution on the sphere. *J. Complexity* **19**, 231–246, 2003, Corrigendum, **20**, 883–884, 2004.

[21] M. Drmota and R. F. Tichy, *Sequences, Discrepancies and Applications.* Lecture Notes in Mathematics, volume 1651. Springer-Verlag, Berlin, 1997.

[22] T. Erber and G. M. Hockney, Equilibrium configurations of $n$ equal charges on a sphere. *J. Phys. A.* **24**, L1369–L1377, 1991.

[23] M. Götz, On the distribution of weighted extremal points on a surface in $\mathbf{R}^d$, $d \geq 3$. *Potential Anal.* **13**(4), 345–359, 2000.

[24] P. J. Grabner, Erdős–Turán type discrepancy bounds. *Monatsh. Math.* **111**, 127–135, 1991.

[25] P. J. Grabner, P. Liardet and R. F. Tichy, Average case analysis of numerical integration. *In: Advances in Multivariate Approximation (Witten-Bommerholz, 1998)* W. Haußmann, K. Jetter and M. Reimer, (eds.), pp. 185–200. Wiley-VCH, Berlin, 1999.

[26] T. C. Hales, A proof of the Kepler conjecture. *Ann. Math. (2)* **162**(3), 1065–1185, 2005.

[27] D. P. Hardin and E. B. Saff, Discretizing manifolds via minimum energy points. *Notices Am. Math. Soc.* **51**(10), 1186–1194, 2004.

[28] D. P. Hardin and E. B. Saff, Minimal Riesz energy point configurations for rectifiable $d$-dimensional manifolds. *Adv. Math.* **193**(1), 174–204, 2005.

[29] J. Korevaar, Fekete extreme points and related problems. *Approximation theory and Function Series (Budapest, 1995).* Bolyai Soc. Math. Stud., volume 5, pp. 35–62. János Bolyai Math. Soc., Budapest, 1996.

[30] A. B. J. Kuijlaars and E. B. Saff, Asymptotics for minimal discrete energy on the sphere. *Trans. Am. Math. Soc.* **350**(2), 523–538, 1998.

[31] L. Kuipers and H. Niederreiter, *Uniform Distribution of Sequences.* Wiley-Interscience, New York, 1974.

[32] N. S. Landkof, *Foundations of Modern Potential Theory.* Springer-Verlag, New York, 1972. Translated from the Russian by A. P. Doohovskoy, Die Grundlehren der mathematischen Wissenschaften, Band 180.

[33] X.-J. Li and J. D. Vaaler, Some trigonometric extremal functions and the Erdős–Turán type inequalities. *Indiana Univ. Math. J.* **48**(1), 183–236, 1999.

[34] H. L. Montgomery, Minimal theta functions. *Glasgow Math. J.* **30**(1), 75–85, 1988.

[35] C. Müller, *Spherical Harmonics.* Lecture Notes in Mathematics, volume 17. Springer-Verlag, Berlin, 1966.

[36] F. J. Narcowich, X. Sun, J. D. Ward and Z. Wu, LeVeque type inequalities and discrepancy estimates for minimal energy configurations on spheres. *J. Approx. Theory* **162**(6), 1256–1278, 2010.

[37] H. Niederreiter, R. F. Tichy and G. Turnwald, An inequality for differences of distribution functions. *Arch. Math. (Basel)* **54**(2), 166–172, 1990.

[38] E. A. Rakhmanov, E. B. Saff and Y. M. Zhou, Minimal discrete energy on the sphere. *Math. Res. Lett.* **1**(6), 647–662, 1994.

[39] E. A. Rakhmanov, E. B. Saff and Y. M. Zhou, Electrons on the sphere. *Computational Methods and Function Theory 1994 (Penang).* Series in Approximations and Decompositions, volume 5, pp. 293–309. World Scientific, River Edge, NJ, 1995.

[40] E. B. Saff and A. B. J. Kuijlaars, Distributing many points on a sphere. *Math. Intelligencer* **19**(1), 5–11, 1997.

[41] P. Sarnak and A. Strömbergsson, Minima of Epstein's zeta function and heights of flat tori. *Invent. Math.* **165**(1), 115–151, 2006.

[42] I. J. Schoenberg, Metric spaces and positive definite functions. *Trans. Am. Math. Soc.* **44**(3), 522–536, 1938.

[43] P. Sjögren, Estimates of mass distributions from their potentials and energies. *Ark. Mat.* **10**, 59–77, 1972.

[44] K. B. Stolarsky, Sums of distances between points on a sphere. *Proc. Am. Math. Soc.* **35**, 547–549, 1972.

[45] K. B. Stolarsky, Sums of distances between points on a sphere. II. *Proc. Am. Math. Soc.* **41**, 575–582, 1973.

[46] R. F. Tichy, A general inequality with applications to the discrepancy of sequences, VII. *Mathematikertreffen Zagreb-Graz (Graz, 1990).* Grazer Math. Ber., volume 313, pp. 65–72. Karl-Franzens-Univ. Graz, Graz, 1991.

[47] G. Wagner, On the product of distances to a point set on a sphere. *J. Aust. Math. Soc. Ser. A* **47**(3), 466–482, 1979.

[48] G. Wagner, On means of distances on the surface of a sphere (lower bounds). *Pacific J. Math.* **144**(2), 389–398, 1990.

[49] G. Wagner, Erdős–Turán inequalities for distance functions on spheres. *Michigan Math. J.* **39**(1), 17–34, 1992.

[50] G. Wagner, On means of distances on the surface of a sphere. II. Upper bounds. *Pacific J. Math.* **154**(2), 381–396, 1992.

# 8

# The cross-correlation measure for families of binary sequences

### Katalin Gyarmati
Eötvös Loránd University, Budapest

### Christian Mauduit
Aix Marseille University, Marseille

### András Sárközy
Eötvös Loránd University, Budapest

*Dedicated to Professor Harald Niederreiter on the occasion
of his 70th birthday.*

## Abstract

Large families of binary sequences of the same length are considered and a new measure, the cross-correlation measure of order $k$, is introduced to study the connection between the sequences belonging to the family. It is shown that this new measure is related to certain other important properties of families of binary sequences. Then the size of the cross-correlation measure is studied. Finally, the cross-correlation measures of two important families of pseudorandom binary sequences are estimated.

## 8.1 Introduction

Pseudorandom binary sequences have many applications, in particular, they play a crucial role in modern cryptography. The pseudorandomness of the individual binary sequences is usually characterized using the notion of linear complexity, and tests based on mathematical statistics ("poker test," "runs test," etc.) are also used. However, these requirements usually study just a single property of the sequence, and they also have other weak points. Thus recently a more comprehensive theory of pseudorandomness of binary sequences has been initiated by Mauduit and Sárközy [31]. They introduced the following notations and definitions.

126

Consider a binary sequence

$$E_N = (e_1, \ldots, e_N) \in \{-1, +1\}^N.$$

Then the *well-distribution measure* of $E_N$ is defined as

$$W(E_N) = \max_{a,b,t} \left| \sum_{j=0}^{t-1} e_{a+jb} \right|$$

where the maximum is taken over all $a, b, t$ with $a, b, t \in \mathbb{N}$, $1 \leq a \leq a + (t-1)b \leq N$, while the *correlation measure of order $k$* of $E_N$ is defined as

$$C_k(E_N) = \max_{M,D} \left| \sum_{n=1}^{M} e_{n+d_1} e_{n+d_2} \cdots e_{n+d_k} \right|$$

where the maximum is taken over all $D = (d_1, \ldots, d_k)$ with nonnegative integers $d_1 < \cdots < d_k$ and $M \in \mathbb{N}$ with $M + d_k \leq N$.

Then $E_N$ is considered a "good" pseudorandom sequence if both of these measures $W(E_N)$ and $C_k(E_N)$ (at least for "small" $k$) are "small" in terms of $N$ (in particular, both are $o(N)$ as $N \to \infty$). Indeed, Cassaigne *et al.* [9] showed later that this terminology is justified since for almost all $E_N \in \{-1, +1\}^N$ both $W(E_N)$ and $C_k(E_N)$ are less than $N^{1/2}(\log N)^c$ (and they are also greater than $\varepsilon N^{1/2}$; see also [5, 6, 25]). It was also shown [31] that the Legendre symbol forms a "good" pseudorandom sequence. Since then, many constructions have been given for binary sequences with strong pseudorandom properties (see, e.g., [10, 11, 16, 17, 27, 29, 34, 36, 37, 39]).

However, these "good" constructions produce only a "few" good sequences while in many applications, for example, in cryptography, one needs *"large" families* of "good" pseudorandom binary sequences. If these sequences are constructed by an algorithm, then we usually speak of a pseudorandom generator, and the algorithm is considered a "good" one if it satisfies the "next bit" test. This approach has certain weak points. Large families consisting of binary sequences which are "good" in terms of the pseudorandom measures defined above have also been constructed, see, for example, [19, 20, 28, 30, 32, 34, 38]. In these constructions it is guaranteed that the individual sequences belonging to the family possess strong pseudorandom properties. However, in many applications it is not enough to know this; it can be much more important to know that the given family has a "rich," "complex" structure, and that there are many "independent" sequences in it. In order to handle this requirement Ahlswede *et al.* [2] (see also [3, 4, 22, 33]) introduced the notion of *family complexity* or briefly *f-complexity* (which can be especially useful in cryptography).

**Definition 8.1** *The $f$-complexity $\Gamma(\mathcal{F})$ of a family $\mathcal{F}$ of binary sequences $E_N \in \{-1, +1\}^N$ is defined as the greatest integer $j$ such that for any specification*

$$e_{i_1} = \varepsilon_1, \ldots, e_{i_j} = \varepsilon_j \ (1 \le i_1 < \cdots < i_j \le N)$$

*(with $\varepsilon_1, \ldots, \varepsilon_j \in \{-1, +1\}$) there is at least one $E_N = (e_1, \ldots, e_N) \in \mathcal{F}$ which satisfies it. The $f$-complexity of $\mathcal{F}$ is denoted by $\Gamma(\mathcal{F})$. (If there is no $j \in \mathbb{N}$ with the property above then we set $\Gamma(\mathcal{F}) = 0$.)*

There are also other properties of families which play an important role in applications. Such a property is the existence of *collisions* in the given family. This notion appears, for example, in [8, 35, 40, 41]; here we will follow Tóth's [40] presentation. Assume that $N \in \mathbb{N}$, $S$ is a given set (e.g., a set of certain polynomials or the set of all the binary sequences of a given length much less than $N$), and to each $s \in S$ we assign a unique binary sequence

$$E_N = E_N(s) = (e_1, \ldots, e_N) \in \{-1, +1\}^N,$$

and let $\mathcal{F} = \mathcal{F}(S)$ denote the family of binary sequences obtained in this way:

$$\mathcal{F} = \mathcal{F}(S) = \{E_N(s) : s \in S\}. \tag{8.1}$$

**Definition 8.2** *If $s \in S$, $s' \in S$, $s \neq s'$ and*

$$E_N(s) = E_N(s'), \tag{8.2}$$

*then (8.2) is said to be a* collision *in $\mathcal{F} = \mathcal{F}(S)$. If there is no collision in $\mathcal{F} = \mathcal{F}(S)$, then $\mathcal{F}$ is said to be* collision free.

In other words, $\mathcal{F} = \mathcal{F}(S)$ is collision free if we have $|\mathcal{F}| = |S|$. An ideally good family of pseudorandom binary sequences is collision free.

There is another related notion appearing in the literature, namely, the notion of *avalanche effect* (see, e.g., [8, 15, 24, 40, 41]); here we will present Tóth's definition.

**Definition 8.3** *If $\mathcal{F}$ is a family of form (8.1), and for any $s \in S$ changing $s$ for any $s' \in S$ with $s' \neq s$ changes "many" elements of $E_N(s)$ (i.e., for $s \neq s'$ many elements of the sequences $E_N(s)$ and $E_N(s')$ are different), then we speak about an* avalanche effect, *and we say that $\mathcal{F} = \mathcal{F}(S)$ possesses the* avalanche property. *If for any $s, s' \in S$, $s \neq s'$ at least $\left(\frac{1}{2} - o(1)\right) N$ elements of $E_N(s)$ and $E_N(s')$ are different then $\mathcal{F}$ is said to possess the* strict avalanche property.

We will also need the following definition.

**Definition 8.4** *If* $N \in \mathbb{N}$, $E_N = (e_1, \ldots, e_N) \in \{-1, +1\}^N$ *and* $E'_N = (e'_1, \ldots, e'_N) \in \{-1, +1\}^N$, *then the* distance $d(E_N, E'_N)$ *between* $E_N$ *and* $E'_N$ *is defined by*

$$d(E_N, E'_N) = \left|\{n : 1 \le n \le N, \ e_n \ne e'_n\}\right|$$

*(so that $d(E_N, E'_N)$ is a variant of the Hamming distance). Moreover, if $\mathcal{F}$ is a family of form (8.1), then the* distance minimum $m(\mathcal{F})$ *of $\mathcal{F}$ is defined by*

$$m(\mathcal{F}) = \min_{\substack{s, s' \in \mathcal{S} \\ s \ne s'}} d\left(E_N(s), E_N(s')\right).$$

*Applying this notion we may say that the family $\mathcal{F}$ in (8.1) is collision free if and only if $m(\mathcal{F}) > 0$, and $\mathcal{F}$ possesses the strict avalanche property if*

$$m(\mathcal{F}) \ge \left(\frac{1}{2} - o(1)\right) N. \tag{8.3}$$

The notions introduced in Definitions 8.3 and 8.4 can also be used when the family $\mathcal{F}$ is not of form (8.1), i.e., no parameter set $\mathcal{S}$ is given. For example, we may say that $\mathcal{F}$ possesses the avalanche property if for any $E_N \in \mathcal{F}$, $E'_N \in \mathcal{F}$, $E_N \ne E'_N$, the sequences $E_N$ and $E'_N$ have many different elements, and the distance minimum can be defined as

$$m(\mathcal{F}) = \min_{\substack{E_N, E'_N \in \mathcal{F} \\ E_N \ne E'_N}} d(E_N, E'_N).$$

We will use these notions in this extended sense.

In this paper our goal is to study a further important property of families of binary sequences. First in Section 8.2 we will introduce a measure called the *cross-correlation measure*, and we will study the connection between this new measure and the other related notions listed above. Then in Section 8.3 we will study the connection between the size of the family and its cross-correlation measure. Finally, in Sections 8.4 and 8.5 we will estimate the cross-correlation of two important families of pseudorandom binary sequences.

## 8.2 The definition of the cross-correlation measure

In Section 8.1 we mentioned $C_k(E_N)$, the correlation measure of order $k$ of the binary sequence $E_N$ which is, perhaps, the most important measure of pseudo-randomness of a *single* binary sequence; in the definition of this measure we consider a fixed sequence and we compare different elements of it (so that this is an *autocorrelation* type quantity). If, instead of a *single* sequence we want to characterize a *family* of sequences, then it is quite natural to compare elements of *different* sequences taken from the family, i.e., to consider a *correlation* type quantity involving different sequences. Thus we suggest the use of the following definition.

**Definition 8.5** *Let* $N \in \mathbb{N}$, $k \in \mathbb{N}$, *and for any* $k$ *binary sequences* $E_N^{(1)}, \ldots, E_N^{(k)}$ *with*

$$E_N^{(i)} = \left( e_1^{(i)}, \ldots, e_N^{(i)} \right) \in \{-1, +1\}^N \text{ (for } i = 1, 2, \ldots, k)$$

*and any* $M \in \mathbb{N}$ *and* $k$-*tuple* $D = (d_1, \ldots, d_k)$ *of nonnegative integers with*

$$0 \leq d_1 \leq \cdots \leq d_k < M + d_k \leq N, \tag{8.4}$$

*write*

$$V_k \left( E_N^{(1)}, \ldots, E_N^{(k)}, M, D \right) = \sum_{n=1}^{M} e_{n+d_1}^{(1)} \cdots e_{n+d_k}^{(k)}. \tag{8.5}$$

*Let*

$$\tilde{C}_k \left( E_N^{(1)}, \ldots, E_N^{(k)} \right) = \max_{M, D} \left| V_k \left( E_N^{(1)}, \ldots, E_N^{(k)}, M, D \right) \right| \tag{8.6}$$

*where the maximum is taken over all* $D = (d_1, \ldots, d_k)$ *and* $M \in \mathbb{N}$ *satisfying* (8.4) *with the additional restriction that if* $E_N^{(i)} = E_N^{(j)}$ *for some* $i \neq j$, *then we must not have* $d_i = d_j$. *Then the* cross-correlation measure of order $k$ *of the family* $\mathcal{F}$ *of binary sequences* $E_N \in \{-1, +1\}^N$ *is defined as*

$$\Phi_k(\mathcal{F}) = \max \tilde{C}_k \left( E_N^{(1)}, \ldots, E_N^{(k)} \right) \tag{8.7}$$

*where the maximum is taken over all* $k$-*tuples of binary sequences* $\left( E_N^{(1)}, \ldots, E_N^{(k)} \right)$ *with*

$$E_N^{(i)} \in \mathcal{F} \text{ for } i = 1, \ldots, k.$$

(Note that other cross-correlation type quantities also occur in [7, 18, 21].)

Then observe first that by the definition of $\tilde{C}_k$, for every $E_N \in \{-1, +1\}^N$ we have

$$\tilde{C}_k (E_N, \ldots, E_N) = C_k(E_N),$$

thus from (8.7) we have the following result.

**Proposition 8.6**  *We have*

$$\Phi_k(\mathcal{F}) \geq \max_{E_N \in \mathcal{F}} C_k(E_N). \tag{8.8}$$

This means that an upper bound for the cross-correlation of order $k$ of the *family* $\mathcal{F}$ is also an upper bound for correlation of order $k$ of *every sequence* $E_N \in \mathcal{F}$. Thus it suffices to estimate $\Phi_k(\mathcal{F})$ : if we have a "good" upper bound for $\Phi_k(\mathcal{F})$, then this guarantees that $\mathcal{F}$ consists of *sequences possessing strong pseudorandom properties*.

In Section 8.1 we said that in the applications it is "important to know that the given family has a rich, complex structure, and that there are many independent sequences in it." Can one use the cross-correlation measure of a family to show that, indeed, this is the case? We will show that already the small cross-correlation measure of order 2 is enough to guarantee that the sequences in the family are far apart (literally).

**Proposition 8.7**  *If* $N \in \mathbb{N}$ *and* $E_N = (e_1, \ldots, e_N) \in \mathcal{F}$, $E_N' = (e_1', \ldots, e_N') \in \mathcal{F}$, $\mathcal{F} \subset \{-1, +1\}^N$, *then we have*

$$\left| d(E_N, E_N') - \frac{N}{2} \right| \leq \frac{1}{2}\tilde{C}_2(E_N, E_N') \leq \frac{1}{2}\Phi_2(\mathcal{F}). \tag{8.9}$$

*Proof.*  Clearly we have

$$\frac{(e_n - e_n')^2}{4} = \begin{cases} 0 & \text{if } e_n = e_n' \\ 1 & \text{if } e_n \neq e_n' \end{cases} \quad \text{for } n = 1, 2, \ldots, N$$

thus

$$d(E_N, E_N') = \sum_{n=1}^{N} \frac{(e_n - e_n')^2}{4} = \frac{N}{2} - \frac{1}{2}\sum_{n=1}^{N} e_n e_n'$$

whence, by (8.5), (8.6) and (8.7),

$$\left| d(E_N, E_N') - \frac{N}{2} \right| = \frac{1}{2}\left| \sum_{n=1}^{N} e_n e_n' \right| \leq \frac{1}{2}\tilde{C}_2(E_N, E_N') \leq \Phi_2(\mathcal{F})$$

which proves (8.9).

If the cross-correlation of order 2 of the family $\mathcal{F} \subseteq \{-1, +1\}^N$ is $o(N)$:

$$\Phi_2(\mathcal{F}) = o(N), \tag{8.10}$$

then it follows from Definition 8.4, (8.9) and (8.10) that

$$m(\mathcal{F}) = \min_{\substack{E_N, E'_N \in \mathcal{F} \\ E_N \neq E'_N}} d(E_N, E'_N) \geq \frac{N}{2} - \frac{1}{2}\Phi_2(\mathcal{F}) = \frac{N}{2} - o(N)$$

so that (8.3) holds. This proves the following result.

**Proposition 8.8** *If $N \in \mathbb{N}$, $\mathcal{F} \subset \{-1, +1\}^N$ and (8.10) holds then the family $\mathcal{F}$ possesses the strict avalanche property.*

So far we have seen that there is a close connection between collision, distance minimum and avalanche property in a family of binary sequences on the one hand and its cross-correlation on the other hand. It remains to see whether there is any connection between the complexity of a family and its cross-correlation. We will show by two examples that these two measures are independent in the sense that a family $\mathcal{F}$ may be "good" concerning its family complexity, i.e., $\Gamma(\mathcal{F})$ is large, but it may be "bad" considering its cross-correlation, i.e., $\Phi_k(\mathcal{F})$ is also large for every small $k$. On the other hand it is also possible that $\mathcal{F}$ can be considered "good" since $\Phi_k(\mathcal{F})$ is small, but that $\mathcal{F}$ can be considered "bad" concerning its small family complexity. (This means that it is not enough to study only one of $\Gamma(\mathcal{F})$ and $\Phi_k(\mathcal{F})$, we have to estimate both of them.)

**Example 8.9** *Let $N \in \mathbb{N}$ and let $\mathcal{F}$ be the set of all the binary sequences of length $N$: $\mathcal{F} = \{-1, +1\}^N$. Then clearly $\Gamma(\mathcal{F})$ is maximal: $\Gamma(\mathcal{F}) = N$. On the other hand, $E_N = (e_1, \ldots, e_N) = (1, \ldots, 1) \in \mathcal{F}$ thus by (8.8), for $k \in \mathbb{N}$, $k \leq N$ we have*

$$\Phi_k(\mathcal{F}) \geq C_k(E_N) = \sum_{n=1}^{N-k+1} e_n e_{n+1} \cdots e_{n+k-1} = \sum_{n=1}^{N-k+1} 1 = N - k + 1$$

*which is also large.*

**Example 8.10** *Consider any family $\mathcal{F}$ of binary sequences of length $N$ with small cross-correlation of order $k$ for any small $k$; for example, we may take $N = p = $ prime and the family $\mathcal{F}_1$ which will be constructed later in Theorem 8.14 and which will satisfy the inequality*

$$\Phi_k(\mathcal{F}_1) < 10kdp^{1/2} \log p$$

*(for any $1 < k < p$). Then for at least half of the sequences $E_p = (e_1, \ldots, e_p) \in \mathcal{F}_1$ either $e_1 = +1$ or $e_1 = -1$ holds; we may assume the first*

*equality. Then let* $\mathcal{F}_1' = \{E_p = (e_1, \ldots, e_p) : e_1 = +1\}$ *so that* $|\mathcal{F}_1'| \geq \frac{|\mathcal{F}_1|}{2}$, *we have*

$$\Phi_k(\mathcal{F}_1') \leq \Phi_k(\mathcal{F}_1) < 10kdp^{1/2}\log p$$

*(which is small) and*

$$\Gamma\left(\mathcal{F}_1'\right) = 0$$

*(which is also small) since there is no* $E_p = (e_1, \ldots, e_p) \in \mathcal{F}_1'$ *satisfying the specification*

$$e_1 = -1.$$

## 8.3 The size of the cross-correlation measure

When we introduce a new pseudorandom measure of sequences or a new family measure, then a question of basic importance is what is the expected size of the new measure, and what is the size that we hope to achieve? In the case of the measures of pseudorandomness of binary *sequences* our starting point was the study of the behavior of a truly random binary sequence of a given length $N$. In the case of families the situation is more complex: usually not only is the length $N$ of the sequences given but also the size of the family $\mathcal{F}$ plays an important role. Our optimal goal is usually to construct a possible large family $\mathcal{F}$ of "good" pseudorandom binary sequences with the property that it possesses the strong avalanche property, i.e., (8.3) holds: $m(\mathcal{F}) \geq \left(\frac{1}{2} - o(1)\right)N$. By Proposition 8.8 this is the case if (8.9) holds: $\Phi_2(\mathcal{F}) = o(N)$. It follows from the results of coding theory [42] that requirement (8.3) can hold for $\mathcal{F}$ with $|\mathcal{F}| > 2^{c_1 N}$ with some $0 < c_1 < 1/2$ (e.g., $c_1 = 0.11$ can be taken) but it is known that there is a $c_2$ such that $c_1 < c_2 < 1/2$ and a family with $|\mathcal{F}| > 2^{c_2 N}$ cannot satisfy (8.3) (e.g., $c_2 = 0.18$ can be taken). If we relax (8.3), then the size of $\mathcal{F}$ may grow. However, one should not forget that the sequences in $\mathcal{F}$ must also possess strong pseudorandom properties; it is not at all easy to combine this requirement with (8.3) and a good lower bound for $|\mathcal{F}|$. In practice, it is quite satisfactory to construct a family $\mathcal{F}$ with $|\mathcal{F}| > \exp\left(N^{c_1}\right)$, $\Phi_k(\mathcal{F}) < N^{c_2}$ (for all small $k$) with some $0 < c_1 < 1$, $c_2 < 1$ (note that by (8.8) it also follows from the upper bound for $\Phi_k(\mathcal{F})$ that every $E_N \in \mathcal{F}$ possesses small correlations of small order). It remains to present constructions for families with these properties. This will be done in Sections 8.4 and 8.5, but first we will study the extremal values of $\Phi_k(\mathcal{F})$. (One also might like to study the behavior of the cross-correlation measures for a truly random family of given size. This seems to be a rather difficult task;

perhaps we will return to this problem in a subsequent paper.) It was shown in [5, 25] that for $N \in \mathbb{N}$, $k \in \mathbb{N}$ we have

$$\min_{E_N \in \{-1,+1\}^N} C_{2k}(E_N) > \left( \frac{1}{2} \left[ \frac{N}{k+1} \right] \right)^{1/2}.$$

By (8.8) the same lower bound can be given for $\Phi_{2k}(\mathcal{F})$. On the other hand, it was shown in [9] that for all $N \in \mathbb{N}$, $k \in \mathbb{N}$, $2k+1 < N$ we have

$$\min_{E_N \in \{-1,+1\}^N} C_{2k+1}(E_N) = 1. \tag{8.11}$$

It is a natural question to ask: what about the extremal values of $\Phi_{2k+1}(\mathcal{F})$? If $E'_N$ denotes the binary sequence of $N$ whose every element is $+1$, $\mathcal{F}$ contains the sequence $E'_N$, and $2k+1 < N$, then by (8.8) we have

$$\Phi_{2k+1}(\mathcal{F}) = C_{2k+1}\left(E'_N\right) = N - 2k.$$

On the other hand, if $2k+1 < N$ then by (8.11) there is a binary sequence $E''_N$ of length $N$ with $C_{2k+1}(E''_N) = 1$. If $\mathcal{F}$ consists of the single sequence $E''_N$ then we have

$$\Phi_{2k+1}(\mathcal{F}) = C_{2k+1}\left(E''_N\right) = 1.$$

But can $\Phi_{2k+1}(\mathcal{F})$ also be small for greater families? We will show that the answer is affirmative if $|\mathcal{F}|$ is "much smaller" than $N$, the length of the sequences in $\mathcal{F}$ (but we do not know what happens in larger families).

**Proposition 8.11** *Let $N \in \mathbb{N}$, $k \in \mathbb{N}$, $2k+1 < N$, $\ell \in \mathbb{N}$ and $\ell < N$. For $i = 1, \ldots, \ell$, define the binary sequence $E_N^{(i)} = \left(e_1^{(i)}, \ldots, e_N^{(i)}\right)$ of length $N$ by*

$$e_n^{(i)} = (-1)^{\left[\frac{n+i}{\ell}\right]} \text{ for } n = 1, \ldots, N,$$

*and let $\mathcal{F}$ be the family $\mathcal{F} = \left\{E_N^{(1)}, \ldots, E_N^{(\ell)}\right\}$. Then we have*

$$\Phi_{2k+1}(\mathcal{F}) \leq 2\ell. \tag{8.12}$$

*Proof.* Using notation (8.5), for any $M$, $1 \leq i_1, \ldots, i_{2k+1} \leq \ell$ and $2k+1$-tuple $D = (d_1, \ldots, d_{2k+1})$ satisfying (8.4) (with $2k+1$ in place of $k$) we have

$$\left| V_{2k+1}\left( E_N^{(i_1)}, \ldots, E_N^{(i_{2k+1})}, M, D \right) \right| = \left| \sum_{n=1}^{M} e_{n+d_1}^{(i_1)} e_{n+d_2}^{(i_2)} \cdots e_{n+d_{2k+1}}^{(i_{2k+1})} \right|$$

$$= \left| \sum_{n=1}^{M} (-1)^{\left[ \frac{n+d_1+i_1}{\ell} \right]} \cdots (-1)^{\left[ \frac{n+d_{2k+1}+i_{2k+1}}{\ell} \right]} \right|$$

$$= \left| \sum_{r=1}^{\ell} \sum_{m=0}^{\left[ \frac{M}{\ell} \right]-1} (-1)^{\left[ \frac{\ell m+r+d_1+i_1}{\ell} \right]} \cdots (-1)^{\left[ \frac{\ell m+r+d_{2k+1}+i_{2k+1}}{\ell} \right]} \right.$$

$$\left. + \sum_{\ell\left[ \frac{M}{\ell} \right]<n\leq M} (-1)^{\left[ \frac{n+d_1+i_1}{\ell} \right]} \cdots (-1)^{\left[ \frac{n+d_{2k+1}+i_{2k+1}}{\ell} \right]} \right|$$

$$\leq \left| \sum_{r=1}^{\ell} \sum_{m=0}^{\left[ \frac{M}{\ell} \right]-1} (-1)^{m+\left[ \frac{r+d_1+i_1}{\ell} \right]} \cdots (-1)^{m+\left[ \frac{r+d_{2k+1}+i_{2k+1}}{\ell} \right]} \right| + \sum_{\ell\left[ \frac{M}{\ell} \right]<n\leq M} 1$$

$$\leq \left| \sum_{r=1}^{\ell} (-1)^{\left[ \frac{r+d_1+i_1}{\ell} \right]+\cdots+\left[ \frac{r+d_{2k+1}+i_{2k+1}}{\ell} \right]} \sum_{m=0}^{\left[ \frac{M}{\ell} \right]-1} (-1)^{(2k+1)m} \right| + \ell$$

$$\leq \left| \sum_{r=1}^{\ell} 1 \right| \left| \sum_{m=0}^{\left[ \frac{M}{\ell} \right]-1} (-1)^{m} \right| + \ell \leq \ell \cdot 1 + \ell = 2\ell. \tag{8.13}$$

Equation (8.12) follows from (8.6), (8.7) and (8.13).

We do not know what happens in larger families.

**Problem 8.12** *Estimate* $\min \Phi_{2k+1}(\mathcal{F})$ *for any fixed* $N, k$ *and* $|\mathcal{F}|$.

## 8.4 A family with small cross-correlation constructed using the Legendre symbol

The first construction for large families of binary sequences with strong pseudorandom properties (in terms of the measures described in Section 8.1) was given by Goubin *et al.* [19] and it used the Legendre symbol (this is, perhaps, still the best construction of this type). They proved the following theorem.

**Theorem 8.13** *If $p$ is a prime number, $f(x) \in \mathbb{F}_p[x]$ has degree $d$ ($d > 0$), $f(x)$ has no multiple zero in $\overline{\mathbb{F}}_p$, and the binary sequence $E_p = E_p(f) = (e_1, \ldots, e_p)$ is defined by*

$$e_n = \begin{cases} \left(\frac{f(n)}{p}\right) & \text{for } (f(n), p) = 1 \\ +1 & \text{for } p \mid f(n) \end{cases} \quad (\text{for } n = 1, 2, \ldots, p) \qquad (8.14)$$

*then we have*

$$W(E_p) < 10dp^{1/2} \log p,$$

*and if either*

(i) $k = 2$,
(ii) *2 is a primitive root modulo $p$ and $k < p$, or*
(iii) *we have*

$$k < \frac{p^{1/d}}{4}, \qquad (8.15)$$

*then*

$$C_k(E_p) < 10kdp^{1/2} \log p$$

*also holds.*

Indeed, this is a combination of Theorems 1 and 2 in [19]. (Note that (8.15) is a corrected form of the inequality appearing in Corollary 2(ii) in [19]; namely, there the exponent of $p$ is $1/4$, while the correct exponent provided by the proof is $1/d$ as in (8.15).)

Let $\mathcal{F}$ denote the family of binary sequences $E_p(f)$ assigned to the polynomials satisfying the conditions in Theorem 8.13. In Sections 8.4 and 8.5 we will show two different ways to modify the definition of the family slightly to give reasonable control over the cross-correlations of the family.

**Theorem 8.14** *Let $d \in \mathbb{N}$, $p$ a prime number, $d < p$, and consider all the irreducible polynomials $f(x) \in \mathbb{F}_p[x]$ of the form*

$$f(x) = x^d + a_2 x^{d-2} + a_3 x^{d-3} + \cdots + a_d \qquad (8.16)$$

*(so that there is no $x^{d-1}$ term) and let $\mathcal{F}_1$ denote the family of the binary sequences $E_p = E_p(f)$ assigned to these polynomials $f$ by the formula (8.14). Then we have*

(i)

$$\Phi_k(\mathcal{F}_1) < 10kdp^{1/2} \log p \qquad (8.17)$$

*for all $k \in \mathbb{N}$, $1 < k < p$, and*
(*ii*) *if $d < p^{1/2}/20 \log p$, then*

$$|\mathcal{F}_1| \geq p^{[d/3]-1}. \tag{8.18}$$

*Proof.* (i) Using the notation in Definition 8.5, we have to estimate

$$\left| V_k \left( E_p^{(1)}, \ldots, E_p^{(k)}, M, D \right) \right| = \left| \sum_{n=1}^{M} e_{n+d_1}^{(1)} \cdots e_{n+d_k}^{(k)} \right|$$

for

$$E_p^{(i)} = E_p^{(i)}(f_i) \in \mathcal{F}_1 \quad (i = 1, 2, \ldots, k)$$

and $M$, $D$ satisfying the conditions in Definition 8.5. Clearly,

$$f_i(n + d_i) \equiv 0 \pmod{p}, \ 1 \leq n \leq M, \ 1 \leq i \leq k$$

has at most $dk$ solutions (in pairs $(n, i)$). Thus defining $\left( \frac{a}{p} \right)$ as 0 for $p \mid a$, we have

$$\begin{aligned}
\left| V_k \left( E_p^{(1)}, \ldots, E_p^{(k)}, M, D \right) \right| &= \left| \sum_{n=1}^{M} e_{n+d_1}^{(1)} \cdots e_{n+d_k}^{(k)} \right| \\
&\leq \left| \sum_{n=1}^{M} \left( \frac{f_1(n+d_1)}{p} \right) \cdots \left( \frac{f_k(n+d_k)}{p} \right) \right| + dk \\
&= \left| \sum_{n=1}^{M} \left( \frac{f_1(n+d_1) \cdots f_k(n+d_k)}{p} \right) \right| + dk.
\end{aligned} \tag{8.19}$$

If for some $1 \leq i < j \leq k$ we have $f_i(x) \neq f_j(x)$, then

$$f_i(x + d_i) \neq f_j(x + d_j) \tag{8.20}$$

since both $f_i$ and $f_j$ are of form (8.16). If $1 \leq i < j \leq k$ and $f_i(x) = f_j(x)$, then by the conditions on $D$ in Definition 8.5 we cannot have $d_i = d_j$, thus again (8.20) holds. Then writing

$$h(x) = f_1(x + d_1) \cdots f_k(x + d_k), \tag{8.21}$$

this polynomial is the product of $k$ distinct monic irreducible polynomials, thus it is square-free. Now we will need the following lemma.

**Lemma 8.15** *If $p$ is a prime number, $\chi$ is a nonprincipal character modulo $p$ of order $t$, $h(x) \in \mathbb{F}_p[x]$ has degree $r$ and it is not of the form $h(x) = cg(x)^t$ with $c \in \mathbb{F}_p$, $g(x) \in \mathbb{F}_p[x]$, and $X, Y$ are real numbers with $0 < Y \le p$, then*

$$\left| \sum_{X < n \le X+Y} \chi(h(n)) \right| < 9rp^{1/2} \log p.$$

*Proof.* This lemma can be derived from Weil's theorem [44] using a method of Vinogradov [43]; see Theorem 2 and Corollary 1 in [31] and Lemma 2 in [3]. (Note that combining Weil's theorem and Vinogradov's inequality with Cochrane's and Peral's result [12], we obtain that the absolute constant 9 in this upper bound can be replaced by $\frac{4}{\pi^2} + o(1)$ for $p \to \infty$, and then the absolute constant 10 in (8.17) in Theorem 8.14 can also be replaced by $\frac{4}{\pi^2} + o(1)$.)

Since the polynomial in (8.21) is square-free, we may use this lemma with the quadratic character

$$\chi(n) = \begin{cases} \left(\dfrac{n}{p}\right) & \text{if } (n, p) = 1 \\ 0 & \text{if } p \mid n, \end{cases}$$

the polynomial $h(x)$ in (8.21) and $t = 2$. Then we get from (8.19) that

$$\left| V_k\left( E_p^{(1)}, \ldots, E_p^{(k)}, M, D \right) \right| < 9kdp^{1/2} \log p. \tag{8.22}$$

Equation (8.17) follows from (8.6), (8.7) and (8.22).

In order to prove (8.18) we need a result of S. D. Cohen [14].

**Lemma 8.16** *Given a prime power $q > 3$ and arbitrary positive integers $n$ and $m \le n/3$, there exists a primitive polynomial $x^n + a_1 x^{n-1} + \cdots + a_n \in \mathbb{F}_q[x]$ with the first $m$ coefficients $a_1, \ldots, a_m$ prescribed in advance, with the exception that there is no primitive cubic over $\mathbb{F}_4$ with zero first coefficient.*

*Proof.* This is Theorem 3 in [14].

Now assume that $d$ satisfies the given condition, and consider two distinct irreducible polynomials $f_1$, $f_2$ of form (8.16). Write

$$E_p(f_1) = \left( e_1^{(1)}, \ldots, e_p^{(1)} \right), \quad E_p(f_2) = \left( e_1^{(2)}, \ldots, e_p^{(2)} \right).$$

Then the proof of (8.19) gives that

$$\left| \sum_{n=1}^{p} e_n^{(1)} e_n^{(2)} \right| \leq \left| \sum_{n=1}^{p} \left( \frac{f_1(n) f_2(n)}{p} \right) \right| + 2d$$
$$< 18 dp^{1/2} \log p + 2d < 20 dp^{1/2} \log p < p$$

thus $E_p(f_1) \neq E_p(f_2)$. It follows that $|\mathcal{F}_1|$ is at least the number of irreducible polynomials of form (8.16). For any fixed

$$a_2 \in \mathbb{F}_p, a_3 \in \mathbb{F}_p, \ldots, a_{[d/3]} \in \mathbb{F}_p, \tag{8.23}$$

there is at least one primitive polynomial $f(x)$ of form (8.16) with these prescribed coefficients (note that $p > 3$ follows from the conditions in the theorem), and these polynomials are also irreducible (since primitive polynomials are irreducible). Since different $a_i$ determine different irreducible polynomials of form (8.16) and distinct polynomials determine sequences $E_p$, then the number of these sequences is at least the number of choices of the $a_i$ in (8.23):

$$|\mathcal{F}_1| \geq p^{[d/3]-1}$$

which proves (8.18).

## 8.5 Another construction

Theorem 8.14 gives a good upper bound for the cross-correlation, and the size of the family $\mathcal{F}_1$ is also large. However, this theorem has a weakness: since no good algorithm is known for constructing "many" irreducible polynomials over $\mathbb{F}_p$ (see [1, 13, 23, 26]), Theorem 8.14 only proves existence and it does not provide an explicit construction. Thus we will now present another construction which will be more explicit, but the price paid for this is that we will be able to control the cross-correlation of order $k$ only if $k = 2$ or $k$ is odd.

**Theorem 8.17** *Let $d \in \mathbb{N}$, $d$ odd, $d < p$, and consider all the polynomials $f(x) \in \mathbb{F}_p[x]$ of the form*

$$f(x) = (x - x_1)(x - x_2) \cdots (x - x_d) \tag{8.24}$$

*where*

$$x_1, x_2, \ldots, x_d \text{ are distinct elements of } \mathbb{F}_p \tag{8.25}$$

*and*

$$x_1 + x_2 + \cdots + x_d = 0. \tag{8.26}$$

*Let $\mathcal{F}_2$ denote the family of the binary sequences $E_p = E_p(f)$ assigned to these polynomials by formula (8.14). Then we have*

(i)

$$\Phi_k(\mathcal{F}_2) < 10kdp^{1/2}\log p \qquad (8.27)$$

*if $k = 2$ or $k$ is odd, and*

(ii)

$$|\mathcal{F}_2| = \frac{1}{d}\binom{p-1}{d-1}. \qquad (8.28)$$

*Proof.* (i) Since a considerable part of the proof is similar to the proof of (8.17) in Theorem 8.14 we will leave some details to the reader.

As in the proof of Theorem 8.14, we have

$$\left| V_k\left(E_p^{(1)}, \ldots, E_p^{(k)}, M, D\right) \right| \le \left| \sum_{n=1}^{M} \left( \frac{f_1(n+d_1)\cdots f_k(n+d_k)}{p} \right) \right| + dk$$

$$(8.29)$$

where $f_1, f_2, \ldots, f_k$ are of form (8.24) (with $x_1, x_2, \ldots, x_d$ satisfying (8.25) and (8.26)). It follows from (8.26) that if $f(x)$ is of this form, $c \in \mathbb{F}_p$ and $c \ne 0$, then $f(x) \ne f(x+c)$, thus by the restriction on the $d_i$ in Definition 8.5, for $k = 2$ we cannot have $f_1(x+d_1) = f_2(x+d_2)$ in the sum in (8.29), and thus the monic polynomial $f_1(x+d_1)f_2(x+d_2)$ is not a square. If $k$ is odd then the degree of the (monic) polynomial $f_1(x+d_1)\cdots f_k(x+d_k)$ is $kd$ which is odd (since both $k$ and $d$ are odd), thus again this polynomial cannot be a square. In both cases we may use Lemma 8.15 to estimate the sum in (8.29), and we get the same upper bound as in the proof of Theorem 8.14 which proves (8.27).

(ii) As in the proof of Theorem 8.14, it follows from (8.24), (8.26) and the proof of (8.27) (with $k = 2$) that for two distinct polynomials $f_1, f_2$ of form (8.24) (with $x_1, x_2, \ldots, x_d$ satisfying (8.25) and (8.26)) we have $E_p(f_1) \ne E_p(f_2)$. Thus $|\mathcal{F}_2|$ is equal to the number of the polynomials $f(x)$ which satisfy (8.24), (8.25) and (8.26). The number of $d$-tuples $x_1, x_2, \ldots, x_d$ satisfying (8.25) and

$$x_1 + x_2 + \cdots + x_d = c$$

is independent of $c$ since there is a bijection between the solutions for different $c$ values (note that $0 < d < p$). Thus the number of solutions of (8.25) and (8.26) is the total number of $d$-tuples satisfying (8.25) divided by $p$: $\frac{1}{p}\binom{p}{d} = \frac{1}{d}\binom{p-1}{d-1}$, and this proves (8.28).

## Acknowledgement

We would like to thank Arne Winterhof and the anonymous referee for their valuable suggestions and comments.

This research is partially supported by ERC-AdG.228005, Hungarian National Foundation for Scientific Research, Grants No. K100291 and NK104183, the János Bolyai Research Fellowship, the Agence Nationale de la Recherche grant ANR-10-BLAN 0103 called MUNUM and the MTA-ELTE Geometric and Algebraic Combinatorics Research Group.

## References

[1]  S. Abrahamyan, M. Alizadeh and M. K. Kyureghyan, Recursive constructions of irreducible polynomials over finite fields. *Finite Fields Appl.* **18**, 738–745, 2012.

[2]  R. Ahlswede, L. H. Khachatrian, C. Mauduit and A. Sárközy, A complexity measure for families of binary sequences. *Period. Math. Hungar.* **46**, 107–118, 2003.

[3]  R. Ahlswede, C. Mauduit and A. Sárközy, Large families of pseudorandom sequences of $k$ symbols and their complexity. I. *General Theory of Information Transfer and Combinatorics*. Lecture Notes in Computer Science, volume 4123, pp. 293–307. Springer, Berlin, 2006.

[4]  R. Ahlswede, C. Mauduit and A. Sárközy, Large families of pseudorandom sequences of $k$ symbols and their complexity. II. *General Theory of Information Transfer and Combinatorics*. Lecture Notes in Computer Science, volume 4123, pp. 308–325. Springer, Berlin, 2006.

[5]  N. Alon, Y. Kohayakawa, C. Mauduit, C. G. Moreira and V. Rödl, Measures of pseudorandomness for finite sequences: minimal values. *Comb. Probab. Comput.* **15** (2005), 1–29, 2005.

[6]  N. Alon, Y. Kohayakawa, C. Mauduit, C. G. Moreira and V. Rödl, Measures of pseudorandomness for finite sequences: typical values. *Proc. London Math. Soc.* **95**, 778–812, 2007.

[7]  V. Anantharam, A technique to study the correlation measures of binary sequences. *Discrete Math.* **308**, 6203–6209, 2008.

[8]  A. Bérczes, J. Ködmön and A. Pethő, A one-way function based on norm form equations. *Period. Math. Hung.* **49**, 1–13, 2004.

[9]  J. Cassaigne, C. Mauduit and A. Sárközy, On finite pseudorandom binary sequences, VII: the measures of pseudorandomness. *Acta Arith.* **103**, 97–118, 2002.

[10]  Z.-X. Chen, Elliptic curve analogue of Legendre sequences. *Monatsh. Math.* **154**, 1–10, 2008.

[11]  Z. Chen, S. Li and G. Xiao, Construction of pseudorandom binary sequences from elliptic curves by using the discrete logarithms. *Sequences and their Applications, SETA 2006*. Lecture Notes in Computer Science, volume 4086, pp. 285–294. Springer, Berlin, 2006.

[12] T. Cochrane and J. C. Peral, An asymptotic formula for a trigonometric sum of Vinogradov. *J. Number Theory* **91**, 1–19, 2001.

[13] S. D. Cohen, The explicit construction of irreducible polynomials over finite fields, *Des. Codes Cryptogr.* **2**, 169–174, 1992.

[14] S. D. Cohen, Primitive polynomials over small fields. In: *Finite Fields and Applications, Seventh International Conference, Toulouse, 2003.* (eds.), G. Mullen *et al.* Lecture Notes in Computer Science, volume 2948, pp. 293–307. Springer, Berlin, 2006.

[15] H. Feistel, W. A. Notz and J. L. Smith, Some cryptographic techniques for machine-to-machine data communications. *Proc. IEEE* **63**, 1545–1554, 1975.

[16] J. Folláth, Construction of pseudorandom binary sequences using additive characters over $GF(2^k)$. *Period. Math. Hung.* **57**, 73–81, 2008.

[17] J. Folláth, Construction of pseudorandom binary sequences using additive characters over $GF(2^k)$. II. *Period. Math. Hung.* **60**, 127–135, 2010.

[18] G. Gong, Character sums and polyphase sequence families with low correlation, discrete fourier transform (DFT), and ambiguty. In: C. Pascale *et al.* (eds.), *Finite Fields and their Applications.* Radon Series on Computational and Applied Mathematics, volume 11, pp. 1–42. de Gruyter, Berlin, 2013.

[19] L. Goubin, C. Mauduit and A. Sárközy, Construction of large families of pseudorandom binary sequences. *J. Number Theory* **106**, 56–69, 2004.

[20] K. Gyarmati, On a family of pseudorandom binary sequences. *Period. Math. Hung.* **49**, 45–63, 2004.

[21] K. Gyarmati, Concatenation of pseudorandom binary sequences. *Period. Math. Hung.* **58**, 99–120, 2009.

[22] K. Gyarmati, On the complexity of a family related to the Legendre symbol. *Period. Math. Hung.* **58**, 209–215, 2009.

[23] S. Huczynska, Existence results for finite field polynomials with special properties. In: P. Charpin *et al.* (eds.), *Finite Fields and their Applications, Character Sums and Polynomials.* Radon Series on Computational and Applied Mathematics, volume 11, pp. 65–87. de Gruyter, Berlin, 2013.

[24] J. Kam and G. Davida, Structured design of substitution-permutation encryption networks, *IEEE Trans. Comput.* **28**, 747–753, 1979.

[25] Y. Kohayakawa, C. Mauduit, C. G. Moreira and V. Rödl, Measures of pseudorandomness for finite sequences: minimum and typical values. *Proceedings WORDS'03.* TUCS Gen. Publ. 27, pp. 159–169. Turku Cent. Comput. Sci., Turku, 2003.

[26] M. K. Kyureghyan, Recurrent methods for constructing irreducible polynomials over $\mathbb{F}_q$ of odd characterics, I, II. *Finite Fields Appl.* **9**, 39–58, 2003; **12**, 357–378, 2006.

[27] H. Liu, New pseudorandom sequences constructed using multiplicative inverses. *Acta Arith.* **125**, 11–19, 2006.

[28] H. N. Liu, A family of pseudorandom binary sequences constructed by the multiplicative inverse. *Acta Arith.* **130**, 167–180, 2007.

[29] H. Liu, New pseudorandom sequences constructed by quadratic residues and Lehmer numbers. *Proc. Am. Math. Soc.* **135**, 1309–1318, 2007.

[30] H. Liu, A large family of pseudorandom binary lattices. *Proc. Am. Math. Soc.* **137**, 793–803, 2009.

[31] C. Mauduit and A. Sárközy, On finite pseudorandom binary sequences I: measures of pseudorandomness, the Legendre symbol. *Acta Arith.* **82**, 365–377, 1997.

[32] C. Mauduit and A. Sárközy, Construction of pseudorandom binary sequences by using the multiplicative inverse. *Acta Math. Hung.* **108**, 239–252, 2005.

[33] C. Mauduit and A. Sárközy, Family complexity and VC-dimension. In: H. Aydinian *et al.* (eds.), *Ahlswede Festschrift.* Lecture Notes in Computer Science, volume 7777, pp. 346–363. Springer, Berlin, 2013.

[34] C. Mauduit, J. Rivat and A. Sárközy, Construction of pseudorandom binary sequences using additive characters. *Monatsh. Math.* **141**, 197–208, 2004.

[35] A. Menezes, P. C. van Oorschot and S. Vanstone, *Handbook of Applied Cryptography.* CRC Press, Boca Raton, FL, 1997.

[36] L. Mérai, A construction of pseudorandom binary sequences using both additive and multiplicative characters. *Acta Arith.* **139**, 241–252, 2009.

[37] L. Mérai, A construction of pseudorandom binary sequences using rational functions. *Unif. Distrib. Theory* **4**, 35–49, 2009.

[38] L. Mérai, Construction of large families of pseudorandom binary sequences. *Ramanujan J.* **18**, 341–349, 2009.

[39] A. Sárközy, A finite pseudorandom binary sequence. *Stud. Sci. Math. Hung.* **38**, 377–384, 2001.

[40] V. Tóth, Collision and avalanche effect in families of pseudorandom binary sequences. *Period. Math. Hung.* **55**, 185–196, 2007.

[41] V. Tóth, The study of collision and avalanche effect in a family of pseudorandom binary sequences. *Period. Math. Hung.* **59**, 1–8, 2009.

[42] M. Tsfasman, S. Vlăduţ and D. Nogin, *Algebraic Geometric Codes: Basic Notions.* Mathematical Surveys and Monographs, volume 139. AMS, Providence, RI, 2007.

[43] I. M. Vinogradov, *Elements of Number Theory.* Dover, 1954.

[44] A. Weil, *Sur les Courbes Algébriques et les Variétés qui s'en Déduisent.* Hermann, Paris, 1948.

# 9

# On an important family of inequalities of Niederreiter involving exponential sums

*Peter Hellekalek*

Salzburg University, Salzburg

*Dedicated to Harald Niederreiter on the occasion of his 70th birthday.*

## Abstract

The Erdős–Turán–Koksma inequality is a fundamental tool for bounding the discrepancy of a sequence in the $s$-dimensional unit cube $[0, 1)^s$, $s \geq 1$, in terms of exponential sums. In an impressive series of papers, Harald Niederreiter established variants of this inequality and proved bounds for the discrepancy for various sequences and point sets, in the context of pseudorandom number generation and in quasi-Monte Carlo methods. These results were an important breakthrough, because they marked the starting point of a thorough theoretical correlation analysis of pseudorandom numbers. Niederreiter's technique also prepared for the study of digital sequences, which are central to modern quasi-Monte Carlo methods.

In this contribution, we present an overview of these concepts and prove a hybrid version of the Erdős–Turán–Koksma inequality, thereby extending a recent result of Niederreiter.

## 9.1 Introduction

This contribution discusses results of Niederreiter [23, 24] that are of fundamental importance in the advancement of two fields, pseudorandom number generation and high-dimensional integration by low-discrepancy point sets. Niederreiter's technique allows assessment of the uniform distribution behavior of certain sequences via exponential sums. It is a major factor in the rapid development of these two fields during the past three decades.

A central question in Monte Carlo and in quasi-Monte Carlo integration is how to approximate the unknown integral $I(f) = \int_{[0,1]^s} f(\mathbf{x}) \, d\mathbf{x}$ of a Riemann-integrable function $f$, $f : [0, 1]^s \rightarrow \mathbb{C}$, by the sample means $S_N(f, \omega) = (1/N) \sum_{n=0}^{N-1} f(\mathbf{x}_n)$, for a sequence $\omega = (\mathbf{x}_n)_{n \geq 0}$ of points in the $s$-dimensional unit cube $[0, 1]^s$.

In the Monte Carlo method, one chooses a finite sequence $\omega = (\mathbf{x}_n)_{n=0}^{N-1}$ of random points, whereas in the quasi-Monte Carlo method the sequence $\omega$ is constructed so that it is as uniformly distributed in $[0, 1]^s$ as possible. In numerical practice, in both cases one employs deterministic sequences, i.e., sequences generated by deterministic algorithms. In the Monte Carlo case, such *pseudorandom* points, as they are called, mimic realizations of a sequence $(\mathbf{X}_n)_{n \geq 0}$ of i.i.d. random variables which are uniformly distributed in $[0, 1]^s$. In the quasi-Monte Carlo setting, so-called low-discrepancy sequences are used. In both cases, one tries to achieve an integration (or quadrature) error $|S_N(f, \omega) - I(f)|$ which is as small as possible.

The function $f$ will vary from one application to another. For this reason it makes sense to choose a class $\mathcal{F} = \{\xi_{\mathbf{k}} : \mathbf{k} \in \Lambda\}$ of test functions $\xi_{\mathbf{k}}$, $\Lambda$ some – usually countable – index set, and to assess, for a given sequence $\omega$, the integration error for all functions in $\mathcal{F}$. If the class $\mathcal{F}$ is appropriately chosen, then this kind of test procedure will allow the quadrature error to be predicted for many other functions $f$. Of course, a given $f$ should be in some sense related to the function class $\mathcal{F}$ in order to get a reliable prediction for the integration error $|S_N(f, \omega) - I(f)|$.[1] As a theoretical background result, the inequality of Koksma and Hlawka tells us that we have to determine how well $\omega$ is uniformly distributed in $[0, 1]^s$ to derive bounds on the integration error (see [5, 19, 26]).

Which function class $\mathcal{F}$ should we choose? In applied statistics, the goodness-of-fit of the empirical distribution of the sample $\omega$ to uniform distribution is often studied with the two-sided Kolmogorov–Smirnov (KS) test statistic (see the monograph by Conover [3]). Let $F_N$ denote the empirical distribution function given by the sample $\omega = (\mathbf{x}_n)_{n=0}^{N-1}$ in $[0, 1]^s$ and let $F$ denote the distribution function of uniform distribution on $[0, 1]^s$. For $\mathbf{t} = (t_1, \ldots, t_s) \in [0, 1]^s$, we have $F_N(\mathbf{t}) = (1/N) \sharp \{n, 0 \leq n \leq N - 1 : \mathbf{x}_n \in \prod_{i=1}^{s} [0, t_i) \}$ and $F(\mathbf{t}) = \prod_{i=1}^{s} t_i$. The two-sided KS test statistic is then defined as the quantity

$$D_N^*(\omega) = \sup_{\mathbf{t} \in [0,1]^s} |F_N(\mathbf{t}) - F(\mathbf{t})| .$$

---

[1] We would like to point out that many tests of randomness are of this form, $f$ being some test statistic and $I(f)$ its expected value.

We note that $D_N^*(\omega)$ is an extremal quadrature error. This is seen as follows. Let $\mathcal{J}^*$ denote the set of those subintervals of $[0,1]^s$ that are anchored at the origin, $\mathcal{J}^* = \{\prod_{i=1}^s [0, t_i), 0 < t_i \leq 1, 1 \leq i \leq s\}$. For a set $M$, let us write $\mathbf{1}_M$ for the indicator function of $M$, $\mathbf{1}_M(x) = 1$ if $x \in M$ and $\mathbf{1}_M(x) = 0$ otherwise. Let $\lambda_s(J)$ stand for the volume of the subinterval $J$ of $[0,1]^s$. Clearly, $\lambda_s(J)$ is equal to the integral $I(\mathbf{1}_J)$. If we put $\mathcal{F} = \{\mathbf{1}_J - \lambda_s(J) : J \in \mathcal{J}^*\}$, then

$$D_N^*(\omega) = \sup_{f \in \mathcal{F}} |S_N(f, \omega)| \, .$$

We will show in Corollary 9.13 how to restrict these suprema to countable classes of intervals. Unfortunately, this does not help with the fact that no efficient algorithm is known to compute $D_N^*(\omega)$ efficiently – i.e., in polynomial time in $s$ – in higher dimensions. For $s = 1$, we refer the reader to Gonzalez *et al.* [7].

The numerical quantity $D_N^*$ is also known in metric number theory as the *star discrepancy*. We will discuss this and related notions in Section 9.2.

Now that we have exhibited a meaningful class $\mathcal{F}$ of test functions, let us return to the Monte Carlo method, in particular to random number generation. Experienced practitioners in the field of stochastic simulation will agree with Compagner [2]: "Monte Carlo results are misleading when correlations hidden in the random numbers and in the simulated system interfere constructively." For this reason, correlation analysis for random numbers is an essential prerequisite for simulation practice. Let $x_0, x_1, \ldots$ be a sequence of uniform random numbers in $[0, 1]$. This is to say, these numbers are realizations of a sequence of independent, identically distributed random variables $X_n, n \geq 0$, each $X_n$ having the uniform distribution in $[0, 1]$; in symbols, $X_n \sim U([0, 1])$. As a consequence of the i.i.d. property of the $X_n$, for every $s \geq 1$, the random vectors $\mathbf{X}_n = (X_{ns}, X_{ns+1}, \ldots, X_{ns+s-1}), n \geq 0$, are i.i.d., $\mathbf{X}_n \sim U([0, 1]^s)$. For realizations $\mathbf{x}_n = (x_{ns}, x_{ns+1}, \ldots, x_{ns+s-1}), n \geq 0$, this means that we expect them to be approximately uniformly distributed in $[0, 1]^s$. If we put $\omega = (\mathbf{x}_n)_{n=0}^{N-1}$, for some sample size $N \in \mathbb{N}$, then the KS test statistic $D_N^*(\omega)$ may be used to measure the deviation of the sample $\omega$ from uniform distribution on $[0, 1]^s$.

In many state-of-the-art pseudorandom number generators, the output sequence $x_0, x_1, \ldots$ in $[0, 1]$ is generated by some map $T$ on a finite state space $\Omega$, $T : \Omega \to \Omega$, usually by letting $z_{n+1} = T(z_n), n \geq 0, z_0 \in \Omega$ some initial value, and then mapping the element $z_n$ to $x_n, n = 0, 1, \ldots$. Unavoidably, such a construction leads to periodic sequences. Further, there will be correlations between consecutive pseudorandom numbers, due to the deterministic generation method. It is the goal of the creator of a pseudorandom number generator to design his algorithm in such a manner that these correlations are so small that they do not show up in the usual statistical tests for random numbers.[2] The correlations between successive pseudorandom numbers $x_0, x_1, \ldots$ can be analyzed in the following way, by a concept called the *serial test* (see Knuth [18]). For a given dimension $s \geq 2$, we construct *nonoverlapping* $s$-tuples $\mathbf{x}_n = (x_{ns}, x_{ns+1}, \ldots, x_{ns+s-1})$, $n \geq 0$, of pseudorandom points and measure the deviation of the sample $\omega = (\mathbf{x}_n)_{n=0}^{N-1}$ in $[0, 1]^s$ from uniform distribution with the KS test statistic $D_N^*(\omega)$, for sample sizes $N$ of our choice. If the values of $D_N^*(\omega)$ are small enough such that an unknowing observer would accept the points $\mathbf{x}_n$ as realizations of i.i.d. random variables $\mathbf{X}_n$, $\mathbf{X}_n \sim U([0, 1]^s)$, then $s$ consecutive pseudorandom numbers in the original sequence $x_0, x_1, \ldots$ are sufficiently uncorrelated. One repeats this assessment for as many dimensions $s$ and sample sizes $N$ as feasible.

This heuristic to assess correlations among successive pseudorandom numbers $x_0, x_1, \ldots$ in $[0, 1]$ via measuring the uniform distribution of the pseudorandom points $\mathbf{x}_0, \mathbf{x}_1, \ldots$ in $[0, 1]^s$ has proven to be highly successful in predicting the behavior of such pseudorandom numbers in stochastic simulation. The performance of full-period point sets $\omega = (\mathbf{x}_n)_{n=0}^{\tau-1}$, $\tau$ denoting the period of the pseudorandom sequence, in various dimensions $s$ seems to be a very reliable indicator to predict the performance of the pseudorandom numbers themselves in various statistical tests and, hence, in many stochastic simulations. We would like to point out that this is an empirical finding and not a mathematically proven fact. We refer the reader to the survey articles [9, 10, 20] for further information.

If the dimension $s$ and the period $\tau$ of the pseudorandom sequence are coprime, then the full-period point set of nonoverlapping $s$-tuples $\mathbf{x}_n$ coincides with the set of *overlapping* $s$-tuples $(x_n, x_{n+1}, \ldots, x_{n+s-1})$, $0 \leq n < \tau$, because in this case the set $\{ns : 0 \leq n < \tau\}$ is a complete residue system modulo $\tau$. For this reason, most estimates for the quantity $D_N^*(\omega)$ are given for the overlapping $s$-tuples in the literature, even if $N < \tau$ (see [23, 24, 32, 35, 36]).

As we have indicated above, no efficient algorithm to compute the KS test statistic $D_N^*(\omega)$ in higher dimensions $s$ is known. For this reason, how is it possible to study $D_N^*(\omega)$ in various dimensions $s \geq 2$? This is where Niederreiter's variants of the Erdős–Turán–Koksma inequality opened the path for a theoretical version of the serial test, see Section 9.2.

---

[2] In this context, we would like to point to the state-of-the-art battery of statistical tests TESTU01 of L'Ecuyer and Simard [21] for testing pseudorandom numbers.

This paper is organized as follows. In Section 9.2 we recall several prerequisites and discuss Niederreiter's fundamental variants of the Erdős–Turán–Koksma inequality. In Section 9.3, we generalize a recent hybrid Erdős–Turán–Koksma inequality due to Niederreiter [29, Theorem 1] by a method using Fourier series.

## 9.2 Concepts

### 9.2.1 Niederreiter's discrepancy bound

Throughout this paper, $b$ denotes a positive integer, $b \geq 2$, and $\mathbf{b} = (b_1, \ldots, b_s)$ stands for a vector of not necessarily distinct integers $b_i \geq 2$, $1 \leq i \leq s$. $\mathbb{N}$ represents the positive integers, and we put $\mathbb{N}_0 = \mathbb{N} \cup \{0\}$.

The underlying space is the $s$-dimensional torus $\mathbb{R}^s/\mathbb{Z}^s$, which will be identified with the half-open unit cube $[0, 1)^s$. The Haar measure on the $s$-torus $[0, 1)^s$ will be denoted by $\lambda_s$. This switch from the closed unit cube $[0, 1]^s$ and Lebesgue measure to the half-open (and compact) unit cube with normalized Haar measure is for the sake of a consistent theory. It does not induce any substantial change in the questions that were discussed in Section 9.1.

We put $e(y) = e^{2\pi i y}$ for $y \in \mathbb{R}$, where i is the imaginary unit. We will use the standard convention that empty sums have value 0 and empty products have value 1.

The trigonometric function system defined below is the classical function system in the theory of uniform distribution of sequences (see [5, 19]).

**Definition 9.1** *Let $k \in \mathbb{Z}$. The $k$th* trigonometric function $e_k$ *is defined as* $e_k : [0, 1) \to \mathbb{C}$, $e_k(x) = e(kx)$. *For* $\mathbf{k} = (k_1, \ldots, k_s) \in \mathbb{Z}^s$, *the* $\mathbf{k}$th *trigono-metric function* $e_{\mathbf{k}}$ *is defined as* $e_{\mathbf{k}} : [0, 1)^s \to \mathbb{C}$, $e_{\mathbf{k}}(\mathbf{x}) = \prod_{i=1}^{s} e(k_i x_i)$, $\mathbf{x} = (x_1, \ldots, x_s) \in [0, 1)^s$. *The* trigonometric function system *in dimension* $s$ *is denoted by* $\mathcal{T}^{(s)} = \{e_{\mathbf{k}} : \mathbf{k} \in \mathbb{Z}^s\}$.

Let $\mathcal{J}$ denote the class of subintervals of $[0, 1)^s$ of the form $\prod_{i=1}^{s}[u_i, v_i)$, $0 \leq u_i < v_i \leq 1$, $1 \leq i \leq s$, and, as in Section 9.1, let $\mathcal{J}^*$ denote the subclass of $\mathcal{J}$ of those intervals anchored at the origin. The so-called extreme discrepancy and the star discrepancy are defined as follows (see [5, 19, 26]).

**Definition 9.2** *Let* $\omega = (\mathbf{x}_n)_{n \geq 0}$ *be a sequence in* $[0, 1)^s$. *The* extreme discrepancy $D_N(\omega)$ *of the first $N$ elements of $\omega$ is defined as*

$$D_N(\omega) = \sup_{J \in \mathcal{J}} |S_N(\mathbf{1}_J - \lambda_s(J), \omega)|.$$

*The* star discrepancy $D_N^*(\omega)$ *of the first $N$ elements of $\omega$ is defined as*

$$D_N^*(\omega) \;=\; \sup_{J \in \mathcal{J}^*} |S_N(\mathbf{1}_J - \lambda_s(J), \omega)| \,.$$

The Erdős–Turán–Koksma inequality gives an upper bound for the discrepancy in terms of exponential sums (see [1, 5]). In its classical version (9.1) below and in all other variants of this inequality, the discrepancy is bounded by the sum of two terms, namely by an error term that stems from approximating the indicator function $\mathbf{1}_J$ of an interval $J$ in $[0, 1)^s$ by a finite Fourier series, and by the sum of weighted exponential sums of the form $|S_N(e_\mathbf{k}, \omega)|$, over a finite domain of indices $\mathbf{k}$. The cut-off index of the Fourier series, i.e., the number $H$ in Theorem 9.3, defines the bound of the finite summation domain. It is intrinsically related to the error term. There is a trade-off, the choice of a higher cut-off index gives a smaller approximation error but a larger summation domain. The same situation appears with other orthonormal bases like the Walsh functions or the **b**-adic functions, which we will introduce below.

We use the convention that the parameters on which the implied constant in a Landau symbol $\mathcal{O}$ depends are written as subscripts of $\mathcal{O}$.

**Theorem 9.3** [5, Theorem 1.21] *Let $\omega$ be a sequence in $[0, 1)^s$ and let $H$ be an arbitrary positive integer. Then*

$$D_N(\omega) = \mathcal{O}_s\left( \frac{1}{H} + \sum_{0 < M(\mathbf{k}) \le H} \frac{1}{r(\mathbf{k})} |S_N(e_\mathbf{k}, \omega)| \right), \qquad (9.1)$$

*where summation is over vectors $\mathbf{k} = (k_1, \ldots, k_s) \in \mathbb{Z}^s$, and*

$$r(\mathbf{k}) = \prod_{i=1}^{s} \max\{1, |k_i|\}, \quad M(\mathbf{k}) = \max_{1 \le i \le s} |k_i|.$$

At present, the implied constant in $\mathcal{O}_s$ equals $(3/2)^s$, which grows exponentially in $s$ (see [5, Theorem 1.21] and [19, p. 116]).

For particular sequences like those appearing in pseudorandom number generation and in quasi-Monte Carlo methods, a considerably smaller constant that grows linearly with the dimension $s$ can be achieved. The following variant of the Erdős–Turán–Koksma inequality due to Niederreiter [23, Lemma 2.2] stood at the beginning of a thorough theoretical analysis of pseudorandom number generators. We present this result in its current version.

**Theorem 9.4** [26, Theorem 3.10] *Let $M \ge 2$ be an integer and let $\mathbf{y}_n \in \mathbb{Z}^s$, $0 \le n < N$. Put $\omega = (\mathbf{x}_n)_{n=0}^{N-1}$, where $\mathbf{x}_n = \{\mathbf{y}_n/M\}$ denotes the fractional part of $\mathbf{y}_n/M$. Then*

$$D_N(\omega) \leq 1 - \left(1 - \frac{1}{M}\right)^s + \sum_{\mathbf{k} \in C_s^*(M)} \frac{1}{r(\mathbf{k}, M)} |S_N(e_{\mathbf{k}}, \omega)|, \qquad (9.2)$$

where $C_s(M) = (-M/2, M/2]^s \cap \mathbb{Z}^s$, $C_s^*(M) = C_s(M) \setminus \{\mathbf{0}\}$, $r(\mathbf{k}, M) = \prod_{i=1}^s r(k_i, M)$, $\mathbf{k} = (k_1, \dots, k_s) \in C_s(M)$, and

$$r(k_i, M) := \begin{cases} 1 & \text{if } k_i = 0 \\ M \sin \pi \frac{|k_i|}{M} & \text{if } k_i \neq 0. \end{cases} \qquad (9.3)$$

In the same monograph Niederreiter [26, Theorem 3.12] gave an estimate of the star discrepancy for point sets $\omega$ where the coordinates of all points have finite digit expansion in some fixed base $b$. This result may be generalized to allow different bases $b_i$, $1 \leq i \leq s$, in every coordinate (see [25, Satz 2]).

It was shown in [8] how to apply the theory of generalized Walsh series, sometimes called Vilenkin–Fourier series, for a concise treatment of discrepancy estimates of the type (9.2). This approach via harmonic analysis yields Theorem 3.10, Corollary 3.11, and Theorem 3.12 of [26] as well as Satz 2 of [25] as corollaries to Theorem 1 of [8].

Recent results of Niederreiter for the discrepancy of *hybrid sequences* are based on a hybrid version of the Erdős–Turán–Koksma inequality (see [29]). Hybrid sequences are sequences of points in the multidimensional unit cube $[0, 1)^s$ where certain coordinates of the points stem from one lower dimensional sequence and the remaining coordinates stem from a second lower dimensional sequence. Usually, the constituent sequences of a hybrid sequence are of different type.

The analysis of the uniformity of hybrid sequences requires new tools. The classical results of the theory of uniform distribution of sequences have to be adapted to the fact that, in a hybrid sequence, several types of arithmetic may be involved. The first hybrid version of the Erdős–Turán–Koksma inequality was established by Niederreiter [29, Theorem 1] and has already found numerous applications (see [6, 27, 28, 29, 30, 31, 33]).

### 9.2.2 Hybrid function systems

For a nonnegative integer $k$, let $k = \sum_{j \geq 0} k_j b^j$, $k_j \in \{0, 1, \dots, b - 1\}$, be the unique $b$-adic representation of $k$ in base $b$. With the exception of at most finitely many indices $j$, the digits $k_j$ are equal to 0.

Every real number $x \in [0, 1)$ has a representation in base $b$ of the form $x = \sum_{j \geq 0} x_j b^{-j-1}$, with digits $x_j \in \{0, 1, \dots, b - 1\}$. If $x$ is a *b-adic rational*, which means that $x = ab^{-g}$, $a$ and $g$ integers, $0 \leq a < b^g$, $g \in \mathbb{N}$, and if $x \neq 0$, then there exist two such representations.

The $b$-adic representation of $x$ is uniquely determined under the condition that $x_j \neq b - 1$ for infinitely many $j$. In the following, we will call this particular representation the *regular ($b$-adic) representation* of $x$.

Let $\mathbb{Z}_b$ denote the compact group of the $b$-adic integers (see [17, 22] for details). Each element $z$ of $\mathbb{Z}_b$ can be expressed uniquely as a formal sum $z = \sum_{j \geq 0} z_j b^j$, with digits $z_j \in \{0, 1, \ldots, b - 1\}$. The set $\mathbb{Z}$ of integers is embedded in $\mathbb{Z}_b$. If $z \in \mathbb{N}_0$, then at most finitely many digits $z_j$ are different from 0. If $z \in \mathbb{Z}$, $z < 0$, then at most finitely many digits $z_j$ are different from $b - 1$. In particular, $-1 = \sum_{j \geq 0} (b - 1) b^j$.

We recall the following concepts from [12, 13, 16].

**Definition 9.5** *The map* $\varphi_b : \mathbb{Z}_b \rightarrow [0, 1)$, *given by* $\varphi_b(\sum_{j \geq 0} z_j b^j) = \sum_{j \geq 0} z_j b^{-j-1}$ (mod 1), *will be called the* $b$-*adic Monna map.*

The restriction of $\varphi_b$ to $\mathbb{N}_0$ is often called the *radical-inverse function* in base $b$. The Monna map is surjective, but not injective. It may be inverted in the following sense.

**Definition 9.6** *We define the* pseudoinverse $\varphi_b^+$ *of the* $b$-*adic Monna map* $\varphi_b$ *by*

$$\varphi_b^+ : [0, 1) \rightarrow \mathbb{Z}_b, \qquad \varphi_b^+ \left( \sum_{j \geq 0} x_j b^{-j-1} \right) = \sum_{j \geq 0} x_j b^j ,$$

*where* $\sum_{j \geq 0} x_j b^{-j-1}$ *stands for the regular* $b$-*adic representation of the element* $x \in [0, 1)$.

The image of $[0, 1)$ under $\varphi_b^+$ is the set $\mathbb{Z}_b \setminus (-\mathbb{N})$. Furthermore, $\varphi_b \circ \varphi_b^+$ is the identity map on $[0, 1)$, and $\varphi_b^+ \circ \varphi_b$ is the identity on $\mathbb{N}_0 \subset \mathbb{Z}_b$. In general, $z \neq \varphi_b^+(\varphi_b(z))$, for $z \in \mathbb{Z}_b$. For example, if $z = -1$, then $\varphi_b^+(\varphi_b(-1)) = \varphi_b^+(0) = 0 \neq -1$.

It was shown [16] that the dual group $\hat{\mathbb{Z}}_b$ can be written in the form $\hat{\mathbb{Z}}_b = \{\chi_k : k \in \mathbb{N}_0\}$, where $\chi_k : \mathbb{Z}_b \rightarrow \{c \in \mathbb{C} : |c| = 1\}$, $\chi_k(\sum_{j \geq 0} z_j b^j) = e(\varphi_b(k)(z_0 + z_1 b + \cdots))$. We note that $\chi_k$ depends only on a finite number of digits of $z$ and, hence, this function is well defined.

As in [13], we employ the function $\varphi_b^+$ to lift the characters $\chi_k$ to the torus.

**Definition 9.7** *For* $k \in \mathbb{N}_0$, *let* $\gamma_k : [0, 1) \rightarrow \{c \in \mathbb{C} : |c| = 1\}$, $\gamma_k(x) = \chi_k(\varphi_b^+(x))$, *denote the* $k$th $b$-*adic function. We put* $\Gamma_b = \{\gamma_k : k \in \mathbb{N}_0\}$ *and call it the* $b$-*adic function system on* $[0, 1)$.

There is an obvious generalization of the preceding notions to the higher dimensional case. Let $\mathbf{b} = (b_1, \ldots, b_s)$ be a vector of not necessarily distinct integers $b_i \geq 2$, let $\mathbf{x} = (x_1, \ldots, x_s) \in [0, 1)^s$, let $\mathbf{z} = (z_1, \ldots, z_s)$ denote an element of the compact product group $\mathbb{Z}_{\mathbf{b}} = \mathbb{Z}_{b_1} \times \cdots \times \mathbb{Z}_{b_s}$ of $\mathbf{b}$-adic integers, and let $\mathbf{k} = (k_1, \ldots, k_s) \in \mathbb{N}_0^s$. We define $\varphi_{\mathbf{b}}(\mathbf{z}) = (\varphi_{b_1}(z_1), \ldots, \varphi_{b_s}(z_s))$, and $\varphi_{\mathbf{b}}^+(\mathbf{x}) = (\varphi_{b_1}^+(x_1), \ldots, \varphi_{b_s}^+(x_s))$.

Let $\chi_{\mathbf{k}}(\mathbf{z}) = \prod_{i=1}^s \chi_{k_i}(z_i)$, where $\chi_{k_i} \in \hat{\mathbb{Z}}_{b_i}$, and define

$$\gamma_{\mathbf{k}}(\mathbf{x}) = \prod_{i=1}^s \gamma_{k_i}(x_i),$$

where $\gamma_{k_i} \in \Gamma_{b_i}$, $1 \leq i \leq s$. Then $\gamma_{\mathbf{k}} = \chi_{\mathbf{k}} \circ \varphi_{\mathbf{b}}^+$. Let $\Gamma_{\mathbf{b}}^{(s)} = \{\gamma_{\mathbf{k}} : \mathbf{k} \in \mathbb{N}_0^s\}$ denote the $\mathbf{b}$-*adic function system* in dimension $s$.

The system $\Gamma_{\mathbf{b}}^{(s)}$ is an orthonormal basis of the Hilbert space $L^2([0, 1)^s)$. A rather elementary proof of this result is given in [16, Theorem 2.12].

For the Walsh functions defined below, we refer the reader to [4, 8, 10] for elementary properties of these functions in the context of uniform distribution of sequences and to [34] for the background in harmonic analysis.

**Definition 9.8** *For $k \in \mathbb{N}_0$, $k = \sum_{j \geq 0} k_j b^j$, and $x \in [0, 1)$, with regular $b$-adic representation $x = \sum_{j \geq 0} x_j b^{-j-1}$, the $k$th Walsh function in base $b$ is defined by $w_k(x) = e((\sum_{j \geq 0} k_j x_j)/b)$. For $\mathbf{k} \in \mathbb{N}_0^s$, $\mathbf{k} = (k_1, \ldots, k_s)$, and $\mathbf{x} \in [0, 1)^s$, $\mathbf{x} = (x_1, \ldots, x_s)$, we define the $\mathbf{k}$th Walsh function $w_{\mathbf{k}}$ in base $\mathbf{b} = (b_1, \ldots, b_s)$ on $[0, 1)^s$ as the following product: $w_{\mathbf{k}}(\mathbf{x}) = \prod_{i=1}^s w_{k_i}(x_i)$, where $w_{k_i}$ denotes the $k_i$th Walsh function in base $b_i$, $1 \leq i \leq s$. The Walsh function system in base $\mathbf{b}$ in dimension $s$ is denoted by $\mathcal{W}_{\mathbf{b}}^{(s)} = \{w_{\mathbf{k}} : \mathbf{k} \in \mathbb{N}_0^s\}$.*

The following notions are based on concepts discussed first in [13, 14, 15], where further details can be found. For given dimensions $s_1$, $s_2$, and $s_3$, with $s_1, s_2, s_3 \in \mathbb{N}_0$, not all equal to 0, put $s = s_1 + s_2 + s_3$, and write a point $\mathbf{y} \in \mathbb{R}^s$ in the form $\mathbf{y} = (\mathbf{y}^{(1)}, \mathbf{y}^{(2)}, \mathbf{y}^{(3)})$ with components $\mathbf{y}^{(j)} \in \mathbb{R}^{s_j}$, $j = 1, 2, 3$. Let us fix two vectors of bases $\mathbf{b}^{(1)} = (b_1, \ldots, b_{s_1})$, and $\mathbf{b}^{(2)} = (b_{s_1+1}, \ldots, b_{s_1+s_2})$, with not necessarily distinct integers $b_i \geq 2$, $1 \leq i \leq s_1 + s_2$, and put $\mathbf{b} = (\mathbf{b}^{(1)}, \mathbf{b}^{(2)})$. Let $\mathbf{k} = (\mathbf{k}^{(1)}, \mathbf{k}^{(2)}, \mathbf{k}^{(3)})$, with components $\mathbf{k}^{(1)} \in \mathbb{N}_0^{s_1}$, $\mathbf{k}^{(2)} \in \mathbb{N}_0^{s_2}$, and $\mathbf{k}^{(3)} \in \mathbb{Z}^{s_3}$, and put $\Lambda = \mathbb{N}_0^{s_1} \times \mathbb{N}_0^{s_2} \times \mathbb{Z}^{s_3}$. The tensor product $\xi_{\mathbf{k}} = w_{\mathbf{k}^{(1)}} \otimes \gamma_{\mathbf{k}^{(2)}} \otimes e_{\mathbf{k}^{(3)}}$, where $w_{\mathbf{k}^{(1)}} \in \mathcal{W}_{\mathbf{b}^{(1)}}^{(s_1)}$, $\gamma_{\mathbf{k}^{(2)}} \in \Gamma_{\mathbf{b}^{(2)}}^{(s_2)}$, and $e_{\mathbf{k}^{(3)}} \in \mathcal{T}^{(s_3)}$, defines a function $\xi_{\mathbf{k}}$ on the $s$-dimensional unit cube,

$$\xi_{\mathbf{k}} : [0, 1)^s \to \mathbb{C}, \quad \xi_{\mathbf{k}}(\mathbf{x}) = w_{\mathbf{k}^{(1)}}(\mathbf{x}^{(1)})\gamma_{\mathbf{k}^{(2)}}(\mathbf{x}^{(2)})e_{\mathbf{k}^{(3)}}(\mathbf{x}^{(3)}),$$

where $\mathbf{x} = (\mathbf{x}^{(1)}, \mathbf{x}^{(2)}, \mathbf{x}^{(3)}) \in [0, 1)^s$. The family of functions

$$\mathcal{F} = \{\xi_{\mathbf{k}} : \mathbf{k} \in \Lambda\} = \mathcal{W}_{\mathbf{b}^{(1)}}^{(s_1)} \otimes \Gamma_{\mathbf{b}^{(2)}}^{(s_2)} \otimes \mathcal{T}^{(s_3)},$$

is called a *hybrid function system* on $[0, 1)^s$.

It follows from [13, Theorem 1, Corollary 4] and the technique exhibited in [16, Theorem 2.12] for nonprime bases $\mathbf{b}$ that such hybrid function systems are an orthonormal basis of $L^2([0, 1)^s)$ and that a Weyl criterion holds.

All of the following results remain valid if we change the order of the factors in the hybrid function system, as will become apparent from the proofs below. In particular, we may select some arbitrary $s_1$ coordinates and treat them with the Walsh system $\mathcal{W}_{\mathbf{b}^{(1)}}^{(s_1)}$, choose $s_2$ of the remaining coordinates for analysis by the $\mathbf{b}^{(2)}$-adic system $\Gamma_{\mathbf{b}^{(2)}}^{(s_2)}$, and employ the trigonometric system $\mathcal{T}^{(s_3)}$ for the remaining $s_3$ coordinates.

For an integrable function $f$ on $[0, 1)^s$, the $\mathbf{k}$th Fourier coefficient of $f$ with respect to the function system $\mathcal{F}$ is defined in the usual manner, as the inner product of $f$ and $\xi_{\mathbf{k}}$ in $L^2([0, 1)^s)$:

$$\hat{f}(\mathbf{k}) = \int_{[0,1)^s} f \overline{\xi_{\mathbf{k}}} \, d\lambda_s, \quad \mathbf{k} \in \Lambda.$$

The reader should notice that this definition encompasses the cases of Walsh, $b$-adic, and classical Fourier coefficients, by letting $s = s_j$, $j = 1, 2, 3$.

We denote the formal Fourier series of $f$ by $s_f$,

$$s_f = \sum_{\mathbf{k} \in \Lambda} \hat{f}(\mathbf{k}) \xi_{\mathbf{k}},$$

where, for the moment, we ignore questions of convergence.

**Definition 9.9** *For a given vector* $\mathbf{b} = (b_1, \dots, b_s)$ *of not necessarily distinct integers* $b_i \geq 2$, $1 \leq i \leq s$, *a* $\mathbf{b}$*-adic elementary interval, or* $\mathbf{b}$*-adic elint for short, is a subinterval* $I_{\mathbf{c},\mathbf{g}}$ *of* $[0, 1)^s$ *of the form*

$$I_{\mathbf{c},\mathbf{g}} = \prod_{i=1}^{s} \left[ \varphi_{b_i}(c_i), \varphi_{b_i}(c_i) + b_i^{-g_i} \right),$$

*where the parameters are subject to the conditions* $\mathbf{g} = (g_1, \dots, g_s) \in \mathbb{N}_0^s$, $\mathbf{c} = (c_1, \dots, c_s) \in \mathbb{N}_0^s$, *and* $0 \leq c_i < b_i^{g_i}$, $1 \leq i \leq s$. *We say that* $I_{\mathbf{c},\mathbf{g}}$ *has resolution* $\mathbf{g}$ *in base* $\mathbf{b}$.

*A* $\mathbf{b}$*-adic interval with resolution* $\mathbf{g} \in \mathbb{N}_0^s$ *in base* $\mathbf{b}$ *is a subinterval of* $[0, 1)^s$ *of the form*

$$\prod_{i=1}^{s} \left[ a_i b_i^{-g_i}, d_i b_i^{-g_i} \right), \quad 0 \leq a_i < d_i \leq b_i^{g_i}, \ a_i, d_i \in \mathbb{N}_0, \ 1 \leq i \leq s \, .$$

For a given resolution $\mathbf{g} \in \mathbb{N}_0^s$ in base $\mathbf{b}$, we define the following domains:

$$\Delta_{\mathbf{b}}(\mathbf{g}) = \left\{ \mathbf{k} = (k_1, \ldots, k_s) \in \mathbb{N}_0^s : 0 \le k_i < b_i^{g_i}, 1 \le i \le s \right\},$$
$$\Delta_{\mathbf{b}}^*(\mathbf{g}) = \Delta_{\mathbf{b}}(\mathbf{g}) \setminus \{\mathbf{0}\},$$

and note that $\Delta_{\mathbf{b}}(\mathbf{0}) = \{\mathbf{0}\}$.

For a given $\mathbf{g} \in \mathbb{N}_0^s$, the family of $\mathbf{b}$-adic elints $\{I_{\mathbf{c},\mathbf{g}} : \mathbf{c} \in \Delta_{\mathbf{b}}(\mathbf{g})\}$ is a partition of $[0, 1)^s$.

We denote the class of all $\mathbf{b}$-adic intervals with resolution $\mathbf{g}$ by $\mathcal{J}_{\mathbf{b},\mathbf{g}}$. The subclass of those $\mathbf{b}$-adic intervals anchored at the origin will be denoted by $\mathcal{J}_{\mathbf{b},\mathbf{g}}^*$. Obviously, both classes are finite. Further, let

$$\mathcal{J}_{\mathbf{b}} = \bigcup_{\mathbf{g} \in \mathbb{N}_0^s} \mathcal{J}_{\mathbf{b},\mathbf{g}}$$

denote the class of all $\mathbf{b}$-adic intervals in $[0, 1)^s$ and put

$$\mathcal{J}_{\mathbf{b}}^* = \bigcup_{\mathbf{g} \in \mathbb{N}_0^s} \mathcal{J}_{\mathbf{b},\mathbf{g}}^*.$$

In the following, we employ several variants of the discrepancy of a sequence $\omega$ in $[0, 1)^s$, by varying the class of (Lebesgue measurable) sets $\mathcal{S}$ that is considered. In order to distinguish between these discrepancies, we introduce the notation

$$d_N(\omega \mid \mathcal{S}) = \sup_{J \in \mathcal{S}} |S_N(\mathbf{1}_J - \lambda_s(J), \omega)|.$$

As a consequence, $D_N(\omega) = d_N(\omega \mid \mathcal{J})$, and $D_N^*(\omega) = d_N(\omega \mid \mathcal{J}^*)$, but it does not seem appropriate in this article to replace the well-established notation $D_N(\omega)$ and $D_N^*(\omega)$ for the "classical" discrepancies.

**Definition 9.10** *Let $\omega$ be a sequence in $[0, 1)^s$, and consider a base $\mathbf{b} = (b_1, \ldots, b_s)$, with not necessarily distinct integers $b_i \ge 2$. Let $\mathbf{g} \in \mathbb{N}_0^s$ be a given resolution in base $\mathbf{b}$. The* discrete extreme discrepancy *of the first $N$ elements of $\omega$, for resolution $\mathbf{g}$ in base $\mathbf{b}$, is defined as*

$$d_N(\omega \mid \mathcal{J}_{\mathbf{b},\mathbf{g}}) = \max_{I \in \mathcal{J}_{\mathbf{b},\mathbf{g}}} |S_N(\mathbf{1}_I - \lambda_s(I), \omega)|.$$

*The* discrete star discrepancy *of the first $N$ elements of $\omega$, for resolution $\mathbf{g}$ in base $\mathbf{b}$, is defined as*

$$d_N(\omega \mid \mathcal{J}_{\mathbf{b},\mathbf{g}}^*) = \max_{I \in \mathcal{J}_{\mathbf{b},\mathbf{g}}^*} |S_N(\mathbf{1}_I - \lambda_s(I), \omega)|.$$

The discrete discrepancies are linked to the usual discrepancies as follows.

**Theorem 9.11** [15, Theorem 5.11] *Let $\omega$ be a sequence in $[0, 1)^s$ and let $\mathbf{b} = (b_1, \ldots, b_s)$ be a vector of $s$ not necessarily distinct integers $b_i \geq 2$. Then, for all $\mathbf{g} = (g_1, \ldots, g_s) \in \mathbb{N}^s$,*

$$d_N(\omega \mid \mathcal{J}_{\mathbf{b},\mathbf{g}}) \leq D_N(\omega) \leq \epsilon_{\mathbf{b}}(\mathbf{g}) + d_N(\omega \mid \mathcal{J}_{\mathbf{b},\mathbf{g}}),$$
$$d_N(\omega \mid \mathcal{J}^*_{\mathbf{b},\mathbf{g}}) \leq D^*_N(\omega) \leq \epsilon^*_{\mathbf{b}}(\mathbf{g}) + d_N(\omega \mid \mathcal{J}^*_{\mathbf{b},\mathbf{g}}),$$

*where the error terms $\epsilon_{\mathbf{b}}(\mathbf{g})$ and $\epsilon^*_{\mathbf{b}}(\mathbf{g})$ are given by*

$$\epsilon_{\mathbf{b}}(\mathbf{g}) = 1 - \prod_{i=1}^{s}(1 - 2b_i^{-g_i}), \quad \epsilon^*_{\mathbf{b}}(\mathbf{g}) = 1 - \prod_{i=1}^{s}(1 - b_i^{-g_i}).$$

**Remark 9.12** An elementary analytic argument shows that $\epsilon_{\mathbf{b}}(\mathbf{g}) \leq 2s\delta_{\mathbf{g}}$, and $\epsilon^*_{\mathbf{b}}(\mathbf{g}) \leq s\delta_{\mathbf{g}}$, where $\delta_{\mathbf{g}} = \max_{1 \leq i \leq s} b_i^{-g_i}$.

**Corollary 9.13** *Theorem 9.11 implies the following discretization:*

$$D_N(\omega) = d_N(\omega \mid \mathcal{J}_{\mathbf{b}}),$$
$$D^*_N(\omega) = d_N(\omega \mid \mathcal{J}^*_{\mathbf{b}}).$$

We recall that a sequence $\omega$ is called uniformly distributed in $[0, 1)^s$ if and only if (see [19])

$$\lim_{N \to \infty} D_N(\omega) = 0. \tag{9.4}$$

The same holds for $D^*_N(\omega)$. It follows from Theorem 9.11, from (9.4) and from the fact that the class of intervals $\mathcal{J}_{\mathbf{b},\mathbf{g}}$ is finite that a sequence $\omega$ is uniformly distributed in $[0, 1)^s$ if and only if

$$\forall I \in \mathcal{J}_{\mathbf{b}} : \lim_{N \to \infty} S_N(\mathbf{1}_I - \lambda_s(I), \omega) = 0.$$

Obviously, this is also true for the class $\mathcal{J}^*_{\mathbf{b},\mathbf{g}}$.

A key ingredient in the theory of uniform distribution of sequences in terms of a Walsh function system $\mathcal{W}^{(s)}_{\mathbf{b}}$ or in terms of a $\mathbf{b}$-adic system $\Gamma^{(s)}_{\mathbf{b}}$ is the study of the Fourier series of indicator functions $\mathbf{1}_I$ of $\mathbf{b}$-adic elints and $\mathbf{b}$-adic intervals $I$. We have the following rather helpful identity, whose first versions were given in [8, 11].

**Lemma 9.14** ([14, Lemma 3.7], [16, Lemma 2.11]) *Let $I_{\mathbf{c},\mathbf{g}}$ be an arbitrary $\mathbf{b}$-adic elint in $[0, 1)^s$ and let $\hat{\mathbf{1}}_{I_{\mathbf{c},\mathbf{g}}}(\mathbf{k})$ denote the $\mathbf{k}$th Fourier coefficient of the*

*indicator function* $1_{I_{c,g}}$ *with respect to the Walsh system* $\mathcal{W}_{\mathbf{b}}^{(s)}$. *Put* $f = 1_{I_{c,g}}$.
*Then* $f = s_f$ *in the space* $L^2([0, 1)^s)$ *and even pointwise equality holds,*

$$\forall \mathbf{x} \in [0, 1)^s : \quad 1_{I_{c,g}}(\mathbf{x}) = \sum_{\mathbf{k} \in \Delta_{\mathbf{b}}(\mathbf{g})} \hat{1}_{I_{c,g}}(\mathbf{k}) w_{\mathbf{k}}(\mathbf{x}).$$

*Further, the same is true for the* **b**-*adic function system* $\Gamma_{\mathbf{b}}^{(s)}$. *In particular*

$$\forall \mathbf{x} \in [0, 1)^s : \quad 1_{I_{c,g}}(\mathbf{x}) = \sum_{\mathbf{k} \in \Delta_{\mathbf{b}}(\mathbf{g})} \hat{1}_{I_{c,g}}(\mathbf{k}) \gamma_{\mathbf{k}}(\mathbf{x}).$$

For an integer base $b \geq 2$, we define the weight functions

$$\rho_b(k) = \begin{cases} 1 & \text{if} \quad k = 0, \\ \dfrac{2}{b^t \sin(\pi \kappa_{t-1}/b)} & \text{if} \quad b^{t-1} \leq k < b^t, \ t \in \mathbb{N}, \end{cases}$$

where the reader should note that the condition $b^{t-1} \leq k < b^t$ implies
the $b$-adic representation $k = \kappa_0 + \kappa_1 b + \cdots + \kappa_{t-1} b^{t-1}$, with digits
$\kappa_j \in \{0, 1, \ldots, b - 1\}$, and with $\kappa_{t-1} \neq 0$, and, for integer vectors $\mathbf{k} = (k_1, \ldots, k_s) \in \mathbb{N}_0^s$,

$$\rho_{\mathbf{b}}(\mathbf{k}) = \prod_{i=1}^{s} \rho_{b_i}(k_i).$$

We also introduce the weights $\rho^*$: $\rho_b^*(0) = 1$, $\rho_b^*(k) = \rho_b(k)/2$ for $k \geq 1$
and, for $\mathbf{k} \in \mathbb{N}_0^s$, $\rho_{\mathbf{b}}^*(\mathbf{k}) = \prod_{i=1}^{s} \rho_{b_i}^*(k_i)$.

**Lemma 9.15** ([8, Lemma 3], [14, Corollary 3.10]) *Let* $I$ *be an arbitrary*
**b**-*adic interval in* $[0, 1)^s$ *with resolution* $\mathbf{g} \in \mathbb{N}_0^s$, $I = \prod_{i=1}^{s}[a_i b_i^{-g_i}, d_i b_i^{-g_i})$,
*with integers* $a_i, d_i, 0 \leq a_i < d_i \leq b_i^{g_i}, 1 \leq i \leq s$, *and put* $f = 1_I - \lambda_s(I)$.
*Then, if* $\hat{f}(\mathbf{k})$ *denotes the* **k***th Fourier coefficient of* $f$ *with respect to the*
*Walsh system* $\mathcal{W}_{\mathbf{b}}^{(s)}$,

$$\forall \mathbf{k} \in \Delta_{\mathbf{b}}^*(\mathbf{g}) : \quad |\hat{f}(\mathbf{k})| \leq \rho_{\mathbf{b}}(\mathbf{k}),$$

*and, if* $I$ *is anchored at the origin, i.e., if all* $a_i$ *are equal to* 0, *then*

$$\forall \mathbf{k} \in \Delta_{\mathbf{b}}^*(\mathbf{g}) : \quad |\hat{f}(\mathbf{k})| \leq \rho_{\mathbf{b}}^*(\mathbf{k}).$$

*Further, the same results hold for the Fourier coefficients* $\hat{f}(\mathbf{k})$ *of* $f$ *with*
*respect to the* **b**-*adic system* $\Gamma_{\mathbf{b}}^{(s)}$.

Lemma 9.15 explains where the weights defined in (9.3) come from. The
pointwise identity in Lemma 9.14 and the bound on the Fourier coefficients in
Lemma 9.15 induce an upper bound for the discrete discrepancies in terms of
Walsh functions,

$$d_N(\omega \mid \mathcal{J}_{\mathbf{b},\mathbf{g}}) = \max_{I \in \mathcal{J}_{\mathbf{b},\mathbf{g}}} |S_N(\mathbf{1}_I - \lambda_s(I), \omega)| \tag{9.5}$$

$$= \max_{I \in \mathcal{J}_{\mathbf{b},\mathbf{g}}} \left| \sum_{\mathbf{k} \in \Delta_{\mathbf{b}}^*(\mathbf{g})} \hat{\mathbf{1}}_{I_{\mathbf{c},\mathbf{g}}}(\mathbf{k}) S_N(w_{\mathbf{k}}, \omega) \right| \leq \sum_{\mathbf{k} \in \Delta_{\mathbf{b}}^*(\mathbf{g})} \rho_{\mathbf{b}}(\mathbf{k}) |S_N(w_{\mathbf{k}}, \omega)|,$$

and in terms of **b**-adic functions,

$$d_N(\omega \mid \mathcal{J}_{\mathbf{b},\mathbf{g}}) \leq \sum_{\mathbf{k} \in \Delta_{\mathbf{b}}^*(\mathbf{g})} \rho_{\mathbf{b}}(\mathbf{k}) |S_N(\gamma_{\mathbf{k}}, \omega)|. \tag{9.6}$$

The bounds for $d_N^*(\omega \mid \mathcal{J}_{\mathbf{b},\mathbf{g}})$ are proved in the same fashion, one replaces the weights $\rho_{\mathbf{b}}(\mathbf{k})$ by $\rho_{\mathbf{b}}^*(\mathbf{k})$.

As a consequence, (9.5), (9.6) and Theorem 9.11 yield the following theorem.

**Theorem 9.16** ([8, 14, 16]) *Let* $\mathbf{b} = (b_1, \ldots, b_s)$ *be a vector of s not necessarily distinct integers* $b_i \geq 2$, *and let* $\mathcal{W}_{\mathbf{b}}^{(s)}$ *denote the Walsh system in base* $\mathbf{b}$. *Then, for every sequence* $\omega$ *in* $[0, 1)^s$,

$$D_N(\omega) \leq \epsilon_{\mathbf{b}}(\mathbf{g}) + \sum_{\mathbf{k} \in \Delta_{\mathbf{b}}^*(\mathbf{g})} \rho_{\mathbf{b}}(\mathbf{k}) |S_N(w_{\mathbf{k}}, \omega)|,$$

*and*

$$D_N^*(\omega) \leq \epsilon_{\mathbf{b}}^*(\mathbf{g}) + \sum_{\mathbf{k} \in \Delta_{\mathbf{b}}^*(\mathbf{g})} \rho_{\mathbf{b}}^*(\mathbf{k}) |S_N(w_{\mathbf{k}}, \omega)|,$$

*Further, the same results hold if we replace* $\mathcal{W}_{\mathbf{b}}^{(s)}$ *by* $\Gamma_{\mathbf{b}}^{(s)}$.

## 9.3 A hybrid Erdős–Turán–Koksma inequality

Suppose that $s_1, s_2, s_3 \in \mathbb{N}_0$, not all equal to 0, and put $s = s_1 + s_2 + s_3$. Let $\mathbf{b}^{(1)} = (b_1, \ldots, b_{s_1})$, and $\mathbf{b}^{(2)} = (b_{s_1+1}, \ldots, b_{s_1+s_2})$ be two vectors of bases, with not necessarily distinct integers $b_i \geq 2$, $1 \leq i \leq s_1 + s_2$, and put $\mathbf{b} = (\mathbf{b}^{(1)}, \mathbf{b}^{(2)})$. Write $\mathbf{J}_{\mathbf{b}}$ for the set of intervals $J$ in $[0, 1)^s$ of the form $J = J^{(1)} \times J^{(2)} \times J^{(3)}$, with component intervals $J^{(1)} \in \mathcal{J}_{\mathbf{b}^{(1)}}$, $J^{(2)} \in \mathcal{J}_{\mathbf{b}^{(2)}}$, and $J^{(3)} \in \mathcal{J}^{(s_3)}$, where $\mathcal{J}^{(s_3)}$ stands for the class of all subintervals of $[0, 1)^{s_3}$. Further, write $\mathbf{J}_{\mathbf{b}}^*$ for the subclass of those intervals in $\mathbf{J}_{\mathbf{b}}$ that are anchored at the origin. As a consequence of Corollary 9.13, we have

$$D_N(\omega) = d_N(\omega \mid \mathbf{J}_{\mathbf{b}}).$$

The identity $D_N^*(\omega) = d_N(\omega \mid \mathbf{J}_{\mathbf{b}}^*)$ for the star discrepancy is straightforward.

Let $\mathbf{g} = (\mathbf{g}^{(1)}, \mathbf{g}^{(2)}) \in \mathbb{N}^{s_1+s_2}$ and consider the class of those intervals in $\mathbf{J_b}$ and in $\mathbf{J_b^*}$ with resolution $\mathbf{g}$ in base $\mathbf{b} = (\mathbf{b}^{(1)}, \mathbf{b}^{(2)})$ in the first two components. Let $\mathbf{J_{b,g}} = \mathcal{J}_{\mathbf{b}^{(1)},\mathbf{g}^{(1)}} \times \mathcal{J}_{\mathbf{b}^{(2)},\mathbf{g}^{(2)}} \times \mathcal{J}^{(s_3)}$ and define the discrete extreme discrepancy by

$$d_N(\omega \mid \mathbf{J_{b,g}}) = \sup_{J \in \mathbf{J_{b,g}}} |S_N(\mathbf{1}_J - \lambda_s(J), \omega)|.$$

The discrete star discrepancy $d_N(\omega \mid \mathbf{J_{b,g}^*})$ is defined accordingly. We note that $\mathcal{J}^{(s_3)}$ is uncountable, hence $d_N(\omega \mid \mathbf{J_{b,g}})$ and $d_N(\omega \mid \mathbf{J_{b,g}^*})$ are not a maximum but a supremum. It is easily seen that the following equivalent to Theorem 9.11 holds.

**Lemma 9.17** *For every sequence $\omega$ in $[0, 1)^s$,*

$$d_N(\omega \mid \mathbf{J_{b,g}}) \le D_N(\omega) \le \epsilon_{\mathbf{b}}(\mathbf{g}) + d_N(\omega \mid \mathbf{J_{b,g}}),$$
$$d_N(\omega \mid \mathbf{J_{b,g}^*}) \le D_N^*(\omega) \le \epsilon_{\mathbf{b}}^*(\mathbf{g}) + d_N(\omega \mid \mathbf{J_{b,g}^*}).$$

*Proof.* The proof is a simple translation of the method of proof of [14, Theorem 3.12] and [15, Theorem 5.11] to our setting. We approximate arbitrary intervals $J = J^{(1)} \times J^{(2)} \times J^{(3)}$ in $\mathbf{J_b}$ by intervals in $\mathbf{J_{b,g}}$. In this approximation, only the first two component intervals are concerned and the third component interval $J^{(3)}$ remains untouched. The number $\epsilon_{\mathbf{b}}(\mathbf{g})$ is an upper bound on the difference of the volumes of the intervals that are involved in the approximation. The same argument applies to $\mathbf{J_b^*}$. $\square$

In order to argue similarly as for the bounds (9.5) and (9.6), we have to deal with the Fourier coefficients of indicator functions $\mathbf{1}_J$ with respect to the trigonometric functions. The following lemma is due to Niederreiter.

**Lemma 9.18** ([29, Lemma 2]) *Let $J$ be an arbitrary subinterval of $[0, 1)$. For every $H \in \mathbb{N}$, there exists a trigonometric polynomial*

$$P_J(x) = \sum_{k=-H}^{H} c_J(k)e_k(x), \quad x \in [0, 1),$$

*with complex coefficients $c_J(k)$, where $c_J(0) = \lambda_1(J)$ and $|c_J(k)| < 1/r(k)$ for $k \ne 0$, such that, for all $x \in [0, 1)$,*

$$|\mathbf{1}_J(x) - P_J(x)| \le \frac{1}{H+1} \sum_{k=-H}^{H} u_J(k)e_k(x),$$

*with complex numbers $u_J(k)$ satisfying $|u_J(k)| \le 1$ for all $k$ and $u_J(0) = 1$.*

From the proof of [29, Theorem 1, p.201] we derive the following.

**Corollary 9.19** *Let $s \geq 1$ and let $J$ be an arbitrary subinterval of $[0, 1)^s$. For every positive integer $H$, there exists a trigonometric polynomial $P_J$,*

$$P_J(\mathbf{x}) = \sum_{\substack{\mathbf{k} \in \mathbb{Z}^s: \\ 0 \leq M(\mathbf{k}) \leq H}} c_J(\mathbf{k})\mathbf{e}_{\mathbf{k}}(\mathbf{x}), \quad \mathbf{x} \in [0, 1)^s \,,$$

*with complex coefficients $c_J(\mathbf{k})$, where $c_J(\mathbf{0}) = \lambda_s(J)$ and $\mid c_J(\mathbf{k}) \mid < 1/r(\mathbf{k})$ for $\mathbf{k} \neq \mathbf{0}$, such that, for all points $\mathbf{x} \in [0, 1[^s$,*

$$\mid \mathbf{1}_J(\mathbf{x}) - P_J(\mathbf{x}) \mid \leq \left(1 + \frac{1}{H+1}\right)^s - 1$$

$$+ \sum_{\substack{\mathbf{k} \in \mathbb{Z}^s: \\ 0 < M(\mathbf{k}) \leq H}} \left(1 + \frac{1}{H+1}\right)^{s-\mathrm{wt}(\mathbf{k})} \frac{1}{(H+1)^{\mathrm{wt}(\mathbf{k})}} u_J(\mathbf{k})\mathbf{e}_{\mathbf{k}}(\mathbf{x}), \quad (9.7)$$

*with complex numbers $u_J(\mathbf{k})$ satisfying $\mid u_J(\mathbf{k}) \mid \leq 1$ for all $\mathbf{k}$ and $u_J(\mathbf{0}) = 1$. Here, $\mathrm{wt}(\mathbf{k})$ denotes the Hamming weight of the vector $\mathbf{k}$, which is to say, the number of nonzero coordinates of $\mathbf{k}$.*

The following theorem generalizes Niederreiter's [29, Theorem 1] hybrid Erdős–Turán–Koksma inequality. In the proof, we employ the Fourier series method that was outlined in Section 9.2.

**Theorem 9.20** *Let $\omega = (\mathbf{x}_n)_{n \geq 0}$ be an arbitrary sequence in $[0, 1)^s$, and let $\mathbf{b}$ and $\mathcal{F} = \mathcal{W}^{(s_1)}_{\mathbf{b}^{(1)}} \otimes \Gamma^{(s_2)}_{\mathbf{b}^{(2)}} \otimes \mathcal{T}^{(s_3)}$ be defined as in Section 9.2. Choose an arbitrary resolution $\mathbf{g} \in \mathbb{N}^{s_1+s_2}$ and an arbitrary cut-off index $H \in \mathbb{N}$ and denote the finite summation domain $\Delta$ defined by $\mathbf{b}$, $\mathbf{g}$, and $H$ by*

$$\Delta = \Delta_{\mathbf{b}^{(1)}}(\mathbf{g}^{(1)}) \times \Delta_{\mathbf{b}^{(2)}}(\mathbf{g}^{(2)}) \times \{\mathbf{k}^{(3)} \in \mathbb{Z}^{s_3} : 0 \leq M(\mathbf{k}^{(3)}) \leq H\}.$$

*Put $\Delta^* = \Delta \setminus \{\mathbf{0}\}$ and write $\rho(\mathbf{k}) = \rho_{\mathbf{b}^{(1)}}(\mathbf{k}^{(1)})\rho_{\mathbf{b}^{(2)}}(\mathbf{k}^{(2)})r(\mathbf{k}^{(3)})^{-1}$, and $\rho^*(\mathbf{k}) = \rho^*_{\mathbf{b}^{(1)}}(\mathbf{k}^{(1)})\rho^*_{\mathbf{b}^{(2)}}(\mathbf{k}^{(2)})r(\mathbf{k}^{(3)})^{-1}$. Then*

$$D_N(\omega) = \mathcal{O}_{s_3}\left(1/H + \epsilon_{\mathbf{b}}(\mathbf{g}) + \sum_{\mathbf{k} \in \Delta^*} \rho(\mathbf{k}) \mid S_N(\xi_{\mathbf{k}}, \omega)\mid\right),$$

$$D_N^*(\omega) \leq \mathcal{O}_{s_3}\left(1/H + \epsilon^*_{\mathbf{b}}(\mathbf{g}) + \sum_{\mathbf{k} \in \Delta^*} \rho^*(\mathbf{k}) \mid S_N(\xi_{\mathbf{k}}, \omega)\mid\right).$$

*Proof.* Let $J = J^{(1)} \times J^{(2)} \times J^{(3)} \in \mathbf{J}_{\mathbf{b},\mathbf{g}} = \mathcal{J}_{\mathbf{b}^{(1)},\mathbf{g}^{(1)}} \times \mathcal{J}_{\mathbf{b}^{(2)},\mathbf{g}^{(2)}} \times \mathcal{J}^{(s_3)}$ be arbitrary. Then

$$S_N(\mathbf{1}_J - \lambda_s(J), \omega) = S_N(\mathbf{1}_{J^{(1)}} \otimes \mathbf{1}_{J^{(2)}} \otimes \mathbf{1}_{J^{(3)}} - \lambda_s(J), \omega)$$

$$= \frac{1}{N} \sum_{n=0}^{N-1} \mathbf{1}_{J^{(1)}}(\mathbf{x}_n^{(1)}) \mathbf{1}_{J^{(2)}}(\mathbf{x}_n^{(2)}) \mathbf{1}_{J^{(3)}}(\mathbf{x}_n^{(3)}) - \lambda_s(J)$$

$$= S_N(\mathbf{1}_{J^{(1)}} \otimes \mathbf{1}_{J^{(2)}} \otimes (\mathbf{1}_{J^{(3)}} - P_{J^{(3)}}), \omega)$$

$$+ S_N(\mathbf{1}_{J^{(1)}} \otimes \mathbf{1}_{J^{(2)}} \otimes P_{J^{(3)}} - \lambda_s(J), \omega)$$

$$= \Sigma_1 + \Sigma_2, \tag{9.8}$$

where $P_{J^{(3)}}$ is a trigonometric polynomial which is given by Corollary 9.19, and $\Sigma_1$ and $\Sigma_2$ denote the two sums in the third line of (9.8).

Let $\omega^{(j)}$ denote the component sequence $(\mathbf{x}_n^{(j)})_{n \geq 0}$, $1 \leq j \leq 3$. From (9.7) and from the proof of Theorem 1 in [29, p. 201] we obtain

$$|\Sigma_1| \leq \frac{1}{N} \sum_{n=0}^{N-1} \left| \mathbf{1}_{J^{(3)}}(\mathbf{x}_n^{(3)}) - P_{J^{(3)}}(\mathbf{x}_n^{(3)}) \right|$$

$$= O_{s_3} \left( \frac{1}{H} + \sum_{\substack{\mathbf{k}^{(3)} \in \mathbb{Z}^{s_3}: \\ 0 < M(\mathbf{k}^{(3)}) \leq H}} r(\mathbf{k}^{(3)})^{-1} \left| S_N(e_{\mathbf{k}^{(3)}}, \omega^{(3)}) \right| \right).$$

In order to estimate $\Sigma_2$, Lemma 9.14 and Corollary 9.19 imply that

$$S_N(\mathbf{1}_{J^{(1)}} \otimes \mathbf{1}_{J^{(2)}} \otimes P_{J^{(3)}}, \omega) =$$

$$\sum_{\substack{\mathbf{k}^{(1)} \in \Delta_{\mathbf{b}^{(1)}}(\mathbf{g}^{(1)}), \\ \mathbf{k}^{(2)} \in \Delta_{\mathbf{b}^{(2)}}(\mathbf{g}^{(2)}), \\ \mathbf{k}^{(3)}: 0 \leq M(\mathbf{k}^{(3)}) \leq H}} \hat{\mathbf{1}}_{J^{(1)}}(\mathbf{k}^{(1)}) \hat{\mathbf{1}}_{J^{(2)}}(\mathbf{k}^{(2)}) c_{J^{(3)}}(\mathbf{k}^{(3)}) S_N(\xi_{\mathbf{k}}, \omega),$$

with $\xi_{\mathbf{k}} = w_{\mathbf{k}^{(1)}} \otimes \gamma_{\mathbf{k}^{(2)}} \otimes e_{\mathbf{k}^{(3)}}$. We note that $\hat{\mathbf{1}}_{J^{(1)}}(\mathbf{0}) \hat{\mathbf{1}}_{J^{(2)}}(\mathbf{0}) c_{J^{(3)}}(\mathbf{0}) = \lambda_s(J)$. From Lemma 9.15 (see also [14, Corollary 3.10]) and from Corollary 9.19 it follows that

$$|\Sigma_2| = \left| S_N(\mathbf{1}_{J^{(1)}} \otimes \mathbf{1}_{J^{(2)}} \otimes P_{J^{(3)}} - \lambda_s(J), \omega) \right|$$

$$\leq \sum_{\mathbf{k} \in \Delta^*} \rho_{\mathbf{b}^{(1)}}(\mathbf{k}^{(1)}) \rho_{\mathbf{b}^{(2)}}(\mathbf{k}^{(2)}) r(\mathbf{k}^{(3)})^{-1} |S_N(\xi_{\mathbf{k}}, \omega)|.$$

The interval $J \in \mathbf{J}_{\mathbf{b},\mathbf{g}}$ was chosen arbitrarily and the bounds for $|\Sigma_1|$ and $|\Sigma_2|$ are independent of $J$. Hence, we have obtained an upper bound for the discrete discrepancy $d_N(\omega \mid \mathbf{J}_{\mathbf{b},\mathbf{g}})$. Lemma 9.17 then implies the bound

$$D_N(\omega) = O_{s_3}\left(\frac{1}{H} + \epsilon_{\mathbf{b}}(\mathbf{g})\right.$$

$$\left. + \sum_{\mathbf{k}\in\Delta^*} \rho_{\mathbf{b}^{(1)}}(\mathbf{k}^{(1)})\rho_{\mathbf{b}^{(2)}}(\mathbf{k}^{(2)})r(\mathbf{k}^{(3)})^{-1}|S_N(\xi_{\mathbf{k}},\omega)|\right).$$

It is an easy exercise to adapt this proof for $D_N^*(\omega)$.  □

**Remark 9.21** The constant terms in the Erdős–Turán–Koksma inequality for the Walsh functions and the **b**-adic functions grow linearly in the dimension $s$ (see [14]). It is an open problem whether we can achieve this growth for the Erdős–Turán–Koksma inequality if the trigonometric function system is involved. We would have to find a better approximation of indicator functions of intervals by trigonometric polynomials than the approximation provided by Lemma 9.18.

## Acknowledgements

The author would like to thank Arne Winterhof (RICAM Linz) for his hospitality in the "Special Semester on Algebra and Number Theory 2013" during which this paper was written, and to the anonymous referee for his helpful comments.

The author is supported by the Austrian Science Fund (FWF): Project F5504-N26, which is a part of the Special Research Program "Quasi-Monte Carlo Methods: Theory and Applications."

## References

[1]  T. Cochrane, On a trigonometric inequality of Vinogradov. *J. Number Theory*, **27**, 9–16, 1987.

[2]  A. Compagner, Operational conditions for random-number generation. *Phys. Rev. E* **52**, 5634–5645, 1995.

[3]  W. L. Conover, *Practical Nonparametric Statistics*, third edition. Wiley, 1999.

[4]  J. Dick and F. Pillichshammer, *Digital Nets and Sequences: Discrepancy Theory and Quasi-Monte Carlo Integration*. Cambridge University Press, Cambridge, 2010.

[5]  M. Drmota and R. F. Tichy, *Sequences, Discrepancies and Applications*. Lecture Notes in Mathematics, volume 1651. Springer, Berlin, 1997.

[6]  D. Gómez-Pérez, R. Hofer and H. Niederreiter, A general discrepancy bound for hybrid sequences involving Halton sequences. *Unif. Distrib. Theory* **8**(1), 31–45, 2013.

[7]  T. Gonzalez, S. Sahni and W. R. Franta, An efficient algorithm for the Kolmogorov–Smirnov and Lilliefors tests. *ACM Trans. Math. Software* **3**, 60–64, 1977.

[8]  P. Hellekalek, General discrepancy estimates: the Walsh function system. *Acta Arith.* **67**, 209–218, 1994.

[9]  P. Hellekalek, Good random number generators are (not so) easy to find. *Math. Comp. Simul.* **46**, 485–505, 1998.

[10] P. Hellekalek, On the assessment of random and quasi-random point sets. In: P. Hellekalek and G. Larcher (eds.), *Random and Quasi-Random Point Sets.* Lecture Notes in Statistics, volume 138, pp. 49–108. Springer, New York, 1998.

[11] P. Hellekalek, A general discrepancy estimate based on $p$-adic arithmetics. *Acta Arith.* **139**, 117–129, 2009.

[12] P. Hellekalek, A notion of diaphony based on $p$-adic arithmetic. *Acta Arith.* **145**, 273–284, 2010.

[13] P. Hellekalek, Hybrid function systems in the theory of uniform distribution of sequences. In: L. Plaskota and H. Woźniakowski (eds.), *Monte Carlo and Quasi-Monte Carlo Methods 2010.* Springer Proceedings in Mathematics and Statistics, volume 25, pp. 435–449. Springer, Berlin, 2012.

[14] P. Hellekalek, A hybrid inequality of Erdős–Turán–Koksma for digital sequences. *Monatsh. Math.* **173**, 55–66, 2014.

[15] P. Hellekalek, The hybrid spectral test: a unifying concept. In: P. Kritzer, H. Niederreiter, F. Pillichshammer and A. Winterhof (eds.), *Uniform Distribution and Quasi-Monte Carlo Methods.* Radon Series in Computational and Applied Mathematics. DeGruyter, Berlin, 2014.

[16] P. Hellekalek and H. Niederreiter, Constructions of uniformly distributed sequences using the $b$-adic method. *Unif. Distrib. Theory* **6**(1), 185–200, 2011.

[17] E. Hewitt and K. A. Ross, *Abstract Harmonic Analysis*, volume 1. Grundlehren der Mathematischen Wissenschaften [Fundamental Principles of Mathematical Sciences], volume 115, second edition. Springer-Verlag, Berlin, 1979.

[18] D. E. Knuth, *The Art of Computer Programming*, volume 2, third edition. Addison-Wesley, Reading, MA, 1998.

[19] L. Kuipers and H. Niederreiter, *Uniform Distribution of Sequences*. John Wiley, New York, 1974. Reprint, Dover Publications, Mineola, NY, 2006.

[20] P. L'Ecuyer, Random number generation. In: J. E. Gentle, W. Haerdle and Y. Mori (eds.), *Handbook of Computational Statistics*, pp. 35–70. Springer, New York, 2004.

[21] P. L'Ecuyer and R. Simard, TestU01: a C library for empirical testing of random number generators. *ACM Trans. Math. Software* **33**(4), Article 22, 2007.

[22] K. Mahler, *p-adic Numbers and their Functions*. Cambridge Tracts in Mathematics, volume 76, second edition. Cambridge University Press, Cambridge, 1981.

[23] H. Niederreiter. Pseudo-random numbers and optimal coefficients. *Adv. Math.* **26**, 99–181, 1977.

[24] H. Niederreiter, Quasi-Monte Carlo methods and pseudo-random numbers. *Bull. Am. Math. Soc.* **84**, 957–1041, 1978.

[25] H. Niederreiter, Pseudozufallszahlen und die Theorie der Gleichverteilung. *Sitzungsber. Österr. Akad. Wiss. Math. Naturwiss Kl. II*, **195**, 109–138, 1986.

[26] H. Niederreiter, *Random Number Generation and Quasi-Monte Carlo Methods.* SIAM, Philadelphia, PA, 1992.

[27] H. Niederreiter, On the discrepancy of some hybrid sequences. *Acta Arith.* **138**, 373–398, 2009.

[28] H. Niederreiter, A discrepancy bound for hybrid sequences involving digital explicit inversive pseudorandom numbers. *Unif. Distrib. Theory* **5**(1), 53–63, 2010.

[29] H. Niederreiter, Further discrepancy bounds and an Erdős–Turán–Koksma inequality for hybrid sequences. *Monatsh. Math.* **161**, 193–222, 2010.

[30] H. Niederreiter, Discrepancy bounds for hybrid sequences involving matrix-method pseudorandom vectors. *Publ. Math. Debrecen* **79**(3–4), 589–603, 2011.

[31] H. Niederreiter, Improved discrepancy bounds for hybrid sequences involving Halton sequences. *Acta Arith.* **155**(1), 71–84, 2012.

[32] H. Niederreiter and I. E. Shparlinski, Recent advances in the theory of nonlinear pseudorandom number generators. In: K.-T. Fang, F. J. Hickernell and H. Niederreiter (eds.), *Monte Carlo and Quasi-Monte Carlo Methods 2000*, pp. 86–102. Springer, New York, 2002.

[33] H. Niederreiter and A. Winterhof, Discrepancy bounds for hybrid sequences involving digital explicit inversive pseudorandom numbers. *Unif. Distrib. Theory* **6**(1), 33–56, 2011.

[34] F. Schipp, W. R. Wade and P. Simon, With the collaboration of J. Pál, *Walsh Series. An Introduction to Dyadic Harmonic Analysis.* Adam Hilger, Bristol, 1990.

[35] A. Topuzoğlu and A. Winterhof, Pseudorandom sequences. *Topics in Geometry, Coding Theory and Cryptography*. Algebra and Applications, volume 6, pp. 135–166. Springer, Dordrecht, 2007.

[36] A. Winterhof, Recent results on recursive nonlinear pseudorandom number generators (invited paper). In: C. Carlet and A. Pott (eds.), *Sequences and their Applications, SETA 2010*. Lecture Notes in Computer Science, volume 6338, pp. 113–124. Springer, Berlin, 2010.

# 10

# Controlling the shape of generating matrices in global function field constructions of digital sequences

*Roswitha Hofer and Isabel Pirsic*
Johannes Kepler University Linz

*This paper is dedicated to H. Niederreiter on the occasion of his
70th birthday.*

## Abstract

Motivated by computational as well as theoretical considerations, we show
how the shape and density of the generating matrices of two optimal constructions of $(t, s)$-sequences and $(u, e, s)$-sequences (the Xing–Niederreiter and
Hofer–Niederreiter sequences) can be controlled by a careful choice of various
parameters. We also present some experimental data to support our assertions
and point out open problems.

## 10.1 Introduction

The usefulness of and need for well-distributed pseudorandom and quasi-
random point sets in very high dimensions has been evidenced by the unbroken
stream of publications and conferences with the topic of Monte Carlo and
quasi-Monte Carlo (MCQMC) methods in scientific computing, most notably
the biannual conference series and proceedings of the same name. Beginning with the well-known Koksma–Hlawka inequality up to the more recent
higher order nets, it became clear that, in particular, applications pertaining to
multivariate numerical integration are an important area covered by MCQMC
methods. Numerous applications in diverse areas of applied mathematics
profit from this fact; often cited are applications in finance, computer aided
visualization and simulations. (The reader is referred to [5], [4], and [18].)

As regards the suitability of even arbitrary point sets for MCQMC methods, the notion of *discrepancy* is well established as a measure for the degree

of equidistribution, which significantly determines, for example, the error of numerical integration. In brief, discrepancy can be defined as measuring the worst case integration error when applied to indicator functions of subintervals of the unit cube. When the coordinates of the intervals are restricted to $b$-adic rationals, we arrive at the notion of $(t, s)$-sequences (in base $b$) [16]; if, furthermore, a different granularity is permitted in different coordinates, we arrive at the recent refinement of $(u, e, s)$-sequences [10, 26].

We review the definitions of these concepts in more detail. In the following, let $b \in \mathbb{N} \setminus \{1\}$; $N, m, s \in \mathbb{N}$; $t, u \in \mathbb{N}_0$ and $q \in \mathbb{N}$ a prime power.

**Definition 10.1** *Given a finite point sequence $\{x_n | n \in \{1, \ldots, N\}\} \in ([0, 1]^s)^N$, its (extremal) discrepancy is defined as*

$$D_N(\{x_n\}) := \sup_{I \subseteq [0,1]^s} \left| \frac{1}{N} \sum_{n=1}^{N} \chi_I(x_n) - \lambda(I) \right|,$$

*where $\chi_I$ is the indicator function of a subinterval $I$ and $\lambda(I)$ is its Lebesque measure.*

*When the additional restriction is introduced that $I$ is anchored at the origin, we get the definition of the* star discrepancy, $D_N^*(\{x_n\})$, *of the same point sequence.*

(For the sake of convenience it is customary to generalize from sets to sequences as this also allows us to include multi-sets.)

**Definition 10.2** *A $b$-ary box (or elementary interval) is a subinterval $I \subseteq [0, 1]^s$ of the form*

$$\prod_{i=1}^{s} \left[ \frac{a_i}{b^{d_i}}, \frac{a_i + 1}{b^{d_i}} \right),$$

*where $a_i, d_i \in \mathbb{N}_0$.*

For the definition of $(t, s)$-sequences and $(u, e, s)$-sequences we make use of the so-called truncation operator. (For the relevance of the truncation operator and further background we refer the reader to, e.g., [24, 25, 28].)

**Definition 10.3** *Let $b \geq 2$ be an integer and let $x = \sum_{j=1}^{\infty} x_j b^{-j}$ be a $b$-adic expansion of $x \in [0, 1]$, where it is allowed that $x_j = b - 1$ for all but finitely many $j$. For every integer $m \geq 1$, the $m$-truncation of $x$ in base $b$ is defined by $[x]_{b,m} = \sum_{j=1}^{m} x_j b^{-j}$ (depending on $x$ via its $b$-adic expansion). For $x \in [0, 1]^s$, the notation $[x]_{b,m}$ means that $m$-truncation is applied to each coordinate of $x$.*

**Definition 10.4** A $(t, s)$-sequence in base $b$ is a sequence $(x_n)_{n\geq 0} \in ([0, 1]^s)^{\mathbb{N}}$, with prescribed b-adic expansions for each coordinate, such that for any $m > t$ and any block of $b^m$ subsequent points $\{y_n | n \in 0, \ldots, b^m - 1\} := \{[x_n]_{b,m} | n = ab^m + n', n' \in \{0, \ldots, b^m - 1\}\}$, $a \in \mathbb{N}_0$, the following holds:

$$\frac{1}{b^m} \sum_{n=0}^{b^m-1} \chi_I(y_n) - b^{t-m} = 0,$$

for any b-ary box $I$ with volume at least $b^{t-m}$.

Any finite sequence $\{y_n\}$ of $b^m$ elements fulfilling the same condition for fixed $m$ is called a $(t, m, s)$-net in base $b$.

If the further restriction is introduced that, given $e = (e_1, \ldots, e_s) \in \mathbb{N}^s$ we have $e_i | d_i$, we speak of a $(u, e, s)$-sequence in base $b$ or $(u, e, m, s)$-net in base $b$, respectively, where $u$ plays the role of $t$. In particular, a $(t, (1, \ldots, 1), s)$-sequence is a $(t, s)$-sequence and vice versa.

For the actual construction of such high-quality point sets the most fruitful framework has long been the *digital method* [16] which utilizes linear maps over a finite field $\mathbb{F}_q$. (Usually in this context the symbol $q$ is used for the base, replacing $b$.)

Set $Z_q := \{0, 1, \ldots, q - 1\} \subset \mathbb{Z}$. Choose

(i) bijections $\psi_r : Z_q \to \mathbb{F}_q$ for all integers $r \geq 0$, satisfying $\psi_r(0) = 0$ for all sufficiently large $r$,
(ii) elements $c_{j,r}^{(i)} \in \mathbb{F}_q$ for $1 \leq i \leq s$, $j \geq 1$, and $r \geq 0$,
(iii) bijections $\lambda_{i,j} : \mathbb{F}_q \to Z_q$ for $1 \leq i \leq s$ and $j \geq 1$.

For the construction of the sequence we make use of the notion of *generating matrices* $C^{(i)} := (c_{j,r}^{(i)})_{j\geq 1, r\geq 0} \in \mathbb{F}_q^{\mathbb{N}\times\mathbb{N}_0}$ for $1 \leq i \leq s$. If these matrices $C^{(1)}, \ldots, C^{(s)}$, in each row of each matrix, contain only finitely many nonzero entries, then we speak of *finite-row generating matrices*. The $i$th coordinate $x_n^{(i)}$ of the $n$th point $x_n = (x_n^{(1)}, \ldots, x_n^{(s)})$ of the sequence is computed as follows. Given an integer $n \geq 0$, let $n = \sum_{r=0}^{\infty} z_r(n)q^r$ be the digit expansion of $n$ in base $q$, with all $z_r(n) \in Z_q$ and $z_r(n) = 0$ for all sufficiently large $r$. Then we form the matrix-vector product over $\mathbb{F}_q$ given by

$$C^{(i)} \cdot \begin{pmatrix} \psi_0(z_0(n)) \\ \psi_1(z_1(n)) \\ \vdots \end{pmatrix} =: \begin{pmatrix} y_{n,1}^{(i)} \\ y_{n,2}^{(i)} \\ \vdots \end{pmatrix}.$$

Finally, we put

$$x_n^{(i)} = \sum_{j=1}^{\infty} \lambda_{i,j}(y_{n,j}^{(i)}) q^{-j}.$$

**Definition 10.5** *A $(t,s)$-sequence in base $q$ constructed by the* digital method *outlined above is called a* digital $(t,s)$-sequence over $\mathbb{F}_q$; *analogously for* $(u, \boldsymbol{e}, s)$-sequences.

**Remark 10.6** Note that when $q = 2$, as is preferred in computer implementations, the construction can be defined more concisely, especially if we assume $0 \in \mathbb{Z}$ always maps to $\bar{0} \in \mathbb{F}_2$ (in both directions). Let $\oplus$ denote the digit-wise binary XOR operation (there are some technicalities here; for simplicity let us assume that in the following we will not encounter complications such as, e.g., two reals with infinite binary expansions adding up to a 2-adic rational). Consider the unique representation $n = \sum_{r=0}^{\infty} 2^{d_r(n)}$ with distinct $d_r(n) \in \mathbb{N}_0$. Then, with a previously chosen sequence $\xi_r \in [0, 1]^s$, $r \geq 0$ we have

$$\boldsymbol{x}_n = \bigoplus_{r=0}^{\infty} \xi_{d_r(n)}.$$

This can be further simplified to an operation only on integers if we impose a maximum precision of, say, $m$ bits in the calculations. Then the $\xi_r$ are 2-adic rationals, i.e., are of the form $\boldsymbol{a}_r / 2^m$, $\boldsymbol{a}_r \in \mathbb{N}_0^s$. Consequently, we get $2^m \boldsymbol{x}_n = \oplus_{r \geq 0} \boldsymbol{a}_{d_r(n)}$. The construction principle in this remark quickly follows from the general matrix approach by choosing $\xi_r := \boldsymbol{y}_{2^r}$ and $\boldsymbol{a}_r := \lfloor 2^m \boldsymbol{y}_{2^r} \rfloor$, respectively, where the $\boldsymbol{y}_r$ are constructed as above via generating matrices. Conversely, the 2-adic expansions of $\xi_{2^r}$, $r \geq 0$, determine the columns of the generating matrices.

The distribution of the sequence $\boldsymbol{x}_0, \boldsymbol{x}_1, \dots$ is mostly determined by the rank structure of the generating matrices. It is well known that the digital method generates a $(t,s)$-sequence in base $q$ if for every integer $m > t$ and all non-negative integers $d_1, \dots, d_s$ such that $d_1 + \cdots + d_s \leq m - t$, the $(m - t) \times m$ matrix over $\mathbb{F}_q$ formed by the row vectors

$$(c_{j,0}^{(i)}, c_{j,1}^{(i)}, \dots, c_{j,m-1}^{(i)}) \in \mathbb{F}_q^m$$

with $1 \leq j \leq d_i$ and $1 \leq i \leq s$ has rank $d_1 + \cdots + d_s$ (see [4, Section 4.4] and [18, Section 4.3]).

Similarly, from [10, Proposition 3] we know that the digital method generates a $(u, \boldsymbol{e}, s)$-sequence in base $q$ with $\boldsymbol{e} = \{e_1, \dots, e_s\} \in \mathbb{N}^s$ if for any integers $d_1, \dots, d_s \geq 0$ with $e_i | d_i$ for $1 \leq i \leq s$ and $d_1 + \cdots + d_s \leq m - u$, the $(d_1 + \cdots + d_s) \times m$ matrix over $\mathbb{F}_q$ formed by the row vectors

$$(c_{j,0}^{(i)}, c_{j,1}^{(i)}, \ldots, c_{j,m-1}^{(i)}) \in \mathbb{F}_q^m$$

with $1 \leq j \leq d_i$ and $1 \leq i \leq s$ has rank $d_1 + \cdots + d_s$.

By these facts, the problem of finding a point set with good distribution properties is translated to a combinatorial problem in linear algebra that is considerably easier to attack. In fact, there exist several more or less closely related approaches called Sobol' [22], Faure [6], Niederreiter [17] and Xing–Niederreiter sequences [28], to mention only the most widely known.

A point of critique that might be brought forward against digital sequences is that, apart from the nontrivial problem of finding good matrices, the construction itself by matrix-vector multiplication is rather costly if compared, for example, to the simple recursive procedures found in pseudorandom number generators. One remedy against this is to utilize a Gray code reordering of the sequence which has the effect that in each step only one vector addition needs to be carried out; this was done, for example, in [1, 3]. Another approach would be to decrease the density of the matrices, i.e., the number of nonzero entries, as much as possible. Apart from a subsequent decrease in the implicit number of operations inside the matrix-vector multiplications, this also has the further advantage of reduced storage size for the generating matrices, a factor which should not be be neglected when dealing with high dimensions. Furthermore, the imposition of constraints on the generating matrices is of theoretical interest, pertaining to the question of how many "bits" of information are needed to obtain a certain degree of "randomness." For practical purposes this could imply that the quasi "canonical form" of the matrices allows a clearer control of the subsequent randomization by scrambling transformations; this is, however, a topic for future research. Then, there is also an implication for Niederreiter–Halton sequences, which are hybrid sequences (meaning here that several digit bases may occur in the dimensions). Within these Niederreiter–Halton sequences the subclass generated by finite-row matrices is of particular interest. (For more details on Niederreiter–Halton sequences and the speciality of finite-row matrices we refer the interested reader to [7, 8, 9, 11, 12, 13].) It is this second approach of density reduction that is the core topic of this paper.

In Section 10.2 we cursorily collect notions, notations and theorems needed from global function fields. We revisit and slightly expand the Xing–Niederreiter construction and the Hofer–Niederreiter construction in Section 10.3. In Section 10.4 we investigate both construction methods with respect to morphological properties of the generating matrices. Finally, in Sections 10.5 and 10.6 we compare the densities of some examples and give a summary.

## 10.2 Global function fields

We refer to the monographs [20, 23] for general background on and basic definitions of global function fields. To give at least a broad idea to the unacquainted reader, we mention the most important notions without much technical detail: a *global function field* is a finite algebraic extension of $\mathbb{F}_q(x)$; the *genus* is a specific measure of its complexity; a *place* can be considered as a prime element in an appropriate associated ring, generalizing the concept of an irreducible polynomial; finally, the *valuation* associated to a place, of an element in the function field, is like the exponent of a prime element in the unique factorization representation.

Let $F$ be a global function field with full constant field $\mathbb{F}_q$ and genus $g$ and let $\mathbb{P}_F$ denote the set of all places of $F$. We write $v_P$ for the *normalized discrete valuation* corresponding to a place $P$. Its *valuation ring* is defined as

$$O_P := \{f \in F : v_P(f) \geq 0\}$$

and has a unique maximal ideal,

$$M_P := \{f \in F : v_P(f) > 0\}.$$

The extension degree of the residue class field $O_P/M_P$ over $\mathbb{F}_q$ is called the *degree* of the place $P$ and is written as $\deg(P)$. If $\deg(P) = 1$ we call $P$ a *rational place*.

A *divisor* $D$ of $F$ is a finite formal sum

$$D = \sum_{P \in \mathbb{P}_F} n_P P$$

with $n_P \in \mathbb{Z}$ for all $P \in \mathbb{P}_F$ and all but finitely many $n_P$ are equal to 0. We also write $n_P = v_P(D)$ and set $\operatorname{supp}(D) = \{P \in \mathbb{P}_F : v_p(D) \neq 0\}$. The *degree* $\deg(D)$ of a divisor $D$ is given by

$$\deg(D) = \sum_{P \in \mathbb{P}_F} n_P \deg(P) = \sum_{P \in \mathbb{P}_F} v_P(D) \deg(P).$$

The *principal divisor* $\operatorname{div}(f)$ of $f \in F^*$ is defined as

$$\operatorname{div}(f) = \sum_{P \in \mathbb{P}_F} v_P(f) P.$$

Note that the degree of a principal divisor $\deg(\operatorname{div}(f))$ with $f \in F^*$ is always 0. Let $D_1, D_2$ be two divisors. We say that $D_1 \leq D_2$ if $v_P(D_1) \leq v_P(D_2)$ for all $P \in \mathbb{P}_F$. Furthermore if $D_1 \geq 0$ (where 0 is the divisor of the empty sum) then we call $D_1$ positive. A very important element of the theory is the *Riemann–Roch space*

$$\mathcal{L}(D) := \{f \in F^* : \mathrm{div}(f) + D \geq 0\} \cup \{0\}$$

associated to a divisor. This is a finite-dimensional vector space over $\mathbb{F}_q$, and we write $\ell(D)$ for its dimension. Obviously, $\mathcal{L}(D_1)$ is a subspace of $\mathcal{L}(D_2)$ whenever $D_1 \leq D_2$. The celebrated *Riemann–Roch theorem* [23, Theorem 1.5.15] ensures that $\ell(D) \geq \deg(D) + 1 - g$, and equality holds whenever $\deg(D) \geq 2g - 1$ (see [23, Theorem 1.5.17]). For a rational place $P$ the *Weierstrass gap theorem* [23, Theorem 1.6.8] says that there are exactly $g$ *gap numbers* $1 = i_1 < \cdots < i_g \leq 2g - 1$, that is, integers $i_j$, $1 \leq j \leq g$, such that $\ell(i_j P) = \ell\big((i_j - 1)P\big)$. Note that for integers $n \geq 2g$ it is clear by the Riemann–Roch theorem that $\ell((n-1)P) = \ell(nP) - 1$.

Finally, choosing a rational place $P \in \mathbb{P}_F$ and a *local parameter sequence* $(z_r)_{r \in \mathbb{Z}} \in F^{\mathbb{Z}}$ at $P$ satisfying $v_P(z_r) = r$ for every $r \in \mathbb{Z}$, then any $f \in F$ has a unique *local expansion* at the place $P$ of the form

$$f = \sum_{r=r_0}^{\infty} \beta_r z_r,$$

where $r_0 \in \mathbb{Z}$ with $v_P(f) \geq r_0$ and $\beta_r \in \mathbb{F}_q$ for all $r \geq r_0$. A description of how to obtain a local expansion can be found in [20, pp. 5–6]. Note that the case of a place $P$ of degree greater than 1 can be reduced to the rational case via a constant field extension of $F$.

For our purposes we deem it convenient to make the nature of the expansion coefficients $\beta_r$ more explicit in the case of a place $P$ that is not rational but of degree $e > 1$. Here, for a local expansion we choose in advance a set $\Gamma = \{\gamma_1, \ldots, \gamma_{q^e}\} \subset F$ of representatives, such that their residue classes with respect to $P$, i.e., the set $\{\gamma_1 + M_P, \ldots, \gamma_{q^e} + M_P\}$, comprise all of the residue class field of $P$, which is isomorphic to $\mathbb{F}_{q^e}$. Any $f \in F$ can then again be uniquely represented as above, with $\beta_r \in \Gamma$ (see [14, Theorem 4.4.1]). In fact, we will specifically use sets $\Gamma$ that are the $\mathbb{F}_q$-linear spans of a basis $\{\Gamma_1, \ldots, \Gamma_e\}$. This ensures that the map from $F$ to $\Gamma^{\mathbb{Z}}$, i.e., from functions to expansion coefficient sequences, is $\mathbb{F}_q$-linear.

## 10.3 Constructions revisited

With the introduction of global function field methods it became possible to achieve the asymptotically best possible quality parameters $t$. The most general and powerful of the four constructions that initiated this line of research was given in [28], where the Xing–Niederreiter construction (sometimes also called the 3rd Niederreiter–Xing construction) was defined.

In subsequent years, Niederreiter, together with different coauthors (Özbudak [19], Mayor [15], Yeo [21]), published many more constructions that either introduced further slight improvements or utilized other aspects of function fields, for example, differentials instead of Riemann–Roch spaces.

Another alternative construction was given by Hofer and Niederreiter [10]. Both, the Xing–Niederreiter construction and the Hofer–Niederreiter construction use as main ingredient one distinct rational place and $s$ further places of arbitrary degree, and both determine the entries of the generating matrices using the coefficients of the local expansions at a place, of certain elements of the function field. The main difference, however, is that the Xing–Niederreiter construction builds the matrices row by row, whereas the Hofer–Niederreiter construction is a column-wise concept.

In the following we give slightly generalized versions of these two constructions with the effect of increasing freedom of choice in the constructions. Thereby, one obtains a higher diversity of the generating matrices that can be exploited to achieve specified morphological properties in Section 10.4.

### 10.3.1 Xing–Niederreiter construction

Xing and Niederreiter introduced the following construction in [28]. Starting with a fixed dimension $s \in \mathbb{N}$, a prime power $q$, and a global function field $F$ with full constant field $\mathbb{F}_q$ and genus $g$, one chooses as main ingredients $s+1$ distinct places $P_\infty, P_1, \ldots, P_s$ of $F$ satisfying $\deg(P_\infty) = 1$ and $\deg(P_i) = e_i \geq 1$ for $1 \leq i \leq s$. Furthermore a positive auxiliary divisor $D$ of $F$ is fixed such that

$$P_\infty \notin \operatorname{supp}(D)$$

and $\deg(D) = 2g$. This auxiliary divisor is an important tool to meet the requirements of the Riemann–Roch theorem.

In a first step, a basis of the vector space $\mathcal{L}(D)$ with dimension $l(D) = g + 1$ is chosen in the following way. By the Riemann–Roch theorem we know $l(D - P_\infty) = g$ and $l(D - (2g+1)P_\infty) = 0$, hence there exist integers $0 = n_0 < n_1 < \cdots < n_g \leq 2g$ such that

$$l(D - n_f P_\infty) = l(D - (n_f + 1)P_\infty) + 1 \text{ for } 0 \leq f \leq g.$$

Now choose $w_f \in \mathcal{L}(D - n_f P_\infty) \setminus \mathcal{L}(D - (n_f + 1)P_\infty)$ to obtain the basis

$$\{w_0, w_1, \ldots, w_g\} \text{ of } \mathcal{L}(D).$$

Note that $v_{P_\infty}(w_f) = n_f$ for every $0 \leq f \leq g$ and $0 = n_0 < n_1 < \cdots < n_g \leq 2g$.

To construct the elements of $F$ that determine the matrix rows, Xing and Niederreiter [28] build a chain of vector spaces

$$\mathcal{L}(D) \subset \mathcal{L}(D + P_i) \subset \mathcal{L}(D + 2P_i) \subset \cdots$$

and step by step add to the set $\{w_0, w_1, \ldots, w_g\}$ basis vectors such that, for each $n \geq 1$, $\{w_0, w_1, \ldots, w_g, k_1^{(i)}, \ldots, k_{ne_i}^{(i)}\}$ forms a basis of $\mathcal{L}(D + nP_i)$. Here, the degree of the auxiliary divisor $D$ ensures that in each step exactly $e_i$ elements are added to the basis.

Next, choose a parameter sequence $(z_r)_{r \geq 0}$ in $F$ satisfying $\nu_{P_\infty}(z_r) = r$, involving the basis $\{w_0, w_1, \ldots, w_g\}$ of $\mathcal{L}(D)$ specified above as follows: set $z_r = w_f$ if $r = n_f$ for some $f \in \{0, 1, \ldots, g\}$, but with no further restrictions for the other elements of the sequence. (Xing and Niederreiter [28] use the special case, where $z_r = z^r$ if $r \notin \{n_0, n_1, \ldots, n_g\}$ with $z$ satisfying $\nu_{P_\infty}(z) = 1$.)

Now the local expansion of $k_j^{(i)}$ at $P_\infty$ is used to determine the $j$th row $c_j^{(i)}$ of $C^{(i)}$ as follows. The local expansion of $k_j^{(i)}$ (note that $\nu_{P_\infty}(k_j^{(i)}) \geq 0$),

$$k_j^{(i)} = \sum_{r=0}^{\infty} a_{j,r}^{(i)} z_r \text{ for } 1 \leq i \leq s \text{ and } j \geq 1, \qquad (10.1)$$

with $a_{j,r}^{(i)} \in \mathbb{F}_q$, gives the preliminary sequence

$$(\widehat{a_{j,n_0}^{(i)}}, a_{j,1}^{(i)}, \ldots, \widehat{a_{j,n_1}^{(i)}}, a_{j,n_1+1}^{(i)}, \ldots, \widehat{a_{j,n_g}^{(i)}}, a_{j,n_g+1}^{(i)}, \ldots) \in \mathbb{F}_q^{\mathbb{N}_0}.$$

After deleting the terms with the hat we arrive at what will be the $j$th row of $C^{(i)}$,

$$c_j^{(i)} = (c_{j,0}^{(i)}, c_{j,1}^{(i)}, \ldots) \in \mathbb{F}_q^{\mathbb{N}_0}.$$

(Note that the deletion of the terms with the hat, which are the coefficients associated to the elements of the basis of $\mathcal{L}(D)$, saves an additive $g + 1$ term in the quality parameters $t$ and $u$.)

As the proofs of [28, Theorems 1 and 2] and [10, Remark 6] actually do not use the special choice $z_r := z^r$, the following theorem still holds with the additional generalization.

**Theorem 10.7** *Given a prime power $q$, a dimension $s \in \mathbb{N}$, let $F$ be a global function field with full constant field $\mathbb{F}_q$ and genus $g$, and with at least one rational place $P_\infty$. Let $D$ be a positive divisor with degree $\deg(D) = 2g$ and $P_\infty \notin \text{supp}(D)$. Finally, let $P_1, \ldots, P_s$ be $s$ distinct places of degrees*

$e_1, \ldots, e_s$ with $P_i \neq P_\infty$ for $1 \leq i \leq s$. Then the matrices $C^{(1)}, \ldots, C^{(s)}$ constructed above generate a $(t, s)$-sequence in base $q$ with

$$t = g + \sum_{i=1}^{s} (e_i - 1)$$

and a $(u, \boldsymbol{e}, s)$-sequence in base $q$ with

$$u = g \text{ and } \boldsymbol{e} = (e_1, \ldots, e_s).$$

The motive for allowing more general $z_r$ is to obtain additional freedom of choice in the construction. Moreover, we also have freedom in the choice of the base systems $(k_j^{(i)})_{j \geq 1}$, $1 \leq i \leq s$. Finally, it would also be possible to vary the set of representatives used for the coefficients in the local expansions, however, we do not take into account these variations here.

**Example 10.8** *Let $F = \mathbb{F}_q(x)$ be the rational function field over $\mathbb{F}_q$ (where $g = 0$), $P_\infty$ the infinite place of $\mathbb{F}_q(x)$, and $P_1, \ldots, P_s$ places corresponding to distinct monic irreducible polynomials $p_1(x), \ldots, p_s(x)$ over $\mathbb{F}_q$ of degrees $e_1, \ldots, e_s$. Then the auxiliary divisor $D$ is zero and, obviously, $w_0 = 1$ forms a basis of $\mathbb{F}_q = \mathcal{L}(D)$. In this case the Xing–Niederreiter construction takes basis vectors $k_{(l-1)e_i+1}^{(i)}, \ldots, k_{(l-1)e_i+e_i}^{(i)}$, for $l \in \mathbb{N}$, from $\mathcal{L}(lP_i) \setminus \mathcal{L}((l-1)P_i)$. These are rational functions with fixed denominator polynomial $(p_i(x))^l$. The numerator polynomials $y_{l,1}^{(i)}(x), \ldots, y_{l,e_i}^{(i)}(x)$ are coprime with $p_i(x)$, have degrees $\leq e_i l$ to ensure $v_{P_\infty}(k_j^{(i)}) \geq 0$, and such that any nontrivial linear combination is not a multiple of $p_i(x)$ in view of the basis property. Using the sequence $z_r = z^r$ with the local parameter $z = 1/x$ at $P_\infty$ we thus arrive via the Xing–Niederreiter construction at generating matrices of generalized Niederreiter sequences. A particular case is the classical Niederreiter construction, where the numerator polynomials $\{y_{l,1}^{(i)}(x), \ldots, y_{l,e_i}^{(i)}(x)\}$ form the set $\{1, x, \ldots, x^{e_i-1}\}$.*

### 10.3.2 Hofer–Niederreiter construction

For this construction we choose a dimension $s \in \mathbb{N}$, a finite field $\mathbb{F}_q$, a global function field $F$ with full constant field $\mathbb{F}_q$ and genus $g$, and $s + 1$ distinct places $P_\infty, P_1, \ldots, P_s$ of $F$ satisfying $\deg(P_\infty) = 1$. The degrees of the places $P_1, \ldots, P_s$ are arbitrary and we set $e_i = \deg(P_i)$ for $1 \leq i \leq s$. In addition to the construction in [10], for each $i \in \{1, \ldots, s\}$ we choose an element $h_i \in F$ satisfying $v_{P_i}(h_i) = 0$.

First we construct a sequence $(y_r)_{r \geq 0}$ of elements of $F$ (employing the Weierstrass gap theorem) that will serve to construct the columns of the matrices. Let $n_0 < n_1 < n_2 < \cdots$ be the pole numbers of $P_\infty$, i.e., the complement $\mathbb{N}_0 \setminus \{i_1, \ldots, i_g\}$ of the $g$ gap numbers of $P_\infty$. Note that $n_r = r + g$ for $r \geq g$. We choose for each integer $r \geq 0$ an element

$$y_r \in \mathcal{L}(n_r P_\infty) \setminus \mathcal{L}((n_r - 1) P_\infty).$$

This entails that $v_P(y_r) \geq 0$ for all places $P \in \mathbb{P}_F \setminus \{P_\infty\}$ and all $r \geq 0$.

Next, for $1 \leq i \leq s$, consider the local expansion of $h_i y_r$ at $P_i$, with the local parameter sequence $(z_k^{(i)})_{k \geq 0}$ in $F$ with $v_{P_i}(z_k^{(i)}) = k$ for every $k \geq 0$. We denote the local expansion by

$$h_i y_r = \sum_{k=0}^{\infty} \beta_{k,r}^{(i)} z_k^{(i)} \qquad \text{with } \beta_{k,r}^{(i)} \in \Gamma^{(i)},$$

where $\Gamma^{(i)}$ is a set of representatives of the residue class field at $P_i$. We assume here that $\Gamma^{(i)}$ is isomorphic to $\mathbb{F}_{q^{e_i}}$ as an $\mathbb{F}_q$-vector space, and that $\Gamma^{(i)}$ is spanned by an ordered basis (as discussed in Section 10.2) so that we can identify $\beta_{k,r}^{(i)}$ with a column vector $\boldsymbol{b}_{k,r}^{(i)} \in \mathbb{F}_q^{e_i}$.

Finally, we construct the generating matrices $C^{(i)}$, $1 \leq i \leq s$, over $\mathbb{F}_q$ column-wise by defining column $r$ ($r \geq 0$) of $C^{(i)}$ as the concatenation of the column vectors $\boldsymbol{b}_{k,r}^{(i)}$, $k \geq 0$. This completes the construction.

**Theorem 10.9** *Given a prime power $q$ and a dimension $s \in \mathbb{N}$, let $F$ be a global function field with full constant field $\mathbb{F}_q$ and genus $g$. Let $P_\infty, P_1, \ldots, P_s$ be $s + 1$ distinct places of $F$ with $\deg(P_\infty) = 1$ and $\deg(P_i) =: e_i$ for $1 \leq i \leq s$. Further, let $h_1, \ldots, h_s$ be elements of $F$ satisfying $v_{P_i}(h_i) = 0$ for $1 \leq i \leq s$. Then the matrices $C^{(1)}, \ldots, C^{(s)}$ constructed above generate a $(u, \boldsymbol{e}, s)$-sequence in base $q$ with*

$$u = g \quad \text{and} \quad \boldsymbol{e} = (e_1, \ldots, e_s)$$

*and also a $(t, s)$-sequence in base $q$ with*

$$t = g + \sum_{i=1}^{s} (e_i - 1).$$

*Proof.* The proof of $u$ and $\boldsymbol{e}$ is carried out quite analogously to [10, Proof of Theorem 1]. One needs to show that for every integer $m > g$ and all $d_1, \ldots, d_s \in \mathbb{N}_0$ such that $e_1 | d_1, \ldots, e_s | d_s$ and $1 \leq d_1 + \cdots + d_s \leq m - g$, the $(d_1 + \cdots + d_s) \times m$ matrix $M$ over $\mathbb{F}_q$ formed by the row vectors

$$(c_{j,0}^{(i)}, c_{j,1}^{(i)}, \ldots, c_{j,m-1}^{(i)}) \in \mathbb{F}_q^m$$

with $1 \leq j \leq d_i$ and $1 \leq i \leq s$ has rank $d_1 + \cdots + d_s$. The full row-rank is ensured by finding a proper bound on the dimension of the kernel of $M$. Let $(v_0, \ldots, v_{m-1})^T \in \mathbb{F}_q^m$ be contained in the kernel of $M$. Hence

$$\sum_{r=0}^{m-1} v_r c_{j,r}^{(i)} = 0 \text{ for } 1 \leq j \leq d_i,\ 1 \leq i \leq s. \tag{10.2}$$

We consider the element $f := \sum_{r=0}^{m-1} v_r y_r$ which is contained in the Riemann–Roch space $\mathcal{L}(n_{m-1} P_\infty)$ with dimension $m$, since the vectors $y_0, \dots, y_{m-1}$ form a basis of $\mathcal{L}(n_{m-1} P_\infty)$. Using the construction principle and (10.2) we derive

$$h_i f = \sum_{r=0}^{m-1} v_r h_i y_r = \sum_{r=0}^{m-1} v_r \sum_{k=0}^{\infty} \beta_{k,r}^{(i)} z_k^{(i)} = \sum_{k=d_i/e_i}^{\infty} \left( \sum_{r=0}^{m-1} v_r \beta_{k,r}^{(i)} \right) z_k^{(i)},$$

with $\beta_{k,r}^{(i)} \in \Gamma^{(i)}$. Note also that $\sum_{r=0}^{m-1} v_r \beta_{k,r}^{(i)} \in \Gamma^{(i)}$, i.e., the right hand side is the unique expansion of $h_i f$ using our chosen representative system $\Gamma^{(i)}$. Thus

$$v_{P_i}(h_i \cdot f) \geq \frac{d_i}{e_i}, \text{ for } 1 \leq i \leq s.$$

From $v_{P_i}(h_i) = 0$ we deduce

$$v_{P_i}(f) \geq \frac{d_i}{e_i}, \text{ for } 1 \leq i \leq s.$$

The rest is done identically to [10, Proof of Theorem 1]. The parameter $t$ follows then from [10, Corollary 1]. $\qquad\qquad\square$

As pointed out in [10], the column-wise concept is simpler than the row-wise concept, since no auxiliary divisor $D$ is needed. Moreover, only one ascending sequence of Riemann–Roch spaces is involved in the construction, whereas the row-wise construction needs $s$ of them. Furthermore, the elements $h_1, \dots, h_s$ bring additional variety to the construction.

**Example 10.10** Let $F = \mathbb{F}_q(x)$ be the rational function field over $\mathbb{F}_q$ (so $g = 0$) and $P_\infty$ the infinite place of $F$. Then the sequence $(y_r(x))_{r \geq 0}$ consists of polynomials in $\mathbb{F}_q[x]$ satisfying $\deg(y_r(x)) = r$ for all $r \geq 0$. Furthermore, in this case the places $P_1, \dots, P_s$ correspond to distinct monic irreducible polynomials $p_1(x), \dots, p_s(x)$ over $\mathbb{F}_q$. Now if the powers of the $p_i(x)$ form the sequence $(z_k^{(i)})_{k \geq 0}$ needed for the local expansions, and $h_1(x), \dots, h_s(x)$ are polynomials over $\mathbb{F}_q$ satisfying $h_i(x)$ coprime with $p_i(x)$ for $1 \leq i \leq s$, then our construction reduces to the one introduced in [8] when it is based on irreducible polynomials.

## 10.4 Designing morphological properties of
## the generating matrices

In this section we will discuss both constructions of the previous section, aiming to obtain special properties of the generating matrices. Specifically, we consider the finite-row property and the upper triangular property. In the following we will define the *length of a row* $(c_0, c_1, \ldots) \in \mathbb{F}_q^{\mathbb{N}_0}$ with only finitely many nonzero entries by $\max \{l \in \mathbb{N} : c_{l-1} \neq 0\}$, with the proviso that the length is 0 if $c_r = 0$ for all $r \geq 0$. Analogously we define the *length of a column* $(c_1, c_2, \ldots) \in \mathbb{F}_q^{\mathbb{N}}$ with only finitely many nonzero entries by $\max \{l \in \mathbb{N} : c_l \neq 0\}$.

### 10.4.1 Special settings within the Hofer–Niederreiter constructions

As worked out in [10], it is relatively easy to obtain finite rows within the column-wise concept. The crucial aspect is to make a sophisticated choice for the sequence $(y_r)_{r \geq 0}$ determining the columns. We use the notation and the construction of Section 10.3.2 with the following additional restriction on the sequence $(y_r)_{r \geq 0}$: in the case where $r > g$ we choose $y_r$ such that

$$y_r \in \mathcal{L} \left( n_r P_\infty - \sum_{i=1}^{s} (l_i + l_{i+1} + \cdots + l_s) \frac{v}{e_i} P_i \right) =: \mathcal{L}(D_r)$$

$$y_r \notin \mathcal{L} \left( n_{r-1} P_\infty - \sum_{i=1}^{s} (l_i + l_{i+1} + \cdots + l_s) \frac{v}{e_i} P_i \right) =: \mathcal{L}(D_r').$$

Here, $v := \mathrm{lcm}(e_1, \ldots, e_s)$, and the integers $l_i \geq 0$ are determined by the following chain of divisions with remainder:

$$r - g = svl_s + w_s, \quad w_s \in \{0, \ldots, sv - 1\},$$
$$w_s = (s-1)vl_{s-1} + w_{s-1}, \quad w_{s-1} \in \{0, \ldots, (s-1)v - 1\},$$
$$\vdots$$
$$w_2 = vl_1 + w_1, \quad w_1 \in \{0, \ldots, v - 1\}.$$

Note that such a $y_r$ exists since $D_r' \leq D_r$, $\deg(D_r) \geq r + g - (r - g) = 2g$ and $\deg(D_r') = \deg(D_r) - 1$, whence $\dim \mathcal{L}(D_r)/\mathcal{L}(D_r') > 0$. This special choice leads to generating matrices $C^{(1)}, \ldots, C^{(s)} \in \mathbb{F}_q^{\mathbb{N} \times \mathbb{N}_0}$.

**Theorem 10.11** *Given a prime power $q$ and a dimension $s \in \mathbb{N}$, let $F$ be a global function field with full constant field $\mathbb{F}_q$ and genus $g$. Let*

$P_\infty, P_1, \ldots, P_s$ be $s + 1$ *distinct places of $F$ with* $\deg(P_\infty) = 1$ *and* $\deg(P_i) =: e_i$ *for* $1 \le i \le s$. *Further, let* $h_1, \ldots, h_s$ *be elements of $F$ satisfying* $v_{P_i}(h_i) = 0$ *for* $1 \le i \le s$. *Then the matrices* $C^{(1)}, \ldots, C^{(s)}$ *constructed above generate a $(u, e, s)$-sequence in base $q$ with*

$$u = g \text{ and } e = (e_1, \ldots, e_s)$$

*and also a $(t, s)$-sequence in base $q$ with*

$$t = g + \sum_{i=1}^{s}(e_i - 1).$$

*Additionally, for all $1 \le i \le s$ and $d \in \mathbb{N}$, the length of the $d$th row of $C^{(i)}$ is at most*

$$g + sv \left\lfloor \frac{d-1}{v} \right\rfloor + iv,$$

*where $v = lcm(e_1, \ldots, e_s)$.*

*Proof.* Since the construction within this section is just a refined choice of the construction in Section 10.3.2, the assertions on the quality parameters $u$, $e$, and $t$ are still valid. The proof on the row lengths follows quite analogously to [10, Proof of Theorem 2] and the fact that $v_{P_i}(h_i \cdot y_r) = v_{P_i}(y_r)$.

We fix $i \in \{1, \ldots, s\}$, $d \in \mathbb{N}$ and let $a \in \mathbb{N}_0$, $w \in \{1, \ldots, v\}$, such that $d = av + w$. We write $C^{(i)} = \left(c_{j,r}^{(i)}\right)_{j \ge 1, r \ge 0} \in \mathbb{F}_q^{\mathbb{N} \times \mathbb{N}_0}$. In order to ensure that for every $r \ge g + sv \lfloor (d-1)/v \rfloor + iv = g + sva + iv$, the entry $c_{d,r}^{(i)}$ in the $d$th row of $C^{(i)}$ is 0, it suffices to show that the $l_1, \ldots, l_s$ in the division chain satisfy $l_i + l_{i+1} + \cdots + l_s \ge a + 1$, whence

$$v_{P_i}(h_i \cdot y_r) = v_{P_i}(y_r) \ge (l_i + l_{i+1} + \cdots + l_s)\frac{v}{e_i} \ge (a+1)v/e_i.$$

(Note that $(a+1)v \ge d$.)

We distinguish two cases of $r - g \ge sva + iv$.

(1) Consider $r - g \ge sva + sv$. In this case we immediately get $l_s \ge a + 1$.
(2) If

$$sva + jv \le r - g < sva + (j+1)v$$

with $i \le j < s$, then $l_s = a$ and $l_j = 1$ and $l_k \ge 0$ for $k \notin \{j, s\}$.

In both cases, $l_i + l_{i+1} + \cdots + l_s \ge a + 1$. $\qquad\square$

### 10.4.2 Special settings in the Xing–Niederreiter constructions

In this section we will see that the tools used to design finite rows in the column-wise concept can also be applied to obtain finite columns in the row-wise concept.

In the Xing–Niederreiter construction we can use any $k^{(i)}_{(l-1)e_i+1}, \ldots, k^{(i)}_{(l-1)e_i+e_i}$ in $\mathcal{L}(D + lP_i) \setminus \mathcal{L}(D + (l-1)P_i)$ that extend a basis of $\mathcal{L}(D + (l-1)P_i)$ to a basis of $\mathcal{L}(D + lP_i)$. In the following we will refine this by making a more specific choice using further subspaces; we take $k^{(i)}_{(l-1)e_i+1}, \ldots, k^{(i)}_{(l-1)e_i+e_i}$ in

$$\mathcal{L}(D + lP_i - ((l-1)e_i + 1)P_\infty) \subseteq \mathcal{L}(D + lP_i)$$

that extend a basis of $\mathcal{L}(D + (l-1)P_i - ((l-1)e_i + 1)P_\infty)$

to a basis of $\mathcal{L}(D + lP_i - ((l-1)e_i + 1)P_\infty)$.

**Lemma 10.12** *Such elements* $k^{(i)}_{(l-1)e_i+1}, \ldots, k^{(i)}_{(l-1)e_i+e_i}$ *exist for every* $l \in \mathbb{N}$ *and* $i \in \{1, \ldots, s\}$. *Furthermore, the set* $\{w_0, w_1, \ldots, w_g, k^{(i)}_1, \ldots, k^{(i)}_{le_i}\}$ *is a basis of* $\mathcal{L}(D + lP_i)$.

*Proof.* Existence follows using the Riemann–Roch theorem, since

$$\deg(D + lP_i - ((l-1)e_i + 1)P_\infty) = 2g + le_i - (l-1)e_i - 1 = 2g + e_i - 1$$

and also

$$\deg(D + (l-1)P_i - ((l-1)e_i + 1)P_\infty) = 2g + (l-1)e_i - (l-1)e_i - 1 = 2g - 1,$$

therefore the dimensions of the corresponding Riemann–Roch spaces are $g + e_i$ and $g$.

To prove the basis property, note that

$$\{w_0, w_1, \ldots, w_g, k^{(i)}_1, \ldots, k^{(i)}_{le_i}\} \subseteq \mathcal{L}(D + lP_i)$$

and $\dim \mathcal{L}(D + lP_i) = g + 1 + le_i$. It remains to prove linear independence of the set. We will prove this by induction.

From the construction of $\{w_0, w_1, \ldots, w_g\}$ linear independence is fulfilled for $l = 0$. The induction step $l \mapsto l + 1$ follows by assuming that an equation

$$\sum_{v=0}^{g} b_v w_v + \sum_{u=1}^{le_i} a_u k^{(i)}_u = \sum_{w=1}^{e_i} c_w k^{(i)}_{le_i+w} =: h_{l+1}, \qquad a_u, b_v, c_w \in \mathbb{F}_q,$$

exists with not all $c_w$ zero, hence by the induction assumption also with not all of the $b_v$ and $a_u$ zero. The left hand side is an element of $\mathcal{L}(D + l P_i) \setminus \{0\}$. But the middle term is an element of

$$\mathcal{L}(D + (l + 1)P_i - (le_i + 1)P_\infty) \setminus \mathcal{L}(D + l P_i - (le_i + 1)P_\infty),$$

hence its valuation at the place $P_i$ is exactly $-(l + 1) - v_{P_i}(D)$. This is a contradiction to $h_{l+1} \in \mathcal{L}(D + l P_i) \setminus \{0\}$, so $\{w_0, w_1, \dots, w_g, k_1^{(i)}, \dots, k_{(l+1)e_i}^{(i)}\}$ are linearly independent over $\mathbb{F}_q$.   $\square$

Construct $C^{(1)}, \dots, C^{(s)}$ via the algorithm of Xing and Niederreiter introduced in Section 10.3.1, now using the special $k_j^{(i)}$ of this section. We show that this choice limits the number of nonzero entries in the columns of the generating matrices.

**Theorem 10.13** *Given a prime power $q$, a dimension $s \in \mathbb{N}$, let $F$ be a global function field with full constant field $\mathbb{F}_q$ and genus $g$, and with at least one rational place $P_\infty$. Let $D$ be a positive divisor with degree $\deg(D) = 2g$ and $P_\infty \notin \operatorname{supp}(D)$. Finally, let $P_1, \dots, P_s$ be $s$ distinct places of degrees $e_1, \dots, e_s$ with $P_i \neq P_\infty$ for $1 \leq i \leq s$. Then the matrices $C^{(1)}, \dots, C^{(s)}$ constructed above generate a $(t, s)$-sequence in base $q$ with*

$$t = g + \sum_{i=1}^{s}(e_i - 1)$$

*and a $(u, e, s)$-sequence in base $q$ with*

$$u = g \text{ and } e = (e_1, \dots, e_s).$$

*Additionally, for $1 \leq i \leq s$, $r \geq 0$, the length of the $r$th column of $C^{(i)}$ is bounded by*

$$r + g + e_i.$$

*Hence, the generating matrix $C^{(i)}$ is nearly an upper triangular matrix, i.e., the subdiagonal bandwidth in the lower triangular part is at most $g + e_i - 1$.*

*Proof.* Lemma 10.12 ensures that the important [28, Lemma 2] is still valid, which states that the system $\{w_0, w_1, \dots, w_g\} \cup \{k_j^{(i)}\}_{1 \leq i \leq s, j \geq 1}$ is linearly independent over $\mathbb{F}_q$. The quality parameter $t$ then follows by identical arguments as in [28] and the parameters $u$ and $e$ follow analogously to [10, Remark 6].

It remains to show the bound on the column lengths. Let $i \in \{1, \dots, s\}$ and $j \geq 1$. By the specific choice of $k_j^{(i)} = k_{ve_i+v'}^{(i)}$ we have $v_{P_\infty}(k_j^{(i)}) \geq (ve_i + 1)$, where $v = \lfloor (j - 1)/e_i \rfloor$, $v' \in \{1, \dots, e_i\}$. Therefore in the local expansions of $k_j^{(i)}$ the coefficients $a_{j,r}^{(i)} = 0$ whenever $r \le ve_1 = e_i \lfloor (j-1)/e_i \rfloor$. Since in the construction $g+1$ coefficients are omitted we have $c_{j,r}^{(i)} = 0$ whenever

$$r + g + 1 \le e_i \left\lfloor \frac{j-1}{e_i} \right\rfloor = j - 1 - e_i \left\{ \frac{j-1}{e_i} \right\}.$$

Hence $c_{j,r}^{(i)} = 0$ whenever

$$j > j - 1 \ge r + g + e_i \ge r + g + 1 + e_i \left\{ \frac{j-1}{e_i} \right\}$$

and the result on the column lengths follows.                                    $\square$

**Remark 10.14** Observe that in the proof above we actually showed more, i.e., that nonzero entries in the lower triangular part only occur in $e_i$ by $e_i$ blocks since the structurally important value $v$ depends on the row index $j$ only via $\lfloor (j-1)/e_i \rfloor$, i.e., the remainder of division by $e_i$ is not involved.

Let us consider Example 10.8 with this more specific choice of the functions $k_{(l-1)e_i+1}^{(i)}, \ldots, k_{(l-1)e_i+e_i}^{(i)}$ required to lie in

$$\mathcal{L}(lP_i - ((l-1)e_i + 1)P_\infty) \setminus \mathcal{L}((l-1)P_i - ((l-1)e_i + 1)P_\infty).$$

Observe that the second space equals $\{0\}$ since the argument is a divisor of negative degree. Again we obtain rational functions with fixed denominator polynomial $(p_i(x))^l$, but now the numerator polynomials need degrees in $\{0, 1, \ldots, e_i-1\}$ and have to be linearly independent. One example would be to choose $y_{l,1}^{(i)}(x) = 1, y_{l,2}^{(i)}(x) = x, \ldots, y_{l,e_i}^{(i)}(x) = x^{e_i-1}$. This setting, together with the local parameter sequence $(1/x^r)_{r \in \mathbb{Z}}$, yields the classical Niederreiter constructions, which are also generated by nearly upper triangular matrices. Hence our refined construction can be interpreted as the "classical case" in the general Xing–Niederreiter construction.

### 10.4.3 A "nearly identity" matrix within the Xing–Niederreiter construction

The classical Niederreiter construction has one more special property, i.e., one generating matrix is the identity matrix. This matrix occurs using the place $P_1$ corresponding to the identity polynomial $p_1(x) = x$ together with the finer choice $k_l^{(1)} \in \mathcal{L}(lP_1 - lP_\infty) \setminus \mathcal{L}((l-1)P_i - lP_\infty)$ so that, for example, $k_l^{(1)} = 1/x^l$. This, together with the local parameter sequence $(1/x^r)_{r \ge 0}$, gives the identity matrix.

We would like to figure out an analog in the Xing–Niederreiter construction that yields the identity matrix (or something close to it) for, say, the first generating matrix. As the example of the classical Niederreiter construction shows, a special correspondence between the sequence $(k_j^{(1)})_{j\geq 1}$ and the local parameter sequence may bring the solution.

Suppose that $P_1$ is a rational place (consequently we will have to restrict ourselves to function fields $F$ with at least two rational places) and $P_1 \notin \text{supp}(D)$. We control the value of $v_{P_\infty}(k_j^{(1)})$ by the two conditions

$$k_j^{(1)} \in \mathcal{L}(D + jP_1 - jP_\infty) \setminus \mathcal{L}(D + (j-1)P_1 - jP_\infty) \qquad (10.3)$$

$$k_j^{(1)} \in \mathcal{L}(D + jP_1 - jP_\infty) \setminus \mathcal{L}(D + jP_1 - (j+1)P_\infty). \qquad (10.4)$$

Note that (10.3) is consistent with the construction introduced at the beginning of Section 10.4.2. The requirement (10.4) imposes further restrictions on the sequence $(k_j^{(1)})_{j\geq 1}$. Again we show existence and basis property.

**Lemma 10.15** *For every $j \in \mathbb{N}$ there exist $k_j^{(1)}$ as above, satisfying $v_{P_\infty}(k_j^{(1)}) = j$ and $v_{P_1}(k_j^{(1)}) = -j$. Also, the set $\{w_0, \ldots, w_g, k_1^{(1)}, \ldots, k_j^{(1)}\}$ is a basis of $\mathcal{L}(D + jP_1)$.*

*Proof.* The claimed value of the valuation at $P_1$, $v_{P_1}(k_j^{(1)}) = -j$, is guaranteed by (10.3) together with $P_1 \notin \text{supp}(D)$; analogously, $v_{P_\infty}(k_j^{(1)}) = j$ by (10.4) together with $P_\infty \notin \text{supp}(D)$. Equation (10.3) implies $\{w_0, \ldots, w_g, k_1^{(1)}, \ldots, k_j^{(1)}\} \subseteq \mathcal{L}(D + jP_1)$ and the fact $v_{P_1}(k_j^{(1)}) = -j$ ensures that $k_1^{(1)}, \ldots, k_j^{(1)}$ are linearly independent. For the linear independence of $\{w_0, \ldots, w_g, k_1^{(1)}, \ldots, k_j^{(1)}\}$ assume that an equation

$$\sum_{v=0}^{g} b_v w_v = \sum_{u=1}^{j} a_u k_u^{(1)} =: h, \qquad a_u, b_v \in \mathbb{F}_q$$

holds, with at least one nonzero $a_u$. Applying the valuation at $P_1$ to $h$ yields a contradiction.

It remains to ensure existence. The degrees of the divisors in (10.3) and (10.4) are $\deg(D + jP_1 - jP_\infty) = 2g$ and $\deg(D + (j-1)P_1 - jP_\infty) = \deg(D + jP_1 - (j+1)P_\infty) = 2g - 1$, respectively. By the Riemann–Roch theorem we infer that $\dim(D + jP_1 - jP_\infty) = g + 1$ and $\dim(D + (j-1)P_1 - jP_\infty) = \dim(D + jP_1 - (j+1)P_\infty) = g$. Since the intersection of the two smaller vector space, contains at least the element 0, we obtain the existence of the requested $k_j^{(1)}$ by

$$q^{g+1} - 2q^g + 1 = (q-2)q^g + 1 > 0,$$

in view of the equation

$$\#((C \setminus A) \cap (C \setminus B)) = \#(C \setminus (A \cup B)) = \#C - (\#A + \#B - \#(A \cap B))$$

for arbitrary sets $A$, $B \subseteq C$.        □

Bearing in mind the restriction $P_1 \notin \mathrm{supp}\,(D)$ and $\deg(P_1) = 1$ we now exploit the freedom with respect to the choice of the parameter sequence for the local expansions.

**Corollary 10.16** *We use the construction of Section 10.4.2 with the restriction $P_1 \notin \mathrm{supp}\,(D)$ and $\deg(P_1) = 1$ and the special sequence $(k_j^{(1)})_{j \geq 1}$ considered above. Furthermore, in the local parameter sequence $(z_r)_{r \geq 0}$ we choose $z_r = k_r^{(1)}$ if $r \notin \{n_0, n_1, \dots, n_g\}$ and $z_{n_f} = w_f$ for $f \in \{0, 1, \dots, g\}$. Then $C^{(1)}$ almost becomes the identity matrix, i.e., if we delete the rows indexed by $n_1, \dots, n_g$ in $C^{(1)}$, we indeed obtain the identity matrix in $\mathbb{F}_q^{\mathbb{N} \times \mathbb{N}_0}$.*

*Proof.* When applying the construction of $C^{(1)}$ in Section 10.3.1 we first need the local expansion of $k_j^{(1)}$. Now if $j \notin \{n_1, \dots, n_g\}$ then $k_j^{(1)} = 1z_j$. For $j \in \{n_1, \dots, n_g\}$ we have $k_j^{(1)} = \sum_{r=0}^{\infty} a_{j,r}^{(1)} z_r$ with possibly many nonzero entries. When building the $j$th row of the matrix $C^{(1)}$ the coefficients associated to the indices $n_0, n_1, \dots, n_g$ have to be omitted. Hence the matrix formed by the rows indexed by $\{j \in \mathbb{N} : j \notin \{n_1, \dots, n_g\}\}$ (in consecutive order) yields the identity matrix.        □

## 10.5 Computational results

We implemented the above constructions on a computer, using the computer algebra systems Magma [2] and Mathematica [27] for the calculations. We will first illustrate the effects of the additional construction parameter requirements on the generating matrices and give some numerical data concerning the densities.

First we present in Figures 10.1 and 10.2 the effect of applying the finite-row mechanism on the generating matrices of the Hofer–Niederreiter sequence. The particular construction parameters for the figures were:

- an optimal binary global function field with $g = 3$,
- using all $s + 1 = 7$ rational places,
- with $m = 64$,
- with designed (and actual) $t$-value 3.

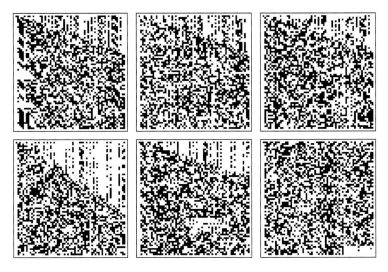

Figure 10.1  Hofer–Niederreiter, no modifications.

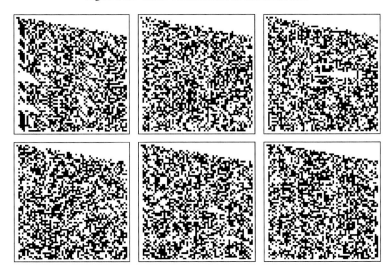

Figure 10.2  Hofer–Niederreiter, with finite rows.

Next, in Figures 10.3, 10.4, and 10.5 we show that we can obtain finite columns and further, even nearly an identity matrix, by exploiting the various degrees of freedom open in the Xing–Niederreiter construction. The particular construction parameters for the figures were:

- an optimal binary global function field with $g = 2$,
- using all $s = 6$ rational places as well as one degree 3 place,

Figure 10.3  Xing–Niederreiter, no modifications.

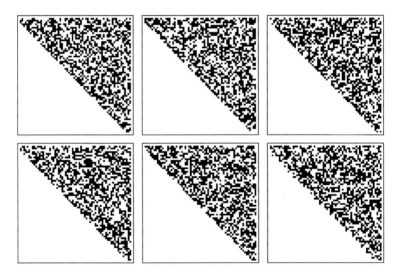

Figure 10.4  Xing–Niederreiter, with finite columns.

- with $m = 64$,
- with designed (and actual) $t$-value 4.

Observe here that the last matrix appears to be more jagged around the diagonals in Figures 10.4 and 10.5 as it actually consists of 3 by 3 blocks, corresponding to the degree of the associated place (cf. Remark 10.14).

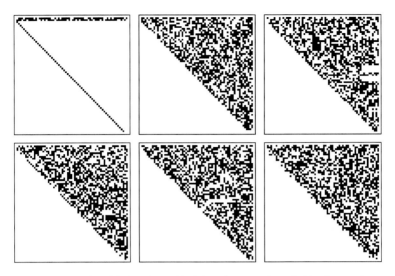

Figure 10.5  Xing–Niederreiter, with finite columns and nearly identity matrix.

The original Xing–Niederreiter matrices appear to be filled with fairly randomly distributed entries so we would expect to obtain a density of, asymptotically, $1/2$. Here we define the density as the ratio of the number of nonzero entries to the total number of entries in initial submatrices. With the use of nearly upper triangular matrices we would expect a further reduction to about $1/4$. Larger values can appear, when genus and maximal place degree increase. Figures 10.6 and 10.7 show the densities of the submatrices versus their size, with respect to the individual matrices in Figures 10.3 and 10.5, as well as the mean densities (thicker line). The graphs support the expectations raised above and illustrate the range of densities that can be obtained by different choices of construction parameter. Observe that although the densities of all matrices other than the first seem to converge more slowly to $1/4$ (if at all), the overall average densities still approach our heuristic bound.

## 10.6  Summary and outlook

We have shown that by paying closer attention to the choices to be made in the construction of digital sequences, here in particular the Hofer–Niederreiter and Xing–Niederreiter constructions, the morphological properties of the generating matrices can be controlled to a great degree. While there may also be computational benefits to having, for example, nearly upper triangular matrices

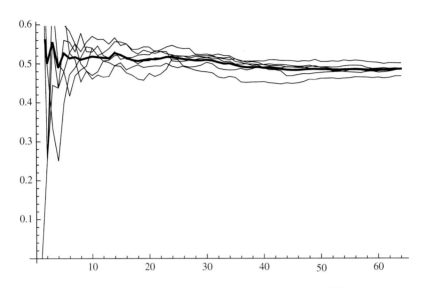

Figure 10.6  Matrix densities of submatrices in Figure 10.3.

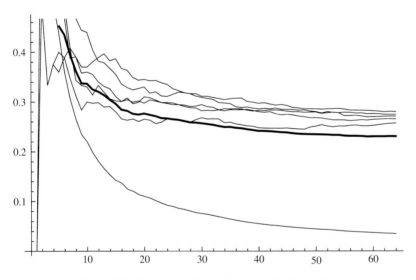

Figure 10.7  Matrix densities of submatrices in Figure 10.5.

instead of a full matrix of more or less random entries, it is the theoretical aspect we would like to emphasize.

The shape of the generating matrices makes it possible to read directly from the matrices several parameters inherent in the global function field employed in the construction, i.e., the genus of the field and the degree of the place associated to a particular coordinate. This mirrors the situation in the classical Niederreiter sequence construction where the degrees of the irreducible polynomials are also apparent in the subdiagonal bandwith of the generating matrices. Furthermore, we can even achieve extreme sparsity in one of the matrices, making it nearly an identity matrix. As a consequence, the projected sequence in this particular dimension is very closely associated to the radical inverse. If we derive finite $(t, m, s)$-nets by the usual procedure of adding equidistant points, we get something very close to the Hammersley sequence as a projection. It is suggestive to consider this shape as something like a canonical or normal form, which would be best suited as a starting point to investigate subsequent scramblings and other transformations.

Finally, a particular open problem we want to point out is whether it is possible in these two constructions to achieve both finite rows and finite columns at the same time. First examples of such generating matrices were given in [9, 11, 12]. The two constructions considered in this paper would need to control the length of the local expansions, in addition to their initial indices. So far, finite expansions of this kind can only be achieved in the field of rational functions or the ring of polynomials over a finite field [7, 8].

## Acknowledgements

R. Hofer was partially sponsored by the FWF projects P24302, P21943 and F5505-N26; I. Pirsic was supported by the FWF projects P23285 and F5511-N26. The projects F5505-N26 and F5511-N26 are part of the Special Research Program "Quasi-Monte Carlo Methods: Theory and Applications."

## References

[1] I. A. Antonov and V. M. Saleev, An effective method for the computation of $\lambda P_\tau$-sequences (Russian). *Zh. Vychisl. Mat. Mat. Fiz.* **19**(1), 243–245, 271, 1979.

[2] W. Bosma, J. Cannon and C. Playoust, The Magma algebra system. I. The user language. *J. Symbol. Comput.* **24**, 235–265, 1997.

[3] P. Bratley, B. L. Fox and H. Niederreiter, Implementation and tests of low-discrepancy sequences. *ACM Trans. Model. Comput. Simul.* **2**(3), 195–213, 1992.

[4] J. Dick and F. Pillichshammer, *Digital Nets and Sequences: Discrepancy Theory and Quasi-Monte Carlo Integration*. Cambridge University Press, Cambridge, 2010.

[5] J. Dick, F. Y. Kuo, G. W. Peters and I. H. Sloan (eds.), *Monte Carlo and Quasi-Monte Carlo Methods 2012*. Springer, Heidelberg, 2013.

[6] H. Faure, Discrépance de suites associées à un système de numération (en dimension $s$). *Acta Arith.* **41**, 337–351, 1982.

[7] R. Hofer, A construction of digital $(0, s)$-sequences involving finite-row generator matrices. *Finite Fields Appl.* **18**, 587–596, 2010.

[8] R. Hofer, A construction of low-discrepancy sequences involving finite-row digital $(t, s)$-sequences. *Monatsh. Math.* **171**, 77–89, 2013.

[9] R. Hofer and G. Larcher, On existence and discrepancy of certain digital Niederreiter–Halton sequences. *Acta Arith.* **141**, 369–394, 2010.

[10] R. Hofer and H. Niederreiter, A construction of $(t, s)$-sequences with finite-row generating matrices using global function fields. *Finite Fields Appl.* **21**, 97–110, 2013.

[11] R. Hofer and G. Pirsic, An explicit construction of finite-row digital $(0, s)$-sequences. *Unif. Distrib. Theory* **6**, 13–30, 2011.

[12] R. Hofer and G. Pirsic, A finite-row scrambling of Niederreiter sequences. In: J. Dick, F. Y. Kuo, G. W. Peter and I. H. Sloan (eds.), *Monte Carlo and Quasi-Monte Carlo Methods 2012*. Springer, Heidelberg, 2013.

[13] R. Hofer, P. Kritzer, G. Larcher and F. Pillichshammer, Distribution properties of generalized van der Corput–Halton sequences and their subsequences. *Int. J. Number Theory* **5**, 719–746, 2009.

[14] H. Koch, *Number Theory: Algebraic Numbers and Functions*. Graduate Studies in Mathematics, volume 24. AMS, Providence, RI, 2000.

[15] D. J. S. Mayor and H. Niederreiter, A new construction of $(t, s)$-sequences and some improved bounds on their quality parameter. *Acta Arith.* **128**, 177–191, 2007.

[16] H. Niederreiter, Point sets and sequences with small discrepancy. *Monatsh. Math.* **104**, 273–337, 1987.

[17] H. Niederreiter, Low-discrepancy and low-dispersion sequences. *J. Number Theory* **30**, 51–70, 1988.

[18] H. Niederreiter, *Random Number Generation and Quasi-Monte Carlo Methods*. CBMS-NSF Regional Conference Series in Applied Mathematics, volume 63. SIAM, Philadelphia, PA, 1992.

[19] H. Niederreiter and F. Özbudak, Low-discrepancy sequences using duality and global function fields. *Acta Arith.* **130**, 79–97, 2007.

[20] H. Niederreiter and C.P. Xing, *Rational Points on Curves over Finite Fields: Theory and Applications*. Cambridge University Press, Cambridge, 2001.

[21] H. Niederreiter and A. S. Yeo, Halton-type sequences from global function fields. *Sci. China Math.* **56**, 1467–1476, 2013.

[22] I. M. Sobol', Distribution of points in a cube and approximate evaluation of integrals (Russian). *Ž. Vyčisl. Mat. Mat. Fiz.* **7**, 784–802, 1967.

[23] H. Stichtenoth, *Algebraic Function Fields and Codes*, second edition. Springer, Berlin, 2009.

[24] S. Tezuka, Polynomial arithmetics of the Halton sequences. *ACM Trans. Model. Comput. Simul.* **3**, 99–107, 1993.

[25] S. Tezuka, *Uniform Random Numbers: Theory and Practice*. Kluwer Academic Publishers, Boston, MA, 1995.

[26] S. Tezuka, On the discrepancy of generalized Niederreiter sequences. *J. Complexity* **29**, 240–247, 2013.

[27] Wolfram Research, Inc., *Mathematica Edition*, Version 8.0. Wolfram Research, Champaign, IL, 2010.

[28] C. P. Xing and H. Niederreiter, A construction of low-discrepancy sequences using global function fields. *Acta Arith.* **73**, 87–102, 1995.

# 11

# Periodic structure of the exponential pseudorandom number generator

*Jonas Kaszián*
RWTH Aachen, Aachen

*Pieter Moree*
Max-Planck-Institut für Mathematik, Bonn

*Igor E. Shparlinski*
University of New South Wales, Sydney

*Dedicated to Harald Niederreiter on the occasion of his*
$\mathbb{F}_2 \times \mathbb{F}_5 \times \mathbb{F}_7$ *birthday.*

## Abstract

We investigate the periodic structure of the exponential pseudorandom number generator obtained from the map $x \mapsto g^x \pmod{p}$ that acts on the set $\{1, \ldots, p-1\}$.

## 11.1 Introduction

### 11.1.1 Motivation and our results

Given a prime $p$ and an integer $g$ with $p \nmid g$ and an initial value $u_0 \in \{1, \ldots, p-1\}$ we consider the sequence $\{u_n\}$ generated recursively by

$$u_n \equiv g^{u_{n-1}} \pmod{p}, \quad 1 \leq u_n \leq p-1, \quad n = 1, 2, \ldots, \quad (11.1)$$

and then, for an integer parameter $k \geq 1$, we consider the sequence of integers $\xi_n^{(k)} \in \{0, \ldots, 2^k - 1\}$ formed by the $k$ least significant bits of $u_n$, $n = 0, 1, \ldots$. This construction is called the *exponential pseudorandom number generator* and has numerous cryptographic applications, see [13, 16, 19, 26, 28, 30] and references therein. Certainly, for the exponential pseudorandom number generator, as for any other pseudorandom number generator, the question of periodicity is of primal interest.

More precisely, the sequence $\{u_n\}$, as any other sequence generated iterations of a function on a finite set, eventually becomes periodic with some *cycle length* $t$. That is, there is some integer $s \geq 0$ such that

$$u_n = u_{n+t}, \qquad n = s, s+1, \ldots. \tag{11.2}$$

We always assume that $t$ is the smallest positive integer with this property. Furthermore, the sequence $u_0, \ldots, u_{s+t-1}$ of length $\ell = s + t$, where $t \geq 1$ and then $s \geq 0$ are chosen to be the smallest possible integers to satisfy (11.2), is called the *trajectory* of $\{u_n\}$ and consists of the *tail* $u_0, \ldots, u_{s-1}$ and the *cycle* $u_s, \ldots, u_{s+t-1}$.

Clearly, we always have $\ell \leq T$ where $T$ is the multiplicative order of $g$ modulo $p$.

Since the sequence $\{u_n\}$ eventually becomes periodic with some cycle length $t$, so does the sequence $\{\xi_n^{(k)}\}$ and its cycle length $\tau_k$ divides $t$.

We further remark that if $g$ is a primitive root modulo $p$, then the map $x \mapsto g^x \pmod{p}$ acts bijectively on the set $\{1, \ldots, p-1\}$ or in other words defines an element of the symmetric group $S_{p-1}$. Therefore, in this case the sequence $\{u_n\}$ is purely periodic, that is, (11.2) holds with $s = 0$. This also means that in this case the sequence $\{\xi_n^{(k)}\}$ is purely periodic.

As usual let $\varphi$ denote Euler's totient function. Recall that there are exactly $\varphi(p-1)$ primitive roots modulo $p$. The above map leads to precisely $\varphi(p-1)$ different elements of $S_{p-1}$. The question is to what extent these $\varphi(p-1)$ permutations represent "generic permutations of $S_{p-1}$." Note that the cardinality $(p-1)!$ of $S_{p-1}$ is vastly larger than $\varphi(p-1)$ which on average behaves as a constant times $p$.

Unfortunately there are essentially no theoretic results about the behavior of either of the sequences $\{u_n\}$ and $\{\xi_n^{(k)}\}$. In fact even the distribution of $t$ has not been properly investigated. If $g$ is a primitive root, which is the most interesting case for cryptographic applications, then heuristically, the periodic behavior of the sequence $\{u_n\}$ can be modeled as a random permutation on the set $\{1, \ldots, p-1\}$; see [1] for a wealth of results about random permutations. For example, by a result of [29] one expects that $t = p^{1+o(1)}$ in this case. If $g$ is not a primitive root it is not clear what the correct statistical model describing the map $x \mapsto g^x \pmod{p}$ should be. Probably, if $g$ is of order $T$ modulo $p$, then one can further reduce the residue $g^x \pmod{p}$ modulo $T$ and consider the associated permutation on the set $\{1, \ldots, T\}$ generated by the map

$$x \mapsto \left( g^x \pmod{p} \right) \pmod{T}.$$

This suggests that in this case one expects $t = T^{1+o(1)}$, but the sequence $\{u_n\}$ is not necessary purely periodic anymore.

For the sequence $\{\xi_n^{(k)}\}$ it is probably natural to expect that $\tau_k = t$ in the overwhelming majority of cases (and for a wide range of values of $k$), but this question has not been properly addressed in the literature.

The only theoretic result here seems to be the bound of [15] relating $t$ and $\tau_k$. First, as in [15, Section 5] we note that there are at most $p2^{-k} + 1$ integers $v \in \{1, \ldots, p-1\}$ with a given string of $k$ least significant bits. Hence, if $2^k < p$ then obviously

$$\tau_k \geq t2^{k-1}/p. \tag{11.3}$$

If $k \leq (1/4 - \varepsilon)r$ for any fixed $\varepsilon > 0$, where $r$ is the bit length of $p$, then it is shown in [15, Section 5] that using bounds of exponential sums one can improve (11.3) to

$$\tau_k \geq c(\varepsilon)t2^{2k}/p, \tag{11.4}$$

where $c(\varepsilon) > 0$ depends only on $\varepsilon > 0$. Clearly the bound (11.4) trivially implies that for $k \geq r/4$ we have

$$\tau_k \geq tp^{-1/2+o(1)}, \tag{11.5}$$

which however is weaker than (11.3) for $k \geq r/2$.

In this paper we use some results of [5] on the concentration of solutions of exponential congruences to sharpen (11.3), (11.4) and (11.5) for $k \geq (3/8 + \varepsilon)r$.

We also use the same method to establish a lower bound for the number of distinct values in the sequence $\{\xi_n^{(k)}\}$. Finally, we also show that for large values of $k$ the modern results on the sum-product problem (see [8]) lead to better estimates.

Our results relate $\tau_k$ and $t$ and are meaningful only when $t$ is sufficiently large. Since no theoretic results about large values of $t$ are known, we study the behavior of $t$ empirically. Our findings are consistent with the map $x \mapsto g^x \pmod{p}$ having a generic cycle structure. In particular, the results of our numerical tests exhibit reasonable agreement with those predicted for random permutations, see [1].

## 11.1.2  Previously known results

Here we briefly review several previously known results about the cycle structure of the map $x \mapsto g^x \pmod{p}$. Essentially only very short cycles, such as fixed points, succumb to the efforts of obtaining rigorous results.

In particular, for an integer $k$ we denote by $N_{p,g}(k)$ the number of $u_0 \in \{1, \ldots, p-1\}$ such that for the sequence (11.1) we have $u_k = u_0$. Note that $N_{p,g}(1)$ is the number of fixed points of the map $x \mapsto g^x \pmod{p}$.

The quantity $N_{p,g}(k)$ for $k = 1, 2, 3$ has recently been studied [3, 4, 12, 18, 21, 22, 23, 27, 31]. Fixed points with various restrictions on $u$ have been considered as well. For example, Cobeli and Zaharescu [12] have shown that

$$\#\{(g, u) \; : \; 1 \le g, u \le p - 1, \; \gcd(u, p - 1) = 1, \; g^u \equiv u \pmod{p}\}$$
$$= \frac{\varphi(p-1)^2}{p-1} + O\left(\tau(p-1)p^{1/2}\log p\right),$$

where $\tau(m)$ is the number of positive integer divisors of $m \ge 1$. Unfortunately, the co-primality condition $\gcd(u, p-1) = 1$ is essential for the method of [12], thus that result does not immediately extend to all $u \in \{1, \ldots, p-1\}$. Several more results and conjectures of similar flavor have been presented by Holden and Moree [23]. Furthermore, an asymptotic formula for the average value $N_{p,g}(1)$ on average over all $p$ and all primitive roots $g \in \{1, \ldots, p-1\}$ is given by Bourgain *et al.* [3, Theorems 13 and 14]:

$$\sum_{p \le Q} \frac{1}{p-1} \sum_{\substack{g=1 \\ g \text{ primitive root}}}^{p-1} N_{p,g}(1) = (A + o(1))\pi(Q)$$

$$\sum_{p \le Q} \frac{1}{p-1} \sum_{g=1}^{p-1} N_{p,g}(1) = (1 + o(1))\pi(Q)$$

as $Q \to \infty$, where

$$A = \prod_{p \text{ prime}} \left(1 - \frac{1}{p(p-1)}\right) = 0.373955\ldots$$

is *Artin's constant* and, as usual, $\pi(Q)$ is the number of primes $p \le Q$. It is also shown in [4, Theorem 11] that

$$\sum_{g=1}^{p-1} N_{p,g}(1) = O(p),$$

however, the conjecture by Holden and Moree [23] that

$$\sum_{g=1}^{p-1} N_{p,g}(1) = (1 + o(1))p \qquad (11.6)$$

remains open. It is known though that

$$\sum_{g=1}^{p-1} N_{p,g}(1) \ge p + O(p^{3/4+o(1)}),$$

see [4, Equation (1.15)]. It is also shown in [4, Section 5.9] that (11.6) may fail only on a very thin set of primes.

It is also known that $N_{p,g}(1) \le \sqrt{2p} + 1/2$ for any $g \in \{1, \ldots, p-1\}$, see [18, Theorem 2].

For $N_{p,g}(2)$, the only known result is the bound

$$N_{p,g}(2) \le C(g)\frac{p}{\log p}$$

of Glebsky and Shparlinski [18, Theorem 3], where $C(g)$ depends on $g$.
Finally, by [18, Theorem 3] we have

$$N_g(3) \le \frac{3}{4}p + \frac{g^{2g+1} + g + 1}{4}$$

(which is certainly a very weak bound).

## 11.2  Preparations

### 11.2.1  Density of points on exponential curves

Let $p$ be a prime and $a$, $b$ and $g$ integers satisfying $p \nmid abg$. Given two intervals $\mathcal{I}$ and $\mathcal{J}$, we denote by $R_{a,b,g,p}(\mathcal{I}, \mathcal{J})$ the number of integer solutions of the system of congruences

$$au \equiv x \pmod{p} \qquad \text{and} \qquad bg^u \equiv y \pmod{p}$$
$$(u, x, y) \in \{1, \ldots, p - 1\} \times \mathcal{I} \times \mathcal{J}.$$

Upper bounds on $R_{1,b,g,p}(\mathcal{I}, \mathcal{J})$ are given in [5, Theorems 23 and 24], which in turn improve and generalize the previous estimates of [9, 10]. We need the following straightforward generalizations of the estimates of [5, Theorems 23 and 24] to an arbitrary $a$ with $p \nmid a$.

**Lemma 11.1** *Suppose that $p \nmid ab$ and that $T$ is the multiplicative order of $g$ modulo $p$. Let $\mathcal{I}$ and $\mathcal{J}$ be two intervals consisting of $K$ and $L$ consecutive integers respectively, where $L \le T$. Then*

$$R_{a,b,g,p}(\mathcal{I}, \mathcal{J}) \le \left(\frac{K}{p^{1/3}L^{1/6}} + 1\right) L^{1/2+o(1)}$$

$$R_{a,b,g,p}(\mathcal{I}, \mathcal{J}) \le \left(\frac{K}{p^{1/8}L^{1/6}} + 1\right) L^{1/3+o(1)}.$$

For intervals $\mathcal{I}$ and $\mathcal{J}$ of the same length, we derive a more explicit form of Lemma 11.1.

**Corollary 11.2** *Assume that $g$ is of multiplicative order $T$ modulo $p$ and that $a$ and $b$ are integers such that $p \nmid ab$. Let $\mathcal{I}$ and $\mathcal{J}$ be two intervals consisting of $H$ consecutive integers respectively, where $H \le T$. Then*

$$R_{a,b,g,p}(\mathcal{I}, \mathcal{J}) \leq H^{o(1)} \begin{cases} H^{1/3} & \text{if } H \leq p^{3/20} \\ H^{7/6} p^{-1/8} & \text{if } p^{3/20} < H \leq p^{3/16} \\ H^{1/2} & \text{if } p^{3/16} < H \leq p^{2/5} \\ H^{4/3} p^{-1/3} & \text{if } p^{2/5} < H. \end{cases}$$

### 11.2.2  Sum-product problem

For a prime $p$, we denote by $\mathbb{F}_p$ the finite field of $p$ elements.

Given a set $\mathcal{A} \subseteq \mathbb{F}_p$ we define the sets

$$2\mathcal{A} = \{a_1 + a_2 \,:\, a_1, a_2 \in \mathcal{A}\} \quad \text{and} \quad \mathcal{A}^2 = \{a_1 \cdot a_2 \,:\, a_1, a_2 \in \mathcal{A}\}.$$

The celebrated result of Bourgain *et al.* [2] asserts that at least one of the cardinalities $\#(\mathcal{A}^2)$ and $\#(2\mathcal{A})$ is always large.

The current state of affairs regarding quantitative versions of this result, due to several authors, has been summarized by Bukh and Tsimerman [8] as follows.

**Lemma 11.3** *For an arbitrary set $\mathcal{A} \subseteq \mathbb{F}_p$, we have*

$$\max \left\{ \# \left( \mathcal{A}^2 \right), \# (2\mathcal{A}) \right\}$$

$$\geq (\#\mathcal{A})^{o(1)} \begin{cases} (\#\mathcal{A})^{12/11} & \text{if } \#\mathcal{A} \leq p^{1/2} \\ (\#\mathcal{A})^{7/6} p^{-1/24} & \text{if } p^{1/2} \leq \#\mathcal{A} \leq p^{35/68} \\ (\#\mathcal{A})^{10/11} p^{1/11} & \text{if } p^{35/68} \leq \#\mathcal{A} \leq p^{13/24} \\ (\#\mathcal{A})^2 p^{-1/2} & \text{if } p^{13/24} \leq \#\mathcal{A} \leq p^{2/3} \\ (\#\mathcal{A})^{1/2} p^{1/2} & \text{if } \#\mathcal{A} \geq p^{2/3}. \end{cases}$$

## 11.3  Main results

### 11.3.1  Period length

For any $k \leq r$ we now obtain an improvement of (11.3).

**Theorem 11.4** *For any $r$-bit prime $p$ and $g$ with $p \nmid g$, we have*

$$\tau_k \geq t p^{o(1)} \begin{cases} (2^k/p)^{1/3} & \text{if } k/r \geq 17/20 \\ 2^{7k/6} p^{-25/24} & \text{if } 17/20 > k/r \geq 13/16 \\ (2^k/p)^{1/2} & \text{if } 13/16 > k/r \geq 3/5 \\ 2^{4k/3} p^{-1} & \text{if } 3/5 > k/r. \end{cases}$$

*Proof.* Recall that we have the divisibility $\tau_k \mid t$ and consider the sequence $u_{s\tau_k}$ for $s = 1, \ldots, t/\tau_k$. By the definition of $\tau_k$, all these numbers end with the same string of $k$ least significant bits. Furthermore, this is also true for $u_{s\tau_k+1} \equiv g^{u_{s\tau_k}} \pmod{p}$. Therefore, there are some integers $\lambda, \mu \in [0, 2^k - 1]$ such that

$$u_{s\tau_k} = 2^k v_s + \lambda \qquad \text{and} \qquad u_{s\tau_k+1} = 2^k w_s + \mu$$

for some integers $v_s, w_s \in [0, 2^{r-k} - 1]$.

Hence, defining $\alpha \in [1, p - 1]$ by the congruence $\alpha 2^k \equiv 1 \pmod{p}$, we see that the residues modulo $p$ of $\alpha u_{s\tau_k}$ and of $\alpha g^{u_{s\tau_k}}$ belong to some intervals of $\mathcal{I}$ and $\mathcal{J}$, respectively, of length $2^{r-k}$ each. Since $t \le T$, where $T$ is the multiplicative order of $g$, for these intervals $\mathcal{I}$ and $\mathcal{J}$ we have

$$t/\tau_k \le R_{\alpha,\alpha,g,p}(\mathcal{I}, \mathcal{J}).$$

Using Corollary 11.2 with $H = 2^{r-k}$, we conclude the proof. $\qquad\square$

Combining Theorem 11.4 with (11.4) and (11.5) we derive the following result.

**Corollary 11.5** *For any $r$-bit prime $p$ and $g$ with $p \nmid g$, we have*

$$\tau_k \ge t p^{o(1)} \begin{cases} (2^k/p)^{1/3} & \text{if } k/r \ge 17/20 \\ 2^{7k/6} p^{-25/24} & \text{if } 17/20 > k/r \ge 13/16 \\ (2^k/p)^{1/2} & \text{if } 13/16 > k/r \ge 3/5 \\ 2^{4k/3} p^{-1} & \text{if } 3/5 > k/r \ge 3/8 \\ p^{-1/2} & \text{if } 3/8 > k/r \ge 1/4 \\ 2^{2k} p^{-1} & \text{if } 1/4 > k/r. \end{cases}$$

### 11.3.2 The number of distinct values

We now obtain a lower bound on the number $v_k(N)$ of distinct values which appear among the elements $\xi_n^{(k)}$, $n = 0, \ldots, N - 1$. Let $\ell = s + t$ be the trajectory length of the sequence $\{u_n\}$, see (11.2).

Note that if $2^k < p$ then the following analog of (11.3) holds:

$$v_k(N) \ge N 2^{k-1}/p. \tag{11.7}$$

In fact for $N = \ell = p^{1+o(1)}$ the bound (11.7) is asymptotically optimal as we obviously have $v_k(N) \le 2^k$. However, for smaller values of $\ell$ we obtain a series of other bounds.

**Theorem 11.6** *For any r-bit prime p and g with $p \nmid g$, we have*

$$
v_k(N) \geq N^{1/2} p^{o(1)} \begin{cases} (2^k/p)^{1/6} & \text{if } 1 \geq k/r \geq 17/20 \\ 2^{7k/12} p^{-25/48} & \text{if } 17/20 > k/r \geq 13/16 \\ (2^k/p)^{1/4} & \text{if } 13/16 > k/r \geq 3/5 \\ 2^{2k/3} p^{-1/2} & \text{if } 3/5 > k/r, \end{cases}
$$

*for all $N \leq \ell$.*

*Proof.* Consider the pairs $(\xi_n^{(k)}, \xi_{n+1}^{(k)})$, $n = 0, \ldots, N - 1$. Then at least one pair $(\lambda, \mu)$ appears at least $N/v_k^2(N)$ times. Since $N \leq \ell < T$, where $T$ is the multiplicative order of $g$, as in the proof of Theorem 11.4 we obtain

$$
N/v_k^2(N) \leq R_{\alpha,\alpha,g,p}(\mathcal{I}, \mathcal{J})
$$

for some intervals $\mathcal{I}$ and $\mathcal{J}$ of length $2^{r-k}$ each and some integer $\alpha \in \{1, \ldots, p - 1\}$. Using Corollary 11.2 with $H = 2^{r-k}$, we conclude the proof.  □

Using the same technique as in [15, Section 5], it is easy to show that any fixed pair $(\lambda, \mu)$ occurs amongst the pairs $(\xi_n^{(k)}, \xi_{n+1}^{(k)})$, $n = 0, \ldots, \ell - 1$, at most $O(p2^{-2k} + p^{1/2}(\log p)^2)$ times. So, we also have

$$
N/v_k^2(N) = O\left(p2^{-2k} + p^{1/2}(\log p)^2\right),
$$

and thus, after simple calculations, we derive the following estimate.

**Corollary 11.7** *For any r-bit prime p and any integer g with $p \nmid g$, we have*

$$
v_k(N) \geq N^{1/2} p^{o(1)} \begin{cases} (2^k/p)^{1/6} & \text{if } k/r \geq 17/20 \\ 2^{7k/12} p^{-25/48} & \text{if } 17/20 > k/r \geq 13/16 \\ (2^k/p)^{1/4} & \text{if } 13/16 > k/r \geq 3/5 \\ 2^{2k/3} p^{-1/2} & \text{if } 3/5 > k/r \geq 3/8 \\ p^{-1/4} & \text{if } 3/8 > k/r \geq 1/4 \\ 2^k p^{-1/2} & \text{if } 1/4 > k/r, \end{cases}
$$

*for all $N \leq \ell$.*

We now obtain a different bound which is stronger than Corollary 11.7 in a wide range of values of $k$ and $\ell$.

**Theorem 11.8** *For any $r$-bit prime $p$ and any integer $g$ with $p \nmid g$, we have*

$$
v_k(N) \geq N^{o(1)} \begin{cases}
N^{6/11}(2^k/p)^{1/2} & \text{if } N \leq p^{1/2} \\
N^{7/12}2^{k/2}p^{-13/24} & \text{if } p^{1/2} < N \leq p^{35/68} \\
N^{5/11}2^{k/2}p^{-9/22} & \text{if } p^{35/68} < N \leq p^{13/24} \\
N2^{k/2}p^{-1} & \text{if } p^{13/24} < N \leq p^{2/3} \\
N^{1/4}2^{k/2}p^{-1/4} & \text{if } N > p^{2/3},
\end{cases}
$$

*for all $N \leq \ell$.*

*Proof.* Consider the set $\mathcal{A} = \{u_n : n = 0, \dots, N - 1\}$. Clearly $\#\mathcal{A} = N$ as the first $N \leq \ell$ elements of the sequence $\{u_n\}$ are pairwise distinct.

Since $u_n = 2^k w_n + \xi_n^{(k)}$ for some integer $w_n \in [0, 2^{r-k} - 1]$, $n = 0, 1, \dots$, we see that

$$\#(2\mathcal{A}) \leq v_k^2(N)2^{r-k+1} \tag{11.8}$$

(even if the addition of the elements of $\mathcal{A}$ is considered in $\mathbb{Z}$ without the reduction modulo $p$).

Furthermore, from the definition of the sequence $\{u_n\}$ we see that

$$\mathcal{A}^2 = \{g^{a_1+a_2} : a_1, a_2 \in A\}$$

(where $g^b$ is computed in $\mathbb{F}_p$), thus we also have

$$\#(\mathcal{A}^2) \leq v_k^2(N)2^{r-k+1}. \tag{11.9}$$

Comparing (11.8) and (11.9) with Lemma 11.3, we conclude the proof.  □

In particular, if $N = p^{1/2+o(1)}$ then Theorem 11.8 improves Corollary 11.7 for $k \geq (41/44 + \varepsilon)r$, with arbitrary $\varepsilon > 0$.

### 11.3.3 Frequency of values

We now give an upper bound on the frequency $V_k(\omega)$ of a given $k$-bit string $\omega$ that appears in the full trajectory $\xi_n^{(k)}$, $n = 0, \dots, \ell - 1$.

More precisely, let $\Omega_k(U)$ be the set of $k$-bit strings $\omega$ for which $V_k(\omega) \geq U$.

**Theorem 11.9** *For any $r$-bit prime $p$ and $g$ with $p \nmid g$, we have*

$$
\#\Omega_k(U) \leq U^{-1}p^{o(1)} \begin{cases}
2^{2k/3}p^{1/3} & \text{if } k/r \geq 17/20 \\
2^{k/6}p^{25/24} & \text{if } 17/20 > k/r \geq 13/16 \\
2^{k/2}p^{1/2} & \text{if } 13/16 > k/r \geq 3/5 \\
2^{-k/3}p & \text{if } 3/5 > k/r.
\end{cases}
$$

*Proof.* Consider the pairs

$$(\xi_n^{(k)}, \xi_{n+1}^{(k)}), \qquad \xi_n^{(k)} \in \Omega_k(U), \ n = 0, \ldots, \ell - 1. \tag{11.10}$$

Clearly, there are

$$W = \sum_{\omega \in \Omega_k(U)} V_k(\omega) \geq \#\Omega_k(U)U$$

such pairs.

Since $\xi_{n+1}^{(k)}$ can take at most $2^k$ possible values, we see that at least one pair $(\omega, \sigma)$ of two $k$-bit strings occurs at least $W/2^k$ times amongst the pairs (11.10). Now, the same argument as used in the proof of Theorem 11.4 implies that

$$W/2^k \leq R_{\alpha,\alpha,g,p}(\mathcal{I}, \mathcal{J})$$

for some intervals $\mathcal{I}$ and $\mathcal{J}$ of lengths $2^{r-k}$ each and some integer $\alpha \in \{1, \ldots, p - 1\}$. Using Corollary 11.2 with $H = 2^{r-k}$, we conclude the proof. $\qquad\square$

Examining the value of $U$ for which the bound of Theorem 11.9 implies that $\#\Omega_k(U) < 1$, we derive the following.

**Corollary 11.10** *For any $r$-bit prime $p$ and $g$ with $p \nmid g$, we have*

$$V_k(\omega) \leq p^{o(1)} \begin{cases} 2^{2k/3}p^{1/3} & \text{if } k/r \geq 17/20 \\ 2^{k/6}p^{25/24} & \text{if } 17/20 > k/r \geq 13/16 \\ 2^{k/2}p^{1/2} & \text{if } 13/16 > k/r \geq 3/5 \\ 2^{-k/3}p & \text{if } 3/5 > k/r. \end{cases}$$

## 11.4 Numerical results on cycles in the exponential map

Here we present results of some numerical tests concerning the cycle structure of the permutation on the set $\{1, \ldots, p - 1\}$ generated by the map $x \mapsto g^x \pmod{p}$.

We use $\mathcal{I}_m$ to denote the dyadic interval $\mathcal{I}_m = [2^{m-1}, 2^m - 1]$.

We test 500 pairs $(p, g)$ of primes $p$ and primitive roots $g$ modulo $p$ selected using a pseudorandom number generator separately taken from each of the intervals $p \in \mathcal{I}_{20}$ and $p \in \mathcal{I}_{22}$ and $p \in \mathcal{I}_{25}$.

We also repeat this for 60 pairs $(p, g)$ in the larger range $p \in \mathcal{I}_{30}$.

Let $L_r(N)$ and $C(N)$ be the length of the $r$th longest cycle and the number of disjoint cycles in a random permutation on $N$ symbols, respectively.

We now recall that by the classical result of Shepp and Lloyd [29] the ratio $\lambda_r(N) = L_r(N)/N$ is expected to be

$$\lambda_r(N) = G_r + o(1),$$

as $N \to \infty$, for some constants $G_r$, $r = 1, 2, \ldots$, explicitly given in [29] via some integral expressions. In particular, we find from [29, Table 1] that

$$G_1 = 0.624329\ldots, \quad G_2 = 0.209580\ldots, \quad G_3 = 0.088316\ldots$$

(we note that values reported in [25] deviate slightly from those of [29], but they agree over the approximations given here). Interestingly, the constants $G_r$ also occur when one considers the size (in terms of number of digits) of the $r$th largest prime factor of an integer $n$, see Knuth and Trabb Pardo [25]. For example, de Bruijn [7] has shown that

$$\sum_{n \leq x} \log P(n) = G_1 x \log x + O(x),$$

with $P(n)$ the largest prime factor of $n$, thus establishing a claim by Dickman. The constant $G_1$ is now known as the Golomb–Dickman constant. For further information and references see the book by Finch [14, Section 5.4].

We also recall that Goncharov [20] has shown that the ratio $\gamma(N) = C(N)/\log N$, is expected to be

$$\gamma(N) = 1 + o(1) \qquad \text{as } N \to \infty.$$

The above asymptotic results can also be found in [1, Section 1.1].

In Table 11.1 we present the average value, over the tested primes $p$ in each group, of the lengths of the 1st, 2nd and 3rd longest cycles normalized by dividing by the size of the set, that is, by $p - 1$.

We also calculated the number of cycles for the above pairs $(p, g)$, normalized by dividing by $\log(p - 1)$, and then we present the average value for each of the ranges.

Table 11.1 *Numbers of connected components*

|  | Range and number of pairs $(p, g)$ | | | |
|---|---|---|---|---|
|  | $\mathcal{I}_{20}$ 500 | $\mathcal{I}_{22}$ 500 | $\mathcal{I}_{25}$ 500 | $\mathcal{I}_{30}$ 60 |
| Average $\lambda_1$ | 0.63946789 | 0.61508766 | 0.63157252 | 0.60441217 |
| Average $\lambda_2$ | 0.19999487 | 0.21687612 | 0.20469932 | 0.21715242 |
| Average $\lambda_3$ | 0.08646438 | 0.08450844 | 0.09092497 | 0.09354165 |
| Average $\gamma$ | 1.03813497 | 1.03324650 | 1.03014896 | 1.05566909 |

We note that we have also tried to compare the length of the smallest cycle with the expected length $e^{-\gamma} \log p$ for a random permutation on $\{1, \ldots, p-1\}$, where $\gamma = 0.5772\ldots$ is the Euler–Mascheroni constant. However, the results are inconclusive and require further tests and investigation.

## 11.5 Comments

It is certainly interesting to study similar questions over arbitrary finite fields, although in this case there is no canonical way to interpret field elements as integer numbers and thus to extract bits from field elements. Probably the most interesting and natural case is the case of binary fields $\mathbb{F}_{2^r}$ of $2^r$ elements with a sufficiently large $r$. First, we use the isomorphism $\mathbb{F}_{2^r} = \mathbb{F}_2(\alpha)$, where $\alpha$ is a root of an irreducible polynomial over $\mathbb{F}_2$ of degree $r$. Now we can represent each element of $\mathbb{F}_{2^r}$ as an $r$-dimensional binary vector of coefficients in the basis $1, \alpha, \ldots, \alpha^{r-1}$, and the bit extraction is now apparent. For example, the proof of [18, Theorem 2] can easily be adjusted to give a square-root bound for the number of fixed points (when we identify elements of $\mathbb{F}_{2^r}$ with $r$-dimensional binary vectors). It is also quite likely that using the results and methods of [11] one can obtain some variants of our results in these settings.

Furthermore, for cryptographic applications, it is also interesting to study the relation between $t$ and $\tau_k$ and, in particular, obtain improvements of Corollaries 11.7 and 11.10 for almost all $p$ and almost all initial values $u_0$. It is quite likely that the method of [6], combined with the ideas of [5], can be used to derive such results.

Finally, we note that exponential maps have also been considered modulo prime powers, see [17, 24]. Although many computational problems, such as the discrete logarithm problem, are easier modulo prime powers, the corresponding exponential pseudorandom number generator does not seem to have any immediate weaknesses.

## Acknowledgements

The authors would like to thank Daniel Panario for useful discussions and references and Arne Winterhof for a careful reading of the manuscript.

This work was finished during a very enjoyable internship of the first author and research stay of the third author at the Max Planck Institute for Mathematics, Bonn. The third author was also supported in part by ARC grants DP110100628 and DP130100237.

# References

[1] R. Arratia, A. D. Barbour and S. Tavaré, *Logarithmic Combinatorial Structures: A Probabilistic Approach.* EMS Monographs in Mathematics. European Mathematical Society, Zürich, 2003.

[2] J. Bourgain, N. Katz and T. Tao, A sum product estimate in finite fields and applications. *Geom. Funct. Anal.* **14**, 27–57, 2004.

[3] J. Bourgain, S. V. Konyagin and I. E. Shparlinski, Product sets of rationals, multiplicative translates of subgroups in residue rings and fixed points of the discrete logarithm. *Int. Math. Res. Notices* **2008**, 1–29, 2008. Corrigendum **2009**, 3146–3147, 2009.

[4] J. Bourgain, S. V. Konyagin and I. E. Shparlinski, Distribution of elements of cosets of small subgroups and applications. *Int. Math. Res. Notices* **2012**, 1968–2009, 2012.

[5] J. Bourgain, M. Z. Garaev, S. V. Konyagin and I. E. Shparlinski, On congruences with products of variables from short intervals and applications. *Proc. Steklov Math. Inst.* **280**, 67–96, 2013.

[6] J. Bourgain, M. Z. Garaev, S. V. Konyagin and I. E. Shparlinski, Multiplicative congruences with variables from short intervals. *J. Anal. Math.*, to appear.

[7] N. G. de Bruijn, On the number of positive integers $\leq x$ and free of prime factors $> y$. *Ned. Acad. Wetensch. Proc. Ser. A.* **54**, 50–60, 1951.

[8] B. Bukh and J. Tsimerman, Sum-product estimates for rational functions. *Proc. London Math. Soc.* **104**, 1–26, 2012.

[9] T. H. Chan and I. E. Shparlinski, On the concentration of points on modular hyperbolas and exponential curves. *Acta Arith.* **142**, 59–66, 2010.

[10] J. Cilleruelo and M. Z. Garaev, Concentration of points on two and three dimensional modular hyperbolas and applications. *Geom. Funct. Anal.* **21**, 892–904, 2011.

[11] J. Cilleruelo and I. E. Shparlinski, Concentration of points on curves in finite fields. *Monatsh. Math.* **171**, 315–327, 2013.

[12] C. Cobeli and A. Zaharescu, An exponential congruence with solutions in primitive roots. *Rev. Roum. Math. Pures Appl.* **44**, 15–22, 1999.

[13] R. R. Farashahi, B. Schoenmakers and A. Sidorenko, Efficient pseudorandom generators based on the DDH assumption. *Public Key Cryptography, PKC 2007.* Lecture Notes in Computer Science, volume 4450, pp. 426–441. Springer-Verlag, Berlin, 2007.

[14] S. R. Finch, *Mathematical Constants.* Encyclopedia of Mathematics and its Applications, volume 94. Cambridge University Press, Cambridge, 2003.

[15] J. B. Friedlander and I. E. Shparlinski, On the distribution of the power generator. *Math. Comp.* **70**, 1575–1589, 2001.

[16] R. Gennaro, An improved pseudo-random generator based on discrete logarithm problem. *J. Cryptol.* **18**, 91–110, 2006.

[17] L. Glebsky, Cycles in repeated exponentiation modulo $p^n$. *Integers* **13**, A66, 2013.

[18] L. Glebsky and I. E. Shparlinski, Short cycles in repeated exponentiation modulo a prime. *Des. Codes Cryptogr.* **56**, 35–42, 2010.

[19] O. Goldreich and V. Rosen, On the security of modular exponentiation with application to the construction of pseudorandom generators. *J. Cryptol.* **16**, 71–93, 2003.

[20] V. Goncharov, Du domaine d'analyse combinatoire. *Bull. Acad. Sci. USSR Ser. Mat. (Izv. Akad. Nauk SSSR)*, **8**, 3–48, 1944.

[21] J. Holden, Fixed points and two cycles of the discrete logarithm. *Algorithmic Number Theory*. Lecture Notes in Computer Science, volume 2369, pp. 405–416. Springer, Berlin, 2002.

[22] J. Holden and P. Moree, New conjectures and results for small cycles of the discrete logarithm. *High Primes and Misdemeanours: Lectures in Honour of the 60th Birthday of Hugh Cowie Williams*. Fields Institute Communications, volume 41, pp. 245–254. American Mathematical Society, Providence, RI, 2004.

[23] J. Holden and P. Moree, Some heuristics and and results for small cycles of the discrete logarithm. *Math. Comp.* **75**, 419–449, 2006.

[24] J. Holden and M. M. Robinson, Counting fixed points, two-cycles, and collisions of the discrete exponential functions using *p*-adic methods. *J. Aust. Math. Soc.* **92**, 163–178, 2012.

[25] D. E. Knuth and L. Trabb Pardo, Analysis of a simple factorization algorithm. *Theor. Comput. Sci.* **3**, 321–348, 1976.

[26] J. C. Lagarias, Pseudorandom number generators in cryptography and number theory. *Proc. Symp. Applied Mathematics*, volume 42, pp. 115–143. American Mathematical Society, Providence, RI, 1990.

[27] M. Levin, C. Pomerance and K. Soundararajan, Fixed points for discrete logarithms. *Algorithmic Number Theory*. Lecture Notes in Computer Science, volume 6197, pp. 6–15. Springer-Verlag, Berlin, 2010.

[28] S. Patel and G. S. Sundaram, An efficient discrete log pseudo random generator. *Advances in Cryptography*. Lecture Notes in Computer Science, volume 1462, pp. 35–44. Springer-Verlag, Berlin, 1999.

[29] L. A. Shepp and S. P. Lloyd, Ordered cycle lengths in a random permutation. *Trans. Am. Math. Soc.* **121**, 340–357, 1966.

[30] H. Shi, S. Jiang and Z. Qin, More efficient DDH pseudorandom generators. *Des. Codes Cryptogr.* **55**, 45–64, 2010.

[31] W. P. Zhang, On a problem of Brizolis (Chinese). *Pure Appl. Math.* **11** (suppl. 1–3).

# 12

# Construction of a rank-1 lattice sequence based on primitive polynomials

*Alexander Keller, Nikolaus Binder and Carsten Wächter*
NVIDIA, Berlin

*Dedicated to Harald Niederreiter on the occasion of his 70th birthday.*

## Abstract

A construction of a rank-1 lattice sequence is introduced. By analogy with the Sobol' sequence, its generator vector is constructed from a sequence of primitive polynomials and is extensible in both the number of dimensions and the number of digits of each component of the generator vector. Some initial numerical evidence is provided by applying the rank-1 lattice sequence to high-dimensional light transport simulation.

## 12.1 Introduction

Niederreiter [19] advocated the use of quasi-Monte Carlo methods in computer graphics. Quasi-Monte Carlo methods, as introduced to a large audience in Niederreiter's seminal book [20], can be considered as the deterministic counterpart of Monte Carlo methods. Based on number theory rather than probability theory, quasi-Monte Carlo methods are exactly reproducible and are simple to parallelize. However, the most important benefit of quasi-Monte Carlo methods is that they can outperform Monte Carlo methods in many practical applications. Besides financial mathematics, light transport simulation in computer graphics is one of the prominent applications of quasi-Monte Carlo methods.

As early as [19], Niederreiter suggested the application of rank-1 lattices to image synthesis. While rank-1 lattices can be generated extremely efficiently, the challenge remains to find a suitable set of parameters that determines the lattice. By constructing a rank-1 lattice sequence suitable for light transport simulation, Niederreiter's visionary idea is put into practice and compared to state-of-the-art quasi-Monte Carlo light transport simulation, as surveyed in [11].

## 12.2 Integro-approximation by rank-1 lattice sequences

Image synthesis can be reduced to the form of an integro-approximation problem [11]

$$h(\vec{y}) := \int_{[0,1)^s} f(\vec{x}, \vec{y})\mathrm{d}\vec{x} = \lim_{n\to\infty} \frac{1}{n}\sum_{i=0}^{n-1} f(\vec{x}_i, \vec{y}) \qquad (12.1)$$

that, given the parameter $\vec{y}$, can be computed efficiently by averaging samples of the function $f$ taken at the points

$$\vec{x}_i := \{\phi_b(i)\vec{g}\} \in (\mathbb{Q}\cap[0,1))^s$$

of a rank-1 lattice sequence [1, 3, 5, 7, 8, 9, 17, 18, 21]. The points $\vec{x}_i$ are the fractional parts (as denoted by the $\{\cdot\}$ operator) of a generator vector $\vec{g}$ multiplied by the radical inverse

$$\phi_b : \mathbb{N}_0 \to \mathbb{Q}\cap[0,1)$$
$$i = \sum_{k=0}^{\infty} a_k(i)b^k \mapsto \sum_{k=0}^{\infty} a_k(i)b^{-k-1} \qquad (12.2)$$

of the index $i \in \mathbb{N}_0$. In fact, the radical inverse results from representing the integer $i$ as digits $a_k$ in base $b$ and mirroring that representation at the decimal point.

### 12.2.1 Some properties of rank-1 lattice sequences

For $0 \le i < b^m$ and $m \in \mathbb{N}_0$ the radical inverse $\phi_b(i)$ enumerates all fractions $\frac{i}{b^m}$ in a permuted order. Consequently, $\phi_b(i)b^m$ is an integer permutation and we have the following properties.

(1) The first $b^m$ points $\vec{x}_i$ of a rank-1 lattice sequence form a rank-1 lattice [20].
(2) In addition, for $n' \le n$, the set of the first $n'$ samples is included in the set of the first $n$ samples, which allows for the progressive approximation of Equation (12.1) without recomputing samples or discarding intermediate results.
(3) For an infinite sequence of unique samples, the components

$$g_j =_b \cdots g_{j,3}g_{j,2}g_{j,1}g_{j,0}$$

of the generator vector $\vec{g} := (g_1, g_2, g_3, \ldots)$ need to be $b$-adic integers, which can be considered sequences of digits $g_{j,k} \in \{0, \ldots, b-1\}$ for $k \in \mathbb{N}_0$. Due to the resulting property

$$\sum_{k=0}^{m} g_{j,k} b^k \equiv \sum_{k=0}^{m-1} g_{j,k} b^k \pmod{b^m}, \tag{12.3}$$

the computation of the points $\vec{x}_i$ becomes practical, because for $n < b^m$ only the $m$ least significant digits are relevant to determine the coordinates $x_{i,j}$.

(4) In fact, partitioning a rank-1 lattice sequence into contiguous blocks of $b^m$ points, each block of points is a shifted copy of the first $b^m$ points and forms a (shifted) rank-1 lattice.

(5) For $m$ digits, there are $b^m - 1$ different, nonzero generator vector components $g_j$ modulo $b^m$. Therefore, components cannot be unique modulo $b^m$ unless the dimension $s < b^m$. Closely related, $g_j \bmod b^m$ results in the same points as produced by $b^m - (g_j \bmod b^m)$.

(6) If $\gcd(g_j, b) = 1$, the set of coordinates $x_{i,j}$ consists of the equidistant samples $\frac{i}{b^m}$ for $0 \le i < b^m$. As a consequence, the resulting lattice is a Latin hypercube sample and projection regular [25, Section 3.5]. Without loss of generality, then one of the components $g_j$ may be chosen as 1.

### 12.2.2 Determining good generator vectors

Although it is known that good rank-1 lattice point sequences exist [8] and that such sequences can be computed very efficiently, determining good generator vectors often involves search and optimization.

Property (12.3) lends itself to methods which extend the generator vectors digit by digit. Digits then can be determined to optimize selected figures of merit [7, 17, 18, 21]. Similar to rank-1 lattices, generator vectors can be constructed component by component (CBC) [1, 3] or the search space may be restricted to generator vectors $\vec{g} = (1, a, a^2, a^3, \ldots)$ in Korobov form [5, 9].

All of these approaches are implemented in the free lattice builder software,[1] which allows for the adaptive construction of rank-1 lattices and evaluation by a collection of figures of merit as documented in [14, 15].

### 12.3 Construction

According to [17, Theorem 3.1.3], a rank-1 lattice sequence is uniformly distributed if the components $g_j$ of its generator vector are linearly independent.

[1] See https://github.com/mungerd/latbuilder/blob/master/README.md.

In analogy to the Sobol' sequence [26], which is a $(t, s)$-sequence [20] in base $b = 2$ constructed from linearly independent sequences of bits of maximum period linear feedback shift registers (LFSR) [16], a family of such infinite sequences of bits may be used to determine uniquely the components of the generator vector.

For the construction in base $b = 2$, we first enumerate the sequence $p_j$ of primitive polynomials over $\mathbb{Z}_2$ in their natural order as implied by considering their coefficients as binary numbers (see Listing 12.3 or [24] for more efficient approaches). Then, the sequence of digits $g_{j,k}$ of the $j$th component of the generator vector is generated as the sequence of bits of the linear feedback shift register generator determined by the $j$th primitive polynomial $p_j$ (see Listing 12.1 and http://en.wikipedia.org/wiki/Linear_feedback_shift_register). This basic construction already results in a uniform rank-1 lattice sequence and is extensible in dimension, because the number of primitive polynomials is unbounded.

In order to make the construction unique, we first require all components $g_j$ to be co-prime to the base $b$. For $b = 2$ this constraint is satisfied by setting all least significant bits $g_{j,0} = 1$, resulting in odd components. Similar to the Sobol' sequence, for each dimension $j$ the sequence $x_{i,j}$ then is a $(0, 1)$-sequence in base $b = 2$ [11, Theorem 2].

*Listing 12.1 C++ routine iterating the linear feedback shift register generator specified by the primitive polynomial until the required number of digits has been produced; the returned sequence of bits may be used as one component of a generator vector and is odd if and only if the initial state* lfsr *is odd*

```
unsigned int LFSR_Bit_Sequence(unsigned int Digits,
    unsigned int PrimitivePolynomial, unsigned int lfsr =
    1)
{
    unsigned int Bit_Sequence = 0;

    for(unsigned int i = 0; i < Digits; ++i)
    {
        unsigned int lsb = lfsr & 1;
        lfsr >>= 1;

        if(lsb == 1) // least significant bit lsb set ?
        {
            lfsr ^= PrimitivePolynomial;
            Bit_Sequence ^= 1 << i;
        }
    }

    return Bit_Sequence;
}
```

*Listing 12.2* NVIDIA CUDA *routine for generating the coordinate* $x_{i,j}$ *of a rank-1 lattice sequence given the generator vector* $\vec{g}$*; note that the radical inverse* $\phi_2$ *is computed using the bit reversal instruction* `brevll`

```
float extensible_lattice(unsigned long long i, unsigned
    int j)
{
    unsigned long long result = __brevll(i) * g[j];

    return int_as_float(0x3F800000 | (result >> (9 + 32))
        ) - 1.0f;
}
```

Second, we require the components to fulfill property (5), which for each integer $m$ and $0 \le j < b^m$ requires the values $g_j$ mod $b^m$ to be unique. Similar to the idea of optimizing the Sobol' sequence [10], this constraint can be satisfied by choosing distinct initial states of the linear feedback shift registers. The construction becomes unique by extending the generator vector component by component, where for each new component the linear feedback shift register is iterated until the uniqueness constraint is fulfilled for the first time. A straightforward implementation of the construction is shown in Listing 12.4.

### 12.3.1 Structure of the generator vector

It is instructive to represent the resulting generator vector

as a bitmap, where each column corresponds to one component $g_j$ and black pixels represent a zero. The bottom scanline represents the least significant bits,

*Listing 12.3 Part one of a C++ program enumerating a requested number s of primitive polynomials; primitive polynomials of degree d are identified by checking whether their associated LFSR has the maximum period of $2^d - 1$*

```
int main(int argc, char *argv[])
{
   const unsigned int s = 128;
   const unsigned int Bits = 31;
   unsigned int p[s], LFSR[s], g[s];
   unsigned int Mask = 3;

   unsigned int Polynomial = 1;

   for(unsigned int j = 0; j < s; ++Polynomial)
   {
      // in order to test primitivity, simply check,
         whether
      // the LFSR defined by the polynomial has full
         period

      unsigned int lfsr = 1;
      unsigned int period = 0;

      do
      {
         unsigned int lsb = lfsr & 1;
         lfsr >>= 1;

         if(lsb == 1)
            lfsr ^= Polynomial;

         ++period;
      } while(lfsr != 1);

      unsigned int Degree_of_Polynomial = 0;

      for(unsigned int k = Polynomial; k; k /= 2)
         ++Degree_of_Polynomial;

      if(period + 1 == 1 << Degree_of_Polynomial)
      {
         p[j] = Polynomial; // now is a primitive
            polynomial
```

which are all one (light gray pixels), because all components are odd. The leftmost column of ones results from the first primitive polynomial. The second line of the equation abstracts the structure of the bitmap, where the nested rectangles indicate the regions of components which are unique modulo $b^m$. This actually can be verified by looking at the bitmap.

*Listing 12.4 Part two of a C++ program, iterating the linear feedback shift register generators to comply with property (5)*

```
            LFSR[j] = 1;
            g[j] = LFSR_Bit_Sequence(Bits, p[j], LFSR[j]);

            // verify property 5 (uniqueness modulo $b^m$)

            bool ok;

            do
            {
                ok = true;

                for(unsigned int k = 0; k < j; ++k)
                    if(((g[k] & Mask) == (g[j] & Mask))
                       || ((g[j] & 1) == 0))
                    {
                        ok = false;

                        // iterate LFSR

                        unsigned int lsb = LFSR[j] & 1;
                        LFSR[j] >>= 1;

                        if(lsb == 1)
                            LFSR[j] ^= p[j];

                        g[j] = LFSR_Bit_Sequence(Bits, p[j],
                            LFSR[j]);

                        break;
                    }
            } while(! ok);

            // component g[j] now fulfills conditions

            if(Mask == 2 * j + 1)
                Mask = (Mask << 1) | 1;

            ++j;
        }
    }

    // use generator vector g[]

    return 0;
}
```

## 12.4  Applications

In a way, rank-1 lattices can connect the world of both quasi-Monte Carlo and Monte Carlo methods, which is best illustrated by a generator vector in Korobov form. While the whole rank-1 lattice may be of low discrepancy, the generator vector $\vec{g}$ is determined by a multiplicative congruential generator as used for computing a stream of pseudorandom numbers.

The construction of rank-1 lattice sequences as introduced in the previous section is similar in the sense that linear feedback shift register generators have often been used for generating sequences of pseudorandom bits. It also resembles the construction of the highly efficient Mersenne twister pseudorandom number generator [23]. With a suitable selection of primitive polynomials, the new construction allows for running of Monte Carlo Markov chains along the dimensions of each point of the rank-1 lattice sequence.

### 12.4.1  Efficient sampling

Compared to low-discrepancy sequences based on generator matrices, like for example the popular Sobol' sequence [26], the generation of rank-1 lattice sequences is more efficient. A highly tuned sequential SIMD implementation of the Sobol' sequence is already 30% slower as compared to a simple rank-1 lattice sequence implementation for the exact same number of samples. Taking into account that the radical inverse needs to be computed only once per rank-1 lattice sequence point for all $s$ dimensions, the Sobol' sequence is 60% slower. A concrete measurement on massively parallel graphics processing units (GPUs) using an implementation as in Listing 12.2 is shown in Table 12.1. Besides computational simplicity, accessing only one component of the generator vector per coordinate, the rank-1 lattice sequences have a clear advantage in terms of memory bandwidth.

Note that rank-1 lattice sequences can easily be partitioned into a multiple of sequences [12], which allows for simple parallelization. The most efficient

Table 12.1 *For the same number of $2^{31}$ samples in 256 dimensions, rank-1 lattice sequences dramatically outperform the Sobol' sequence with respect to generation speed on massively parallel graphics processing units (GPUs)*

| GPU | Sobol' sequence | Rank-1 lattice sequence | Speedup |
| --- | --- | --- | --- |
| NVIDIA Tesla K40 | 108162.26 ms | 5366.55 ms | 20.15 × |
| NVIDIA GTX480 | 187194.16 ms | 6291.79 ms | 29.75 × |

methods of randomization [22] are a Cranley–Patterson rotation [2] or even a simple `xor`-scrambling [4, 13].

## 12.4.2  Light transport simulation

The new construction has been tested in the context of light transport simulation. As described in [11, Section 4.4], a hybrid sequence is used for image synthesis, where the first two dimensions of the Sobol' sequence are used to sample the image plane [6] and then light transport paths are determined by the rank-1 lattice sequence [11, Section 3.1.5]. The simulations in Figure 12.1 are based on a state-of-the-art bidirectional path tracer using multiple importance sampling, and demonstrate that the new construction

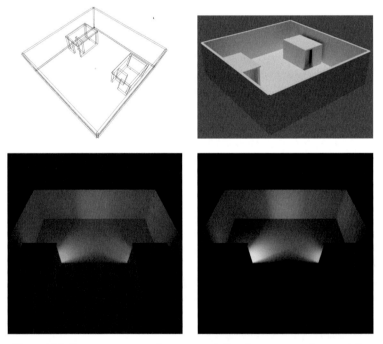

Figure 12.1 Convergence test using a more challenging test scene (top row) for light transport simulation. Inside each of the two symmetric rooms one polyhedral light source illuminates the inner court through the slit between the blockers in the door frame. The two images in the bottom row have been rendered using the same number of $2^{18}$ samples per pixel, where the left image used the Sobol' sequence, while the right image used the new rank-1 lattice construction to sample light transport paths. The images cannot be distinguished with respect to quality, however, sampling using the rank-1 lattice sequence algorithm is simpler and much more efficient. (See color plate.)

achieves at least the same quality as simulations based on the Sobol' sequence.

The new construction has also been tested in NVIDIA's iray® path tracing based GPU renderer. In this light transport simulation process, the lower dimensions are more important and therefore property (5) efficiently reduces transient correlation effects that are visible using rank-1 lattice sequences that do not comply with the uniqueness constraints of the construction described in Section 12.3. A visual comparison with the original Sobol' sequence is shown in Figure 12.2 for a relatively small number of samples. So far, we have not

Figure 12.2 A visual comparison of the new methodology (top) and the Sobol' sequence (bottom) in NVIDIA iray® . While the Sobol' sequence exposes the typical structured artifacts in the form of rectangular patterns (for example on the open book), the rank-1 lattice sequence exposes more noise (for example in the highlight on the glass of the petrol lamp). Both images have been rendered using only 64 path space samples per pixel. (See color plate.)

explored how different mappings of the components of the generator vector to dimensions of the simulation algorithm affect image quality.

## 12.5 Conclusion

In the domain of light transport simulation the new construction of a rank-1 lattice sequence based on primitive polynomials can compete with sampling methods based on generator matrices. For the same number of samples, the new rank-1 lattice sequence achieves an image quality comparable with that resulting from sampling with the Sobol' sequence. However, memory bandwidth, register pressure, and performance of sampling are much more favorable as compared to the Sobol' sequence.

Whether the new construction is of low discrepancy is an open question. First indicators need to be explored by measuring figures of merit, for example as provided by the lattice builder tool [14, 15]. Further numerical experiments include the application to Genz's test functions [25] and financial mathematics.

## References

[1]  R. Cools, F. Kuo and D. Nuyens, Constructing embedded lattice rules for multivariate integration. *SIAM J. Sci. Comput.* **28**, 2162–2188, 2006.

[2]  R. Cranley and T. Patterson, Randomization of number theoretic methods for multiple integration. *SIAM J. Numer. Anal.* **13**, 904–914, 1976.

[3]  J. Dick, F. Pillichshammer and B. Waterhouse, The construction of good extensible rank-1 lattices. *Math. Comp.* **77**, 2345–2373, 2008.

[4]  I. Friedel and A. Keller, Fast generation of randomized low-discrepancy point sets. In: H. Niederreiter, K. Fang and F. Hickernell (eds.), *Monte Carlo and Quasi-Monte Carlo Methods 2000*, pp. 257–273. Springer, 2002.

[5]  H. Gill and C. Lemieux, Searching for extensible Korobov rules. *J. Complexity* **23**, 603–613, 2007.

[6]  L. Grünschloß, M. Raab and A. Keller, Enumerating quasi-Monte Carlo point sequences in elementary intervals. In: L. Plaskota and H. Woźniakowski (eds.), *Monte Carlo and Quasi-Monte Carlo Methods 2010*, pp. 399–408. Springer, 2012.

[7]  F. Hickernell and H. Hong, Computing multivariate normal probabilities using rank-1 lattice sequences. In: G. Golub, S. Lui, F. Luk and R. Plemmons (eds.), *Proceedings of the Workshop on Scientific Computing in Hong Kong*, pp. 209–215. Springer Verlag, Singapore, 1997.

[8]  F. Hickernell and H. Niederreiter, The existence of good extensible rank-1 lattices. *J. Complexity* **19**, 286–300, 2003.

[9]  F. Hickernell, H. Hong, P. L'Ecuyer and C. Lemieux, Extensible lattice sequences for quasi-Monte Carlo quadrature. *SIAM J. Sci. Comput.* **22**, 1117–1138, 2001.

[10] S. Joe and F. Kuo, Constructing Sobol' sequences with better two-dimensional projections. *SIAM J. Sci. Comput.* **30**(5), 2635–2654, 2008.

[11] A. Keller, Quasi-Monte Carlo image synthesis in a nutshell. In: J. Dick, F. Kuo, G. Peters and I. Sloan (eds.), *Monte Carlo and Quasi-Monte Carlo Methods 2012*, pp. 203–238. Springer, Heidelberg, 2013.

[12] A. Keller and L. Grünschloß, Parallel quasi-Monte Carlo integration by partitioning low discrepancy sequences. In: L. Plaskota and H. Woźniakowski (eds.), *Monte Carlo and Quasi-Monte Carlo Methods 2010*, pp. 487–498. Springer, 2012.

[13] T. Kollig and A. Keller, Efficient multidimensional sampling (Proc. Eurographics 2002). *Comput. Graphics Forum* **21**(3), 557–563, 2002.

[14] P. L'Ecuyer and D. Munger, Constructing adapted lattice rules using problem-dependent criteria. *Proceedings of the 2012 Winter Simulation Conference*, pp. 373–384. IEEE Press, New York, 2012.

[15] P. L'Ecuyer and D. Munger, Latticebuilder: a general software tool for constructing rank-1 lattice rules. *ACM Trans. Math. Software*, submitted 2012.

[16] R. Lidl and H. Niederreiter, *Introduction to Finite Fields and their Applications*. Cambridge University Press, Cambridge, 1986.

[17] E. Maize, *Contributions to the theory of error reduction in quasi-Monte Carlo methods*. PhD Thesis, Claremont Graduate School, 1980.

[18] E. Maize, J. Sepikas and J. Spanier, Accelerating the convergence of lattice methods by importance sampling-based transformations. In: L. Plaskota and H. Woźniakowski (eds.), *Monte Carlo and Quasi-Monte Carlo Methods 2010*, pp. 557–572. Springer, 2012.

[19] H. Niederreiter, Quasirandom sampling in computer graphics. *Proceedings 3rd International Seminar on Digital Image Processing in Medicine, Remote Sensing and Visualization of Information (Riga, Latvia)*, pp. 29–34, 1992.

[20] H. Niederreiter, *Random Number Generation and Quasi-Monte Carlo Methods*. SIAM, Philadelphia, PA, 1992.

[21] H. Niederreiter and F. Pillichshammer, Construction algorithms for good extensible lattice rules. *Construct. Approx* **30**, 361–393, 2009.

[22] A. Owen, Monte Carlo extension of quasi-Monte Carlo. *Proceedings of the 1998 Winter Simulation Conference*, pp. 571–577. IEEE Press, New York, 1998.

[23] M. Saito and M. Matsumoto, SIMD-oriented fast Mersenne twister: a 128-bit pseudorandom number generator. In: A. Keller, S. Heinrich and H. Niederreiter (eds.), *Monte Carlo and Quasi-Monte Carlo Methods 2006*, pp. 607–622. Springer, 2007.

[24] N. Saxena and E. McCluskey, Primitive polynomial generation algorithms implementation and performance analysis. Technical Report, Stanford University, Center for Reliable Computing (CRC), TR 04-03, 2004.

[25] I. Sloan and S. Joe, *Lattice Methods for Multiple Integration*. Clarendon Press, Oxford, 1994.

[26] I. Sobol', On the Distribution of points in a cube and the approximate evaluation of integrals. *Zh. Vychisl. Mat. Mat. Fiz.* **7**(4), 784–802, 1967.

# 13

# A quasi-Monte Carlo method for the coagulation equation

*Christian Lécot*
Université de Savoie, Le Bourget-du-Lac

*Ali Tarhini*
Université Libanaise, Nabatieh

*Dedicated to Harald Niederreiter on the occasion of his 70th birthday.*

## Abstract

We propose a quasi-Monte Carlo algorithm for the simulation of the continuous coagulation equation. The mass distribution is approximated by a finite number $N$ of numerical particles. Time is discretized and quasi-random points are used at every time step to determine whether each particle is undergoing a coagulation. Convergence of the scheme is proved when $N$ goes to infinity, if the particles are relabeled according to their increasing mass at each time step. Numerical tests show that the computed solutions are in good agreement with analytical solutions, when available. Moreover, the error of the QMC algorithm is smaller than the error given by a standard Monte Carlo scheme using the same time step and number $N$ of numerical particles.

## 13.1  Introduction

Coagulation models have applications in many domains of science, technology and engineering: aerosol dynamics, nanoparticle generation, crystallization, precipitation, granulation, polymerization, combustion processes, food processes, pollutant formation in flames, microbial systems. The survey paper [1] lists fields of applications. In his seminal work [40, 41], M. von Smoluchowski determined the rate of coagulation of particles in a static dispersing medium where only binary collisions occur. The changes in the number of particles of different sizes were described by an infinite system of nonlinear ordinary differential equations. H. Müller was the first to rewrite these equations in terms of an integrodifferential equation for the time evolution of the particle size density function [29]. This method gives us the coagulation equation for the density $c(x, t)$ of particles of size (or mass) $x$ at times $t$:

$$\frac{\partial c}{\partial t}(x, t) = \frac{1}{2} \int_0^x K(x - y, y)c(x - y, t)c(y, t)dy$$
$$- \int_0^{+\infty} K(x, y)c(x, t)c(y, t)dy, \quad x > 0, \ t > 0, \quad (13.1)$$

with the initial condition $c(x, 0) = c_0(x)$, which is a given nonnegative function. Here $K(x, y)$ is the coagulation kernel describing the rate (or the probability) of formation of a particle of size $x + y$ by coagulation of two particles of size $x$ and $y$. The function is assumed to be nonnegative and symmetric. The integrals on the right hand side of (13.1) represent, respectively, (a) production of particles of size $x$ by coagulation of particles of size $x - y$ and $y < x$, and (b) disappearance of particles of size $x$, due to their coagulation with particles of size $y > 0$. For an integer $m \geq 0$, the order $m$ moment is defined by:

$$M_m(t) := \int_0^{+\infty} x^m c(x, t)dx, \quad t \geq 0.$$

The total number of particles $M_0(t)$ can decrease by coagulation, while the total mass of particles $M_1(t)$ remains constant if there is no gelation (formation of particles of infinite size by coagulation, see [1]).

Theorems concerning the existence, uniqueness, boundedness and positiveness of the solutions of the coagulation equation have been established [5, 7, 27]. Exact solutions to the coagulation equation have also been derived [5, 26, 36, 37, 38]; they are valid for specific kernels and initial data. So it is a key issue to obtain accurate numerical solutions. Many algorithms have been devised for the numerical approximation of the coagulation equation. We refer to [4, 10, 13, 15, 16, 25] for deterministic strategies; comparisons of various approaches are presented in [17, 42]. Following the article by D. T. Gillepsie [11], several Monte Carlo (MC) methods were developed concurrently. In [8] the density was approximated using a particle system with variable particle number. In contrast to this *direct simulation* scheme, where the particles represent the number density, a *mass flow* scheme was first introduced in [2] (for the Smoluchowski equation): in this case, the particles represent the mass density so that the number of particles is kept constant throughout the simulation. A mass flow scheme for the coagulation equation was presented in [9]. Another *constant number* MC method was proposed in [39]. The series of papers [45, 46, 47, 48] introduced another dichotomy: *time-driven* versus *event-driven* MC. In time-driven simulations a time step is chosen, then all possible events are implemented within that step; in event-driven MC, first an event is selected to occur, and time is advanced by an appropriate increment (no explicit time step is needed). It was found [48] that event-driven methods generally provide better accuracy than time-driven methods; nevertheless, time-driven algorithms are more suitable in cases where the coagulation equation is to be solved within a larger process simulator that performs explicit integration in time. Among recent articles, Zou *et al.* [49] proposed a coupling between deterministic and stochastic approaches and Wei and Kruis [43] introduced an implementation of Monte Carlo simulation of coagulation using graphic processing units.

The aim of the present work is to analyze a new time-driven and constant number MC method with improved accuracy. This is attained by replacing the pseudorandom numbers used in the simulation by quasi-random numbers, which are deterministic points having a *low discrepancy*. We refer to the monographs [3, 6, 14, 33] for basic notations and concepts of low-discrepancy point sets. We first denote $I := [0, 1)$. Let $s \geq 1$ be a fixed dimension and denote by $\lambda_s$ the $s$-dimensional Lebesgue measure. For a set $U = \{\mathbf{u}_0, \ldots, \mathbf{u}_{N-1}\}$ of $N$ points in the $s$-dimensional unit cube $I^s$ and for a Borel set $B \subset I^s$ we define the *local discrepancy* by

$$D_N(B, U) := \frac{1}{N} \sum_{0 \leq k < N} 1_B(\mathbf{u}_k) - \lambda_s(B),$$

where $1_B$ denotes the indicator function of $B$. The *discrepancy* of $U$ is defined as $D_N(U) := \sup_J |D_N(J, U)|$, the supremum being taken over all subintervals $J \subset I^s$. The *star discrepancy* of $U$ is $D_N^\star(U) := \sup_{J^\star} |D_N(J^\star, U)|$, where $J^\star$ runs through all subintervals of $I^s$ with a vertex at the origin.

Quasi-Monte Carlo (QMC) methods are deterministic versions of MC methods, where the random samples are replaced by low-discrepancy point sets. In the context of numerical integration, the QMC method outperforms the MC method for many types of integrals [28]. Surveys on QMC schemes are presented in [30, 33]. The most powerful current methods of constructing low-discrepancy point sets are based on the theory of $(t, m, s)$-nets and $(t, s)$-sequences. This theory was developed by H. Niederreiter [31, 32] and surveys of the constructions are presented in [18, 34, 35]. For an integer $b \geq 2$, an *elementary interval in base $b$* is an interval $\prod_{i=1}^{s}[a_i b^{-d_i}, (a_i + 1)b^{-d_i})$, with integers $d_i \geq 0$ and $0 \leq a_i < b^{d_i}$ for $1 \leq i \leq s$. If $0 \leq t \leq m$ are integers, a $(t, m, s)$-net in base $b$ is a point set $U$ consisting of $b^m$ points in $I^s$ such that $D_N(J, U) = 0$ for every elementary interval $J$ in base $b$ with measure $b^{t-m}$. If $b \geq 2$ and $t \geq 0$ are integers, a sequence $\mathbf{u}_0, \mathbf{u}_1, \ldots$ of points in $I^s$ is a

$(t,s)$-*sequence in base $b$* if, for all integers $n \geq 0$ and $m > t$, the points $\mathbf{u}_p$ with $nb^m \leq p < (n+1)b^m$ form a $(t,m,s)$-net in base $b$.

There are QMC methods not only for numerical integration, but also for various other numerical problems. For many MC schemes, it is possible to develop corresponding QMC algorithms. We have proposed QMC methods for the Boltzmann equation [20], for Smoluchowski's coagulation equation [23] and for the general aerosol dynamic equation [22]. In these approaches, it was seen that it is convenient to take special measures in order to benefit from the greater uniformity of quasi-random points: at every time step, the simulation particles are renumbered according to position or to size. A QMC method for the numerical simulation of the continuous coagulation equation was proposed in [24], where a convergence theorem is stated without a proof; the proof as well as detailed analysis is given here.

The remainder of this paper is organized as follows. In Section 13.2, the QMC algorithm is presented. In Section 13.3 we establish a convergence result for the method. Section 13.4 is devoted to computational experiments. We finish the paper with a short conclusion.

## 13.2 The quasi-Monte Carlo algorithm

Let $M_1 := M_1(0)$; by multiplying Equation (13.1) by $x/M_1$ and introducing the mass density function $f(x,t) := xc(x,t)/M_1$, we obtain the *mass flow equation*: for $x > 0$, $t > 0$,

$$\frac{\partial f}{\partial t}(x,t) = \int_0^x \widetilde{K}(x-y,y)f(x-y,t)f(y,t)\mathrm{d}y$$
$$- \int_0^{+\infty} \widetilde{K}(x,y)f(x,t)f(y,t)\mathrm{d}y, \qquad (13.2)$$

where $\widetilde{K}$ is the modified coagulation kernel, $\widetilde{K}(x,y) := M_1 K(x,y)/y$. We denote by $f_0(x) := xc_0(x)/M_1$ the initial data. By integrating Equation (13.2) between 0 and $+\infty$, we obtain:

$$\frac{\mathrm{d}}{\mathrm{d}t}\int_0^{+\infty} f(x,t)\mathrm{d}x = 0.$$

This equation means that the total mass of particles in the system is conserved. Since

$$\int_0^{+\infty} f(x,t)\mathrm{d}x = \int_0^{+\infty} f_0(x)\mathrm{d}x = 1, \qquad (13.3)$$

$f(x, t)$ is a probability density function. We want to introduce a weak formulation of Equation (13.2), so we define a set of test functions. Let us denote $\mathbb{R}_+^* := (0, +\infty)$; a function $\sigma : \mathbb{R}_+^* \to \mathbb{R}_+$ is said to be *simple* if its image is a finite point set of $\mathbb{R}_+$; we denote by $\mathcal{S}(\mathbb{R}_+^*)$ the set of all measurable simple functions on $\mathbb{R}_+^*$. By multiplying Equation (13.2) by a simple function $\sigma \in \mathcal{S}(\mathbb{R}_+^*)$ and by using integration over $\mathbb{R}_+^*$, we obtain a weak formulation of the mass flow equation:

$$\frac{\mathrm{d}}{\mathrm{d}t} \int_0^{+\infty} f(x, t)\sigma(x)\mathrm{d}x =$$
$$\int_0^{+\infty} \int_0^{+\infty} \widetilde{K}(x, y) f(x, t) f(y, t)(\sigma(x + y) - \sigma(x))\mathrm{d}y\mathrm{d}x. \qquad (13.4)$$

We suppose that $\widetilde{K}$ is bounded and we set $\widetilde{K}^\infty := \sup_{x,y>0} \widetilde{K}(x, y)$. In the remainder of the paper, we use the following notation: if $1 \le r < s$ are integers, $\pi_r^s$ denotes the projection defined by $\pi_r^s(x_1, \ldots, x_s) := (x_1, \ldots, x_r)$. For $b \ge 2$ and $m \ge 1$, we put $N := b^m$: this is the number of numerical particles used for the simulation. We need a low-discrepancy sequence for the time evolution: $U = \{\mathbf{u}_0, \mathbf{u}_1, \ldots\} \subset I^3$. For $n \in \mathbb{N}$, we write $U^n := \{\mathbf{u}_p : nN \le p < (n + 1)N\}$. We make the following assumptions:

(H1)  $U$ is a $(t, 3)$-sequence in base $b$ (for some $t \ge 0$),
(H2)  $\pi_2^3(U^n)$ is a $(0, m, 2)$-net in base $b$.

### 13.2.1  Initialization

We choose a set $X^0 = \{x_0^0, \ldots, x_{N-1}^0\} \subset \mathbb{R}_+^*$ of $N$ particles such that the initial mass probability $f_0(x)\mathrm{d}x$ is approximated by the probability distribution:

$$f^0(x) := \frac{1}{N} \sum_{0 \le k < N} \delta(x - x_k^0),$$

where $\delta(x - \xi)$ is the Dirac delta measure concentrated at a point $\xi \in \mathbb{R}_+^*$. This reverts to generating $N$ samples from the density function $f_0$, which can be done by the *inversion method*:

$$x_k^0 := F_0^{-1}\left(\frac{2k + 1}{2N}\right), \quad 0 \le k < N,$$

where $F_0$ is the cumulative distribution function, $F_0(x) := \int_0^x f_0(y)\mathrm{d}y$.

### 13.2.2 Time discretization

A fixed time step $\Delta t$ is chosen such that $\Delta t \widetilde{K}^{\infty} < 1$. We set $t_n := n\Delta t$ and we denote $f_n(x) := f(x, t_n)$: it follows from Equation (13.3) that $f_n$ is a probability density function. We suppose that a set $X^n = \{x_0^n, \ldots, x_{N-1}^n\} \subset \mathbb{R}_+^*$ of $N$ particles has been computed so that

$$f^n(x) := \frac{1}{N} \sum_{0 \leq k < N} \delta(x - x_k^n)$$

approximates (in a certain sense to be made precise below, see Section 13.3) the exact mass probability $f_n(x)dx$. Then the approximation of the solution at time $t_{n+1}$ is calculated as follows.

**1. Renumbering the particles** Particles are relabeled at the beginning of the time step by increasing mass:

$$x_0^n \leq x_1^n \leq \cdots \leq x_{N-1}^n. \tag{13.5}$$

This type of sorting was first introduced in [19]. It guarantees theoretical convergence: since the process can be described by a series of numerical integration, the sorting reverts to minimizing the amplitude of the jumps of the function to be integrated.

**2. Coagulation** We define an auxiliary probability measure $g^{n+1}$ using an explicit Euler discretization scheme in time for Equation (13.4):

$$\frac{1}{\Delta t} \left( \int_0^{+\infty} g^{n+1}(x)\sigma(x) - \int_0^{+\infty} f^n(x)\sigma(x) \right) =$$
$$\int_0^{+\infty} \int_0^{+\infty} \widetilde{K}(x, y) f^n(x) f^n(y)(\sigma(x+y) - \sigma(x)), \tag{13.6}$$

that is, replacing $f^n(x)$ with its expression,

$$\int_0^{+\infty} g^{n+1}(x)\sigma(x) = \frac{1}{N} \sum_{0 \leq k < N} \left( 1 - \frac{\Delta t}{N} \sum_{0 \leq \ell < N} \widetilde{K}(x_k^n, x_\ell^n) \right) \sigma(x_k^n)$$
$$+ \frac{\Delta t}{N} \sum_{0 \leq k < N} \left( \frac{1}{N} \sum_{0 \leq \ell < N} \widetilde{K}(x_k^n, x_\ell^n)\sigma(x_k^n + x_\ell^n) \right). \tag{13.7}$$

The measure $g^{n+1}$ certainly approximates $f_{n+1}(x)dx$, but it is not a sum of Dirac delta measures, like $f^n$. We recover this kind of approximation if we use a QMC quadrature rule. Let $R_{k,\ell} := [k/N, (k+1)/N) \times [\ell/N, (\ell+1)/N) \subset I^2$ be an elementary interval in base $b$. We denote by $1_{R_{k,\ell}}$ its indicator function. Let $\chi_{k,\ell}^n$ be the indicator function of the interval $I_{k,\ell}^n := [0, \Delta t \widetilde{K}(x_k^n, x_\ell^n))$. Corresponding to $\sigma \in \mathcal{S}(\mathbb{R}_+^*)$ we have the indicator function:

$$C_\sigma^{n+1}(\mathbf{u}) := \sum_{0 \leq k, \ell < N} 1_{R_{k,\ell}}(u_1, u_2) \left( \left(1 - \chi_{k,\ell}^n(u_3)\right) \sigma(x_k^n) \right.$$
$$\left. + \chi_{k,\ell}^n(u_3) \sigma(x_k^n + x_\ell^n) \right)$$

(for $\mathbf{u} = (u_1, u_2, u_3) \in I^3$), such that

$$\int_0^{+\infty} g^{n+1}(x) \sigma(x) = \int_{I^3} C_\sigma^{n+1}(\mathbf{u}) d\mathbf{u}. \tag{13.8}$$

We determine $f^{n+1}$ by performing a QMC quadrature in $I^3$:

$$\int_0^{+\infty} f^{n+1}(x) \sigma(x) = \frac{1}{N} \sum_{nN \leq p < (n+1)N} C_\sigma^{n+1}(\mathbf{u}_p), \quad \sigma \in \mathcal{S}(\mathbb{R}_+^*).$$

It is possible to summarize the calculation on a time step as follows. If $u \in [0, 1)$, let $k(u) := \lfloor Nu \rfloor$. Then, for every $p$ with $nN \leq p < (n+1)N$, we have:

$$x_{k(u_{p,1})}^{n+1} = \begin{cases} x_{k(u_{p,1})}^n + x_{k(u_{p,2})}^n & \text{if } u_{p,3} < \Delta t \widetilde{K}(x_{k(u_{p,1})}^n, x_{k(u_{p,2})}^n) \\ x_{k(u_{p,1})}^n & \text{otherwise.} \end{cases} \tag{13.9}$$

A rewording of the step from time $t_n$ to time $t_{n+1}$ is as follows. For every $p$, the numbers $u_{p,1}$ and $u_{p,2}$ select particles; the particle $k(u_{p,1})$, with size $x_{k(u_{p,1})}^n$, has for coagulation partner the particle $k(u_{p,2})$, with size $x_{k(u_{p,2})}^n$, and the coagulation probability is $p_c := \Delta t \widetilde{K}(x_{k(u_{p,1})}^n, x_{k(u_{p,2})}^n)$. Then $u_{p,3}$ is used to select an event: (a) if $0 \leq u_{p,3} < p_c$, particle $k(u_{p,1})$ performs coagulation with particle $k(u_{p,2})$ and they give birth to a particle of size $x_{k(u_{p,1})}^n + x_{k(u_{p,2})}^n$, and (b) if $p_c \leq u_{p,3} < 1$, no coagulation occurs.

## 13.3 Convergence analysis

In this section, we prove a convergence result for the QMC scheme previously described. In order to do that, we need to adapt the basic tools of QMC methods to the new algorithm.

Let $f$ be a probability density function on $\mathbb{R}_+^*$. If $z > 0$, we denote by $\sigma_z$ the indicator function of $(0, z)$. We define the *local discrepancy* of the set $X = \{x_k : 0 \leq k < N\} \subset \mathbb{R}_+^*$ relative to $f$ by:

$$D_N(z, X; f) := \frac{1}{N} \sum_{0 \le k < N} \sigma_z(x_k) - \int_0^{+\infty} \sigma_z(x) f(x) \mathrm{d}x.$$

The *star discrepancy* of $X$ relative to $f$ is defined as:

$$D_N^\star(X; f) := \sup_{z > 0} |D_N(z, X; f)|.$$

The *error* of the present QMC scheme at time $t_n$ is defined to be the star discrepancy of $X^n$ relative to $f_n$. The concept of variation of function in the sense of Hardy and Krause can be extended to a function $\phi$ defined on $\mathbb{R}_+^{*s}$ and is denoted by $V(\phi)$. A fundamental tool for the error analysis of QMC methods is the generalized Koksma–Hlawka inequality. This inequality can be proved using arguments similar to those of S. K. Zaremba [44].

**Proposition 13.1** *Let $f$ be a probability density function over $\mathbb{R}_+^*$. If $\phi$ has bounded variation $V(\phi)$ on $\mathbb{R}_+^*$, then for any set $X = \{x_k : 0 \le k < N\} \subset \mathbb{R}_+^*$, we have:*

$$\left| \frac{1}{N} \sum_{0 \le k < N} \phi(x_k) - \int_0^{+\infty} \phi(x) f(x) \mathrm{d}x \right| \le V(\phi) D_N^\star(X; f).$$

The following result is an adaptation of Lemma 1 of [21] and is similarly established.

**Lemma 13.2** *Let $\phi : \mathbb{R}_+^{*s} \to \mathbb{R}$ be a function of bounded variation $V(\phi)$ in the sense of Hardy and Krause. For $1 \le i \le s$, let $0 < z_{0,i} < z_{1,i} < \cdots < z_{n_i,i}$. For $\mathbf{m} = (m_1, \cdots, m_s) \in \mathbb{N}^s$ such that $\forall i \ 0 \le m_i < n_i$, let $I_{\mathbf{m}} := \prod_{i=1}^s [z_{m_i,i}, z_{m_i+1,i}]$ and $\mathbf{x_m}, \mathbf{y_m} \in I_{\mathbf{m}}$. Then*

$$\sum_{\mathbf{m} < \mathbf{n}} |\phi(\mathbf{x_m}) - \phi(\mathbf{y_m})| \le V(\phi) \prod_{i=1}^s n_i \sum_{j=1}^s \frac{1}{n_j}.$$

As written before, we analyze the accuracy of the QMC method by using the local discrepancy $D_N(z, X^n; f_n)$ of the set $X^n = \{x_0^n, \dots, x_{N-1}^n\}$ relative to the exact solution $f_n$ at time $t_n$. We finally prove an upper bound for the star discrepancy $D_N^\star(X^n; f_n)$ of $X^n$ relative to $f_n$. We introduce the following intermediate terms.

- The *local truncation error*:

$$\varepsilon_z^n := \frac{1}{\Delta t} \int_0^{+\infty} (f_{n+1}(x) - f_n(x)) \sigma_z(x) \mathrm{d}x$$
$$- \int_0^{+\infty} \int_0^{+\infty} \widetilde{K}(x, y) f_n(x) f_n(y) (\sigma_z(x + y) - \sigma_z(x)) \mathrm{d}y \mathrm{d}x.$$

- The *additional error*:

$$e_z^n := \int_0^{+\infty} \int_0^{+\infty} \widetilde{K}(x, y) f^n(x) f^n(y)(\sigma_z(x + y) - \sigma_z(x))$$

$$- \int_0^{+\infty} \int_0^{+\infty} \widetilde{K}(x, y) f_n(x) f_n(y)(\sigma_z(x + y) - \sigma_z(x)) \mathrm{d}y\mathrm{d}x.$$

- The *QMC integration error*:

$$d_z^n := \frac{1}{N} \sum_{nN \le p < (n+1)N} C_{\sigma_z}^{n+1}(\mathbf{u}_p) - \int_{I^3} C_{\sigma_z}^{n+1}(\mathbf{u})\mathrm{d}\mathbf{u}.$$

We have the recurrence formula:

$$D_N(z, X^{n+1}; f_{n+1}) = D_N(z, X^n; f_n) - \Delta t \varepsilon_z^n + \Delta t e_z^n + d_z^n. \qquad (13.10)$$

The local truncation error is bounded as follows.

**Lemma 13.3** *We suppose (a) for every $x > 0$, the function $t \to f(x, t)$ is twice continuously differentiable over $(0, T)$ and (b) $f, \frac{\partial f}{\partial t}, \frac{\partial^2 f}{\partial t^2}$ are integrable over $\mathbb{R}_+^* \times (0, T)$. Then, for $t_{n+1} \le T$,*

$$|\varepsilon_z^n| \le \int_0^{+\infty} \int_{t_n}^{t_{n+1}} \left| \frac{\partial^2 f}{\partial t^2}(x, t) \right| \mathrm{d}t\mathrm{d}x.$$

For the additional error, we have the following bound.

**Lemma 13.4** *We suppose (a) for every $y > 0$, the function $\widetilde{K}(\cdot, y) :$ $x \in (0, +\infty) \mapsto \widetilde{K}(x, y)$ is of bounded variation $V(\widetilde{K}(\cdot, y))$ and $\sup_{y>0} V(\widetilde{K}(\cdot, y)) < +\infty$, and (b) for every $x > 0$, the function $\widetilde{K}(x, \cdot) :$ $y \in (0, +\infty) \mapsto \widetilde{K}(x, y)$ is of bounded variation $V(\widetilde{K}(x, \cdot))$ and $\sup_{x>0} V(\widetilde{K}(x, \cdot)) < +\infty$. Then*

$$|e_z^n| \le \left( \sup_{y>0} V(\widetilde{K}(\cdot, y)) + \sup_{x>0} V(\widetilde{K}(x, \cdot)) + 3\widetilde{K}^\infty \right) D_N^\star(X^n; f_n).$$

*Proof.* We define:

$$\xi_1^n(x) := \frac{1}{N} \sum_{0 \le \ell < N} \widetilde{K}(x, x_\ell^n)(\sigma_z(x + x_\ell^n) - \sigma_z(x)), \quad x > 0,$$

$$\xi_2^n(y) := \int_0^{+\infty} \widetilde{K}(x, y) f_n(x)(\sigma_z(x + y) - \sigma_z(x))\mathrm{d}x, \quad y > 0.$$

Then $e_z^n$ may be written as $e_z^n = e_{z,1}^n + e_{z,2}^n$, with

$$e_{z,1}^n := \frac{1}{N} \sum_{0 \le k < N} \xi_1^n(x_k^n) - \int_0^{+\infty} \xi_1^n(x) f_n(x) dx,$$

$$e_{z,2}^n := \frac{1}{N} \sum_{0 \le \ell < N} \xi_2^n(x_\ell^n) - \int_0^{+\infty} \xi_2^n(y) f_n(y) dy.$$

Using the generalized Koksma–Hlawka inequality (Proposition 13.1), we obtain:

$$|e_{z,1}^n| \le V(\xi_1^n) D_N^\star(X^n; f_n) \quad \text{and} \quad |e_{z,2}^n| \le V(\xi_2^n) D_N^\star(X^n; f_n).$$

If we define: $\xi(x, y) := \widetilde{K}(x, y)(\sigma_z(x + y) - \sigma_z(x))$, then

$$\xi_1^n(x) = \frac{1}{N} \sum_{0 \le \ell < N} \xi(x, x_\ell^n) \quad \text{and} \quad \xi_2^n(y) = \int_0^{+\infty} \xi(x, y) f_n(x) dx.$$

In addition,

$$V(\xi_1^n) \le \sup_{y>0} V(\xi(\cdot, y)) \le \sup_{y>0} V(\widetilde{K}(\cdot, y)) + 2\widetilde{K}^\infty$$

$$V(\xi_2^n) \le \sup_{x>0} V(\xi(x, \cdot)) \le \sup_{x>0} V(\widetilde{K}(x, \cdot)) + \widetilde{K}^\infty.$$

Hence, the desired result follows. □

For bounding the QMC integration error $d_z^n$, we use the following result of H. Niederreiter, see Lemma 3.4 of [31].

**Lemma 13.5** *Let $X$ be a $(t, m, s)$-net in base $b$. For every elementary interval $J' \subset I^{s-1}$ and for every $x_s \in \bar{I}$, we have: $\left|D_{b^m}(J' \times [0, x_s), X)\right| \le b^{t-m}$.*

**Lemma 13.6** *We suppose that $\widetilde{K}$ is of bounded variation in the sense of Hardy and Krause; then*

$$|d_z^n| \le (2 + c_K \Delta t) \frac{1}{b^{\lfloor (m-t)/3 \rfloor}},$$

*where $c_K := 4V(\widetilde{K}) + 3\widetilde{K}^\infty$.*

*Proof.* The function $C_{\sigma_z}^{n+1}$ is the indicator function of some subset $P_z^n$ of $I^3$, thus $d_z^n = D_N(P_z^n, U^n)$. We have $P_z^n = (P_{z,0}^n \setminus \widetilde{P}_{z,1}^n) \cup P_{z,2}^n = \left(P_{z,0}^n \setminus P_{z,1}^n\right) \cup P_{z,2}^n$, where

$$P_{z,0}^n := \bigcup_{\substack{0 \le k,\ell < N \\ x_k^n < z}} R_{k,\ell} \times I, \quad \widetilde{P}_{z,1}^n := \bigcup_{\substack{0 \le k,\ell < N \\ x_k^n < z}} R_{k,\ell} \times (I_{k,\ell}^n \cup J_k^n),$$

$$P_{z,1}^n := \bigcup_{\substack{0 \le k,\ell < N \\ x_k^n < z}} R_{k,\ell} \times I_{k,\ell}^n, \quad P_{z,2}^n := \bigcup_{\substack{0 \le k,\ell < N \\ x_k^n + x_\ell^n < z}} R_{k,\ell} \times I_{k,\ell}^n.$$

Since $P_{z,1}^n \subset P_{z,0}^n$ and $\left( P_{z,0}^n \setminus P_{z,1}^n \right) \cap P_{z,2}^n = \emptyset$, $d_z^n$ can be split up:

$$d_z^n := D_N(P_{z,0}^n, U^n) - D_N(P_{z,1}^n, U^n) + D_N(P_{z,2}^n, U^n).$$

Using hypothesis (H2), we have $D_N(P_{z,0}^n, U^n) = D_N(\pi_2^3(P_{z,0}^n), \pi_2^3(U^n)) = 0$, because $\pi_2^3(P_{z,0}^n)$ is a disjoint union of elementary intervals in base $b$, of measure $b^{-m}$, and $\pi_2^3(U^n)$ is a $(0, m, 2)$-net in base $b$.

In order to obtain upper bounds for the other terms, we introduce the following functions. For $(u_1, u_2) \in I^2$,

$$\kappa_{z,1}^n(u_1, u_2) := \sum_{0 \le k,\ell < N} 1_{R_{k,\ell}}(u_1, u_2) \widetilde{K}(x_k^n, x_\ell^n) \sigma_z(x_k^n),$$

$$\kappa_{z,2}^n(u_1, u_2) := \sum_{0 \le k,\ell < N} 1_{R_{k,\ell}}(u_1, u_2) \widetilde{K}(x_k^n, x_\ell^n) \sigma_z(x_k^n + x_\ell^n).$$

Then we have, for $\alpha = 1, 2$, $P_{z,\alpha}^n = \{ \mathbf{u} \in I^3 : u_3 < \Delta t \kappa_{z,\alpha}^n(u_1, u_2) \}$. Let $d_1, d_2$ be integers (to be determined hereafter) such that $d_1 + d_2 \le m - t$. For $(a_1, a_2) \in \mathbb{N}^2$ with $0 \le a_1 < b^{d_1}$, $0 \le a_2 < b^{d_2}$, we define:

$$R'_{a_1, a_2} := [a_1 b^{-d_1}, (a_1 + 1) b^{-d_1}) \times [a_2 b^{-d_2}, (a_2 + 1) b^{-d_2})$$

and, for $\alpha = 1, 2$,

$$\underline{P}_{z,\alpha}^n := \bigcup_{\substack{0 \le a_1 < b^{d_1} \\ 0 \le a_2 < b^{d_2}}} R'_{a_1, a_2} \times [0, \Delta t \inf_{R'_{a_1, a_2}} \kappa_{z,\alpha}^n),$$

$$\overline{P}_{z,\alpha}^n := \bigcup_{\substack{0 \le a_1 < b^{d_1} \\ 0 \le a_2 < b^{d_2}}} R'_{a_1, a_2} \times [0, \Delta t \sup_{R'_{a_1, a_2}} \kappa_{z,\alpha}^n),$$

$$\partial P_{z,\alpha}^n := \bigcup_{\substack{0 \le a_1 < b^{d_1} \\ 0 \le a_2 < b^{d_2}}} R'_{a_1, a_2} \times [\Delta t \inf_{R'_{a_1, a_2}} \kappa_{z,\alpha}^n, \Delta t \sup_{R'_{a_1, a_2}} \kappa_{z,\alpha}^n).$$

It is easy to check that, for $\alpha = 1, 2$, $\underline{P}_{z,\alpha}^n \subset P_{z,\alpha}^n \subset \overline{P}_{z,\alpha}^n$ and $\overline{P}_{z,\alpha}^n \setminus \underline{P}_{z,\alpha}^n \subset \partial P_{z,\alpha}^n$. Consequently, we have for $\alpha = 1, 2$:

$$D_N(\underline{P}_{z,\alpha}^n, U^n) - \lambda_3(\partial P_{z,\alpha}^n) \le D_N(P_{z,\alpha}^n, U^n) \le D_N(\overline{P}_{z,\alpha}^n, U^n) + \lambda_3(\partial P_{z,\alpha}^n).$$

Since the $R'_{a_1,a_2}$ are disjoint elementary intervals in base $b$, then using hypothesis (H1) and Lemma 13.5 allows to obtain the following bounds, for $\alpha = 1, 2$,

$$|D_N(\underline{P}^n_{z,\alpha}, U^n)| \leq b^{d_1+d_2+t-m} \quad \text{and} \quad |D_N(\overline{P}^n_{z,\alpha}, U^n)| \leq b^{d_1+d_2+t-m}.$$

In addition,

$$\lambda_3(\partial P^n_{z,\alpha}) = \frac{\Delta t}{b^{d_1+d_2}} \sum_{\substack{0 \leq a_1 < b^{d_1} \\ 0 \leq a_2 < b^{d_2}}} \left( \sup_{R'_{a_1,a_2}} \kappa^n_{z,\alpha} - \inf_{R'_{a_1,a_2}} \kappa^n_{z,\alpha} \right).$$

We introduce the following functions. For $x, y > 0$,

$$\xi_{z,1}(x, y) := \widetilde{K}(x, y)\sigma_z(x) \quad \text{and} \quad \xi_{z,2}(x, y) := \widetilde{K}(x, y)\sigma_z(x + y).$$

Then, for $\alpha = 1, 2$, we have $\kappa^n_{z,\alpha}(u_1, u_2) = \xi_{z,\alpha}(x^n_{k(u_1)}, x^n_{k(u_2)})$. We define the following sets:

$$E^n_{a_1,a_2} := \left[ x^n_{a_1 b^{m-d_1}}, x^n_{(a_1+1)b^{m-d_1}-1} \right] \times \left[ x^n_{a_2 b^{m-d_2}}, x^n_{(a_2+1)b^{m-d_2}-1} \right].$$

Since the particles are reordered by increasing size, we have:

$$(u_1, u_2) \in R'_{a_1,a_2} \implies (x^n_{k(u_1)}, x^n_{k(u_2)}) \in E^n_{a_1,a_2}.$$

Consequently, for $\alpha = 1, 2$, we have:

$$\sup_{R'_{a_1,a_2}} \kappa^n_{z,\alpha} - \inf_{R'_{a_1,a_2}} \kappa^n_{z,\alpha} \leq \sup_{E^n_{a_1,a_2}} \xi_{z,\alpha} - \inf_{E^n_{a_1,a_2}} \xi_{z,\alpha}.$$

Using Lemma 13.2, we obtain:

$$\sum_{\substack{0 \leq a_1 < b^{d_1} \\ 0 \leq a_2 < b^{d_2}}} \left( \sup_{E^n_{a_1,a_2}} \xi_{z,1} - \inf_{E^n_{a_1,a_2}} \xi_{z,1} \right) \leq V(\widetilde{K})(b^{d_1} + b^{d_2}) + \widetilde{K}^\infty b^{d_2},$$

$$\sum_{\substack{0 \leq a_1 < b^{d_1} \\ 0 \leq a_2 < b^{d_2}}} \left( \sup_{E^n_{a_1,a_2}} \xi_{z,2} - \inf_{E^n_{a_1,a_2}} \xi_{z,2} \right) \leq (V(\widetilde{K}) + \widetilde{K}^\infty)(b^{d_1} + b^{d_2}).$$

Finally, we have:

$$|d^n_z| \leq 2b^{d_1+d_2+t-m} + \frac{\Delta t}{b^{d_1+d_2}} \left( 2V(\widetilde{K})(b^{d_1} + b^{d_2}) + \widetilde{K}^\infty(b^{d_1} + 2b^{d_2}) \right).$$

By choosing $d_1 = d_2 = \lfloor (m - t)/3 \rfloor$, the conclusion follows.  $\square$

By combining the results of Lemmas 13.3, 13.4 and 13.6, we obtain an upper bound for the error $D_N^\star(X^n; f_n)$. Let us remark that the hypotheses of Lemma 13.4 are satisfied if the hypotheses of Lemma 13.6 are satisfied.

**Proposition 13.7** *We suppose:*

- *for every $x > 0$, the function $t \to f(x, t)$ is twice continuously differentiable over $(0, T)$ and $f$, $\frac{\partial f}{\partial t}$, $\frac{\partial^2 f}{\partial t^2}$ are integrable over $\mathbb{R}_+^* \times (0, T)$,*
- *$\widetilde{K}$ is of bounded variation in the sense of Hardy and Krause.*

*Then*

$$D_N^\star(X^n; f_n) \leq e^{ct_n} D_N^\star(X^0; f_0) + \Delta t \int_0^{+\infty} \int_0^{t_n} e^{c(t_n-t)} \left| \frac{\partial^2 f}{\partial t^2}(x, t) \right| dt \, dx$$

$$+ \left( \frac{2}{\Delta t} + c_K \right) \frac{1}{b^{\lfloor (m-t)/3 \rfloor}} \frac{e^{ct_n} - 1}{c},$$

*where $c_K := 4V(\widetilde{K}) + 3\widetilde{K}^\infty$ and $c := \sup_{x>0} V(\widetilde{K}(x, .)) + \sup_{y>0} V(\widetilde{K}(\cdot, y)) + 3\widetilde{K}^\infty$.*

*Proof.* By using Equation (13.10), we obtain:

$$D_N^*(X^{n+1}; f_{n+1}) \leq (1 + c\Delta t)D_N^*(X^n; f_n)$$

$$+ \Delta t \int_0^{+\infty} \int_{t_n}^{t_{n+1}} \left| \frac{\partial^2 f}{\partial t^2}(x, t) \right| dt \, dx + \frac{2 + c_K \Delta t}{b^{\lfloor (m-t)/3 \rfloor}}.$$

The desired result follows by induction. ☐

**Remark 13.8** It should be noted that the upper bound is of order $\mathcal{O}(1/N^{1/3})$, which is worse than the length of the confidence interval of MC methods. As a matter of fact, numerical experiments show that the QMC method converges faster than the corresponding MC scheme, see Section 13.4. In addition, the upper bound increases when $\Delta t$ decreases, but that was never observed in the numerical experiments.

This shows that the previous result should be improved. The QMC approach, if not thoroughly justified, may lead to efficient schemes, as shown in the next section.

## 13.4 Numerical results

In this section, the efficiency of the QMC scheme is tested. To do this, approximate solutions are compared with exact solutions in some cases where analytical solutions are available. The QMC solutions are also compared with those given by the MC scheme adapted from the algorithm described in [2]. It has long been recognized that three particular kernels $K(x, y)$ are mathematically tractable: $K_0(x, y) = 1$, $K_1(x, y) = x + y$, $K_2(x, y) = xy$. The results of numerical experiments with a constant kernel are reported in [24]. On the other hand it is well known that when $K_2$ is used, gelation occurs in finite time and the total mass is not conserved: this is beyond the scope of the present study. So only $K_1$ is considered here. For all QMC calculations, the low-discrepancy sequence used is a $(0, 3)$-sequence in base 3 of H. Niederreiter [32].

We consider the coagulation kernel $K_1(x, y) = x + y$. With initial condition $c_0(x) := e^{-x}$, the exact solution of Equation (13.1), as calculated in [36], is given by, for $x > 0$, $t \geq 0$,

$$c(x, t) = \frac{1}{x\sqrt{1 - e^{-t}}} \exp\left(-x(2 - e^{-t}) - t\right) I_1\left(2x\sqrt{1 - e^{-t}}\right), \quad (13.11)$$

where $I_1$ is the modified Bessel function of the first kind of order one [12]. We perform the simulation up to $T = 1.0$ with $N$ particles ($N$ varying from $3^4$ to $3^{12}$) and $P$ time steps (varying from $2 \times 100$ to $2^4 \times 100$).

In order to reduce scatter, we compute the *averaged discrepancy* defined as:

$$D_{N,P} := \frac{1}{100} \sum_{h=1}^{100} D_N^*(X^{hp}, f_{hp}),$$

where $p = P/100$.

Figure 13.1 presents log-log plots of the variation of $D_{N,P}$ for different values of $N$ and $P$, for both methods (MC and QMC).

For a given number of particles and time step, the QMC scheme always gives smaller errors than the MC scheme; the improvement is effective for small time steps, and this gain is more apparent when both discretization parameters (the number $N$ of particles and the number $P$ of time steps) are large. For instance, if $P = 1600$, the error of the QMC scheme with $3^{11}$ particles is smaller than the error of the MC scheme using $3^{12}$ particles. Similar comments are made in [24] for the constant kernel case.

## 13.5 Conclusion

In this paper, we analyzed a QMC algorithm for the approximation of the continuous coagulation equation. A sample of $N$ numerical particles was used

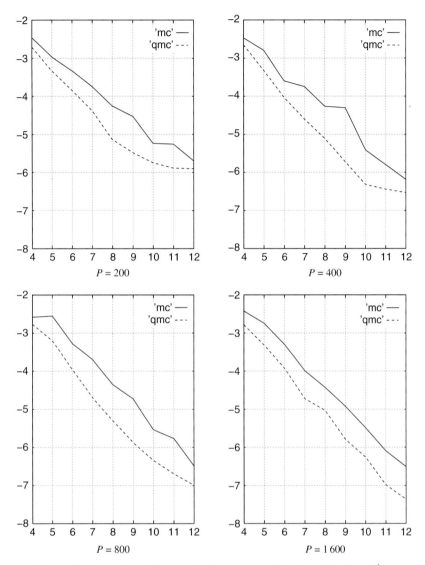

Figure 13.1 Linear kernel: averaged discrepancy as a function of $N$ (from $3^4$ to $3^{12}$) for $P$ between 200 and 1600. Log-log plots of MC (solid lines) versus QMC (dotted lines).

to simulate the behavior of the system as a whole. Time was discretized into $P$ time steps and the mass density was approximated by a sum of $N$ Dirac measures. Particles were renumbered by increasing mass at each time step. Quasi-random points were used to change the particle masses according to the dynamics of the equation.

Convergence of the scheme was proved when the number $N$ of numerical particles tends to infinity: the deterministic error of the QMC scheme at a discrete time $t_n$ is defined to be the star discrepancy of the set of discrete masses relative to the exact mass distribution. The accuracy of the algorithm was assessed through comparison with exact values, in a test case where analytical results are available. Moreover, comparisons with MC results were performed. The numerical approximations given by the QMC scheme converge to the exact results when the number $P$ of time steps and the number $N$ of numerical particles tend to infinity. This is verified by computing the star discrepancy mentioned above. The various errors given by the QMC simulation are always smaller than the corresponding errors given by the MC method using the same $N$ and $P$.

The numerical examples given here or in [24] show that the QMC algorithm performs better than the upper bound of $\mathcal{O}(1/N^{1/3})$ would suggest. This shows that it may be possible to improve the bound given here; it would be an interesting problem for future work.

The accuracy of the QMC approach is stated here on an academic test problem, where an exact solution is available. An application of the scheme to the simulation of the drop size distribution in a spray is described in [24]. Validation of the QMC method will be pursued in other realistic settings.

# References

[1] D. J. Aldous, Deterministic and stochastic models for coalescence (aggregation and coagulation): a review of the mean-field theory for probabilists. *Bernoulli* **5**, 3–48, 1999.

[2] H. Babovsky, On a Monte Carlo scheme for Smoluchowski's coagulation equation. *Monte Carlo Methods Appl.* **5**, 1–18, 1999.

[3] J. Dick and F. Pillichshammer, *Digital Nets and Sequences*. Cambridge University Press, Cambridge, 2010.

[4] R. B. Diemer and J. H. Olson, A moment methodology for coagulation and breakage problems: part 2 – moment models and distribution reconstruction. *Chem. Eng. Sci.* **57**, 2211–2228, 2002.

[5] R. L. Drake, A general mathematical survey of the coagulation equation. In: G. M. Hidy and J. R. Brock (eds.), *Topics in Current Aerosol Research, Part 2*, pp. 201–376. Pergamon Press, Oxford, 1972.

[6] M. Drmota and R. F. Tichy, *Sequences, Discrepancies and Applications*. Springer-Verlag, Berlin, 1997.

[7] P. B. Dubovskiĭ, *Mathematical Theory of Coagulation*. Lecture Notes 23. Global Analysis Research Center, Seoul National University, 1994.

[8] A. Eibeck and W. Wagner, An efficient stochastic algorithm for studying coagulation dynamics and gelation phenomena. *SIAM J. Sci. Comput.* **22**, 802–821, 2000.

[9] A. Eibeck and W. Wagner, Stochastic particle approximations for Smoluchowski's coagulation equation. *Ann. Appl. Prob.* **11**, 1137–1165, 2001.

[10] F. Filbet and P. Laurençot, Numerical simulation of the Smoluchowski coagulation equation. *SIAM J. Sci. Comput.* **25**, 2004–2028, 2004.

[11] D. T. Gillespie, An exact method for numerically simulating the stochastic coalescence process in a cloud. *J. Atmos. Sci.* **32**, 1977–1989, 1975.

[12] I. S. Gradshteyn and I. M. Ryzhik, *Table of Integrals, Series and Products*, seventh edition. Academic Press, San Diego, CA, 2007.

[13] M. Kostoglou, Extended cell average technique for the solution of coagulation equation. *J. Colloid Interface Sci.* **306**, 72–81, 2007.

[14] L. Kuipers and H. Niederreiter, *Uniform Distribution of Sequences*. John Wiley & Sons, New York, 1974.

[15] S. Kumar and D. Ramkrishna, On the solution of population balance equation by discretization – I. A fixed pivot technique. *Chem. Eng. Sci.* **51**, 1311–1332, 1996.

[16] J. Kumar, M. Peglow, G. Warnecke, S. Heinrich and L. Mörl, Improved accuracy and convergence of discretized population balance for aggregation: the cell average technique. *Chem. Eng. Sci.* **61**, 3327–3342, 2006.

[17] J. Kumar, G. Warnecke, M. Peglow and S. Heinrich, Comparison of numerical methods for solving population balance equations incorporating aggregation and breakage. *Powder Technol.* **189**, 218–229, 2009.

[18] G. Larcher, Digital point sets: analysis and application. In: P. Hellekalek and G. Larcher (eds.), *Random and Quasi-Random Point Sets*. Lecture Notes in Statistics, volume 138, pp. 167–222. Springer-Verlag, Berlin, 1998.

[19] C. Lécot, A direct simulation Monte Carlo scheme and uniformly distributed sequences for solving the Boltzmann equation. *Computing* **41**, 41–57, 1989.

[20] C. Lécot, A quasi-Monte Carlo method for the Boltzmann equation. *Math. Comput.* **56**, 621–644, 1991.

[21] C. Lécot, Error bounds for quasi-Monte Carlo integration with nets. *Math. Comput.* **65**, 179–187, 1996.

[22] C. Lécot and A. Tarhini, A quasi-stochastic simulation of the general dynamics equation for aerosols. *Monte Carlo Methods Appl.* **13**, 369–388, 2007.

[23] C. Lécot and W. Wagner, A quasi-Monte Carlo scheme for Smoluchowski's coagulation equation. *Math. Comput.* **73**, 1953–1966, 2004.

[24] C. Lécot, M. Tembely, A. Soucemarianadin and A. Tarhini, Numerical simulation of the drop size distribution in a spray. In: L. Plaskota and H. Wozniakowski (eds.), *Monte Carlo and Quasi-Monte Carlo Methods 2010*, pp. 503–517. Springer-Verlag, Berlin, 2012.

[25] G. Madras and B. J. McCoy, Numerical and similarity solutions for reversible population balance equations with size-dependent rates. *J. Colloid Interface Sci.* **246**, 356–365, 2002.

[26] Z. A. Melzak, The effect of coalescence in certain collision processes. *Q. J. Appl. Math.* **11**, 231–234, 1953.

[27] Z. A. Melzak, A scalar transport equation. *Trans. Am. Math. Soc.* **85**, 547–560, 1957.

[28] W. J. Morokoff and R. E. Caflisch, Quasi-Monte Carlo integration. *J. Comput. Phys.* **122**, 218–230, 1995.

[29] H. Müller, Zur allgemeinen Theorie der raschen Koagulation. *Kolloidchem. Beih.* **27**, 223–250, 1928.

[30] H. Niederreiter, Quasi-Monte Carlo methods and pseudo-random numbers. *Bull. Am. Math. Soc.* **84**, 957–1041, 1978.

[31] H. Niederreiter, Point sets and sequences with small discrepancy. *Monatsh. Math.* **104**, 273–337, 1987.

[32] H. Niederreiter, Low-discrepancy and low-dispersion sequences. *J. Number Theory* **30**, 51–70, 1988.

[33] H. Niederreiter, *Random Number Generation and Quasi-Monte Carlo Methods.* SIAM, Philadelphia, PA, 1992.

[34] H. Niederreiter, Constructions of $(t, m, s)$-nets. In: H. Niederreiter and J. Spanier (eds.), *Monte Carlo and Quasi-Monte Carlo Methods 1998*, pp. 70–85. Springer-Verlag, Berlin, 2000.

[35] H. Niederreiter, Constructions of $(t, m, s)$-nets and $(t, s)$-sequences. *Finite Fields Appl.* **11**, 578–600, 2005.

[36] T. E. Ramabhadran, T. W. Peterson and J. H. Seinfeld, Dynamics of aerosol coagulation and condensation. *AIChE J.* **22**, 840–851, 1976.

[37] T. E. W. Schumann, Theoretical aspects of the size distribution of fog particles. *Q. J. R. Meteorol. Soc.* **66**, 195–207, 1940.

[38] W. T. Scott, Analytic studies of cloud droplet coalescence I. *J. Atmos. Sci.* **25**, 54–65, 1968.

[39] M. Smith and T. Matsoukas, Constant-number Monte Carlo simulation of population balances. *Chem. Eng. Sci.* **53**, 1777–1786, 1998.

[40] M. von Smoluchowski, Versuch einer mathematischen Theorie der Koagulationskinetik kolloider Lösungen. *Z. Phys. Chem.* **92**, 129–168, 1916.

[41] M. von Smoluchowski, Drei Vorträge über Diffusion, Brownsche Molekularbewegung und Koagulation von Kolloidteilchen. *Phys. Z.* **17**, 557–599, 1916.

[42] M. Sommer, F. Stenger, W. Peukert and N. J. Wagner, Agglomeration and breakage of nanoparticles in stirred media mills – a comparison of different methods and models. *Chem. Eng. Sci.* **61**, 135–148, 2006.

[43] J. Wei and F. E. Kruis, GPU-accelerated Monte Carlo simulation of particle coagulation based on the inverse method. *J. Comput. Phys.* **249**, 67–79, 2013.

[44] S. K. Zaremba, Some applications of multidimensional integration by parts. *Ann. Pol. Math.* **21**, 85–96, 1968.

[45] H. Zhao and C. Zheng, A new event-driven constant-volume method for solution of the time evolution of particle size distribution. *J. Comput. Phys.* **228**, 1412–1428, 2009.

[46] H. Zhao and C. Zheng, Correcting the multi-Monte Carlo method for particle coagulation. *Powder Technol.* **193**, 120–123, 2009.

[47] H. Zhao, C. Zheng and M. Xu, Multi-Monte Carlo method for particle coagulation: description and validation. *Appl. Math. Comput.* **167**, 1383–1399, 2005.

[48] H. Zhao, A. Maisels, T. Matsoukas and C. Zheng, Analysis of four Monte Carlo methods for the solution of population balances in dispersed systems. *Powder Technol.* **173**, 38–50, 2007.

[49] Y. Zou, M. E. Kavousanakis, I. G. Kevrekidis and R. O. Fox, Coarse-grained computation for particle coagulation and sintering processes by linking quadrature method of moments with Monte Carlo. *J. Comput. Phys.* **229**, 5299–5314, 2010.

# 14

## Asymptotic formulas for partitions with bounded multiplicity

*Pierre Liardet[1] and Alain Thomas*
Aix Marseille University, Marseille

*Dedicated to Professor Harald Niederreiter on the occasion
of his 70th birthday.*

## Abstract

Let $U = (u_k)_k$ be an increasing sequence of positive integers, let $q$ be an integer such that $q \geq 2$ and let $M_{U,q}(n)$ denote the number of partitions of $n$ into elements from $U$ with at most $q - 1$ possible repetitions. In this paper we investigate the behavior of $M_{U,q}(n)$ with $U$, verifying a general condition introduced in 1954 by K. F. Roth and G. Szekeres in their study of the case $q = 2$ and the unrestricted case ($q = \infty$).

## 14.1 Introduction

Let $U$ denote an increasing sequence in $\mathbb{N}$, the set of positive integers, or equivalently a subset of $\mathbb{N}$ indexed in increasing order $u_1 < u_2 < u_3 < \cdots$ and let $q$ be an integer larger than or equal to 2 or $q = \infty$. A partition of a natural number $n$ in parts (or summands) from $U$ with multiplicity at most $q - 1$ is given by a sum $n = \sum_k e_k u_k$ identified by the sequence of integers $(e_k)_{k \geq 1}$ satisfying $0 \leq e_k < q$ for all indices $k$. In this sum the integer $e_k$ is called the multiplicity of the part $u_k$. Let $M_{U,q}(n)$ denote the number of such partitions of $n$ (and $M_{U,q}(0) = 1$ by convention). The partition problem with multiplicity at most $q - 1$ consists in finding an accurate asymptotic estimate of $M_{U,q}(n)$. In a seminal paper G. H. Hardy and S. Ramanujan [14] introduced the so-called *circle method* to obtain a sharp asymptotic formula for $p(n) := M_{\mathbb{N},\infty}(n)$ (unrestricted partitions problem) leading in particular to

---

[1] It is with deep regret that we announce the first author passed away on August 29, 2014.

$$p(n) \sim \frac{e^{\pi\sqrt{2n/3}}}{4n\sqrt{3}} \qquad (14.1)$$

as $n$ tends to $\infty$. This circle method was perfected by H. Rademacher [22], who obtained a convergent series for $p(n)$, and was revisited subsequently by E. M. Wright [29] who derived an asymptotic formula of the number of unrestricted partitions $p^{(k)}(n)$ of $n$ into $k$th powers. In fact, the order of magnitude of $p^{(k)}(n)$ was previously announced by Hardy and Ramanujan [14] in order to illustrate a wide range of applications of the circle method and, among many other possible results that the authors could obtain, an asymptotic formula for the number $q(n)$ of partitions of $n$ into unequal parts was given (restricted partitions problem), but without proof. A weak form of this result is given by the following equivalence

$$q(n) \sim \frac{1}{4 \cdot 3^{1/4} n^{3/4}} e^{\pi\sqrt{n/3}} \qquad (n \to \infty). \qquad (14.2)$$

Later, modifying Hardy–Ramanujan–Rademacher's method, L.-K. Hua [15] obtained an equality for $q(n)$, while A. E. Ingham [16] established a general Tauberian theorem and derived naturally an asymptotic behavior for $M_{U,\infty}(n)$ for subsets $U$ of $\mathbb{N}$ subject to certain conditions including $\sum_{k,\, u_k \leq t} 1 = Bt^{\beta} + R(t)$ with $B > 0$, $\beta > 0$ and $\int_0^t \frac{R(x)}{x} dx = b \log t + c + o(1)$ with constants $b$ and $c$. Ingham's result includes Hardy and Ramanujan's formula (14.1), and also the cases of partitions into $k$th powers or partitions into $k_1$th and $k_2$th powers. Moreover, it covers the case of restricted partitions but with additional hypothesis on $U$. Later on, Pennington [20] gave more applications of Ingham's method, including the unrestricted partitions into powers of a fixed integer $r > 1$, previously studied by K. Mahler [17] and completed by N. G. de Bruijn [4].

Let $F_{U,q}(\cdot)$ be the generating series of the function $M_{U,q}(\cdot)$. The identity

$$F_{U,q}(w) := \sum_{n=0}^{\infty} M_{U,q}(n) w^n = \prod_{k=1}^{\infty} (1 + w^{u_k} + \cdots + w^{(q-1)u_k}) \qquad (14.3)$$

holds formally; but the power series has radius of convergence 1 and its sum is equal to the infinite product. Starting from Cauchy's formula

$$M_{U,q}(n) = \frac{1}{2i\pi} \int_{\mathcal{C}} \frac{F_{U,q}(w)}{w^{n+1}} dw \qquad (14.4)$$

for $q = 2$, where the path of integration is any circle with center at the origin and radius $r < 1$, K. F. Roth and G. Szekeres [26] extended Ingham's result by means of the *saddle-point method*. They considered subsets $U$ verifying the following conditions which are much less restrictive than those of Ingham:

$$\text{(I)} \quad s = \lim_{k \to \infty} \frac{\log u_k}{\log k} < \infty,$$

$$\text{(II)} \quad \lim_{k \to \infty} \left( \inf_\alpha \left\{ (\log k)^{-1} \sum_{\nu=1}^{k} \|u_\nu \alpha\|^2 \right\} \right) = \infty,$$

where the lower bound is taken over all $\alpha$ satisfying $\frac{1}{2u_k} < \alpha \le \frac{1}{2}$ and $\|x\|$ denotes the distance of $x$ from the nearest integer. They also applied their argument, after a slight modification, to the case $q = \infty$. Examples of sequences $U$ satisfying conditions (I) and (II) are the sequence $(p_k)_k$ of prime numbers, the polynomial sequences $(P(k))_k$ which are monotonic increasing with polynomial $P$ such that $P(0) \ge 1$, $P(\mathbb{N}) \subset \mathbb{N}$ and the $P(k)$ have no common divisor except 1, and the sequences $P(p_k)$ which are polynomials in prime numbers with the natural constraints as above (see [26]).

The partition problem has stimulated much research involving both various elementary or analytic tools and new families of subsets $U$ but most studies are concerned with restricted or unrestricted partitions; see for example [1, 2, 3, 7, 8, 9, 18, 19, 24, 25, 27].

In 1956, M. Dutta [6], motivated by recent investigations in statistical physics [28], exploited the Tauberian approach initiated in [14] to derive the approximate value $\log M_{\mathbb{N},q}(n) \sim \pi \sqrt{\frac{2}{3} n \frac{q-1}{q}}$ for any integer $q \ge 2$. P. Hagis [13], going back to the circle method of Hardy–Ramanujan–Rademacher, obtained a convergent series representation of $M_{\mathbb{N},q}(n)$ and the asymptotic formula

$$M_{\mathbb{N},q}(n) = \frac{1}{2 \cdot 6^{1/4}} \frac{(q-1)^{1/4}}{(q(n + \frac{q-1}{24}))^{3/4}} \exp \left( \pi \sqrt{\left( \frac{2}{3} n + \frac{q-1}{24} \right) \frac{q-1}{q}} \right)$$
$$\times \left( 1 + \mathcal{O}\left( \frac{1}{\sqrt{n}} \right) \right). \tag{14.5}$$

Recently, some authors have paid attention to the case of partitions into powers of a fixed integer $a > 1$ with finite multiplicity $q$ [5, 21, 23]. Their results reveal that the asymptotic behavior of $M_{(a^k)_k,q}(n)$ is oscillating and takes values depending on the arithmetical structure of $n$ with respect to $q$ and $a$, so that one cannot expect an asymptotic formula. In [10] the authors proved in a general setting that the limit $\lim_{n \to \infty, n \in S} \log M_{V,q}(n) / \log n$ exists for some subset $S$ of $\mathbb{N}$ (depending on $V$ and $q$) with natural density 1. Here $V := (v_k)_k$ is a Pisot scale, in other words $v_0 = 1$ and

$$v_k = c(\theta)\theta^k + \gamma_k \tag{14.6}$$

where $\theta$ is a positive Pisot number, $c(\theta) \in \mathbb{Q}(\theta)$ and $\lim_k \gamma_k = 0$ (in fact the series $\sum_{k=0}^{\infty} |\gamma_k|$ converges anyway, see the comment following [10, Lemma 18]). It is interesting to notice that such a sequence does not satisfy condition (II). This is clear for the particular Pisot sequences $V = (a^k)_k$. The general case follows from the fact that the series $\sum_{k=0}^{\infty} \|\theta^{-m} v_k\|^2$ converge for any integer $m$.

The present paper is based on the work of K. F. Roth and G. Szekeres where only the cases $q = 2$ and $q = \infty$ are considered. Our goal is to obtain an asymptotic expansion of $M_{U,q}$, without restriction on the multiplicity $q$, for sequences $U$ subject to (I) and (II). The following remarks are interesting for clarifying some aspects of condition (II).

**Remarks 14.1**

(1) The condition $\frac{1}{2u_k} < \alpha \le \frac{1}{2}$ in (II) can be replaced by $\frac{1}{2u_k} < \|\alpha\|$. Notice that the sum $\sum_{v=1}^{k} \|u_v \alpha\|^2$ can easily be estimated from $U$ in the case $\|\alpha\| \le \frac{1}{2u_k}$ because this sum is exactly $\|\alpha\|^2 \sum_{v=1}^{k} u_v{}^2$.

(2) If condition (I) (respectively (II)) is satisfied for $U$ it is also satisfied for any $V \subset \mathbb{N}$ such that the symmetric difference $U \cup V \setminus U \cap V$ is finite. Notice also that the same property is verified by the negation of (II).

(3) Obviously, the sequence of multiples of a given integer $a > 1$ does not verify (II). More generally, if there exists an integer $a \ge 2$ such that

$$\text{card}\{\ell \, ; \, 1 \le \ell \le k \text{ and } u_\ell \not\equiv 0 \, (\text{mod } a)\} \in \mathcal{O}(\log k),$$

then condition (II) is not satisfied.

(4) In contrast, recall that it was proved in [26] that under condition (I) with $s < 3/2$, the sequence $U$ verifies (II) if the following condition holds.

> **(II$'$)** There exist constants $k_0$ and $c > 0$ such that, for all integers $k, \ell$ satisfying $k > k_0$ and $1 < \ell \le 18u_k/k$, at least $c\ell^2(\log k)^2$ of the integers $u_1, u_2, \dots, u_k$ are not divisible by $\ell$.

Clearly (II$'$) is satisfied by any increasing sequence $(u_k)_k$ of integers such that $u_k \in \mathcal{O}(k)$ and for any integer $m \ge 2$ the sequence $k \mapsto u_k \, (\text{mod } m)$ is uniformly distributed in $\mathbb{Z}/m\mathbb{Z}$. In particular, the sequence $U$ defined by $u_k = \lfloor \alpha k + \beta \rfloor$ with $\alpha$ irrational, $\alpha > 1$ and $\beta \ge 0$, verifies conditions (I) and (II). $\square$

Our main theorem (Theorem 14.2) is given in Section 14.2. It is accompanied with an important lemma which specifies the asymptotic values of certain constants of the theorem according to the parameter $n$. Due to (I), the polynomial sequences $(k^s)_{k \geq 1}$ play a fundamental role in the computations. For this reason, these sequences are studied independently in order to obtain more accurate asymptotic formulas (Theorem 14.7 and Corollary 14.8).

Section 14.3 is devoted to the proof of Theorem 14.2. The classical saddle-point method (see [11]) applied to (14.4) gives for the radius $r$ of the circle $\mathcal{C}$ the solution of the saddle-point equation

$$x \frac{F'_{U,q}(x)}{F_{U,q}(x)} = n \quad (0 < x < 1).$$

By upward convexity, this solution is unique. Equivalently, after setting $r = e^{-\eta}$ and a straightforward calculation, the saddle-point equation takes the form (14.9).

## 14.2 Asymptotic expansion of $M_{U,q}$

### 14.2.1 The main theorem

To state our result we need to introduce the polynomials

$$S_0(x) = \sum_{h=0}^{q-1} x^h \quad \text{and} \quad S_k(x) = \sum_{h=0}^{q-1} (q-1-h)^k x^h \quad (k \in \mathbb{N}).$$

As in [26], we shall use the notation

$$\sum_{k}{}_{U}^{\ell} f(u)$$

to simplify $\sum_{v=k}^{\ell} f(u_v)$. In particular, for any $y > 0$,

$$\log F_{U,q}(e^{-y}) = \sum_{1}{}_{U}^{\infty} \log(S_0(e^{-yu})). \tag{14.7}$$

The following theorem can be viewed as a generalization of Hagis's asymptotic behavior in (14.5) but the price to pay under condition (I) and (II) is the introduction of functions of $n$ involved in the asymptotic estimation (14.8) and expressed by rather intricate formulas.

**Theorem 14.2** *For any integer $m \geq 2$,*

$$M_{U,q}(n) = (2\pi A_2)^{-1/2} \exp\left[n\eta + \log(F_{U,q}(e^{-\eta}))\right]$$

$$\times \left\{ 1 + \sum_{\rho=1}^{m-2} D_\rho + \mathcal{O}(n^{-(m-1)(s+1)^{-1}+\delta}) \right\}, \tag{14.8}$$

*where $\delta$ is an arbitrarily small positive number, $\eta = \eta(n)$ is the unique solution of*

$$n = \sum_{1}^{\infty}{}_U \frac{u S_1(e^{\eta u})}{S_0(e^{\eta u})}, \tag{14.9}$$

*and $D_\rho$ ($\rho = 1, 2, \ldots, m-1$) is given by*

$$D_\rho = D_\rho(n) = A_2^{-6\rho} \sum_{\mu_1=2}^{\infty}{}' \cdots \sum_{\mu_{5\rho}=2}^{\infty}{}' d_{\mu_1, \mu_2, \ldots, \mu_{5\rho}} A_{\mu_1} A_{\mu_2} \ldots A_{\mu_{5\rho}}, \tag{14.10}$$

*the summation being subject to*

$$\mu_1 + \mu_2 + \cdots + \mu_{5\rho} = 12\rho, \tag{14.11}$$

*where the d are certain numerical coefficients, and*

$$A_\mu = A_\mu(n) = \sum_{1}^{\infty}{}_U g_\mu(e^{\eta u}) \left( \frac{u}{S_0(e^{\eta u})} \right)^\mu. \tag{14.12}$$

*Here $g_\mu$ is a certain polynomial (given in Lemma 14.9(ii)) of highest degree term $x^{(q-1)\mu-1}$ and, in particular, $g_1 = S_1$ and $g_2 = S_0 S_2 - S_1^2$. The asymptotic behaviors of $\eta(n)$, $A_\mu(n)$ and $D_\rho(n)$ are specified in Lemma 14.3, Corollary 14.4 and Corollary 14.5.*

The definitions of $\eta$, $D_\rho$ and $A_\mu$ in (14.9), (14.10) and (14.12) correspond to those given in [26] with $q = 2$. Notice that, since the generating function $F_{U,q}(z)$ has nonnegative coefficients, the inequality $M_{U,q}(n) \leq \exp(n\eta + \log(F_{U,q}(e^{-\eta})))$ holds. Theorem 14.2 shows that in regard to the exponential term, this bound is the best one. Another main consequence of Theorem 14.2 is to assert that, for a large class of sequences $U$ growing polynomially, the number of partitions $M_{U,q}(n)$ has effectively an asymptotic behavior. The counterpart of this general result is a formula (14.8) involving rather complicated functions.

## 14.2.2 Preparatory lemmas

It is relevant to compute asymptotic estimations of $\log \eta(n)$, $\log A_\mu(n)$ and consequently $\log D_\rho(n)$ with respect to $\log n$. They are given by the next lemma and its corollaries which are analogous to Lemma 1 and its corollary in [26] but where $q = 2$. In fact the formulas are identical (independent of $q$) but we need a proof.

**Lemma 14.3** *With the definition of Theorem 14.2 one has the following.*

*(i)* $0 < \eta(n) < +\infty$, $A_1(n) = n$ *and* $A_2(n) > 0$.
*(ii)* $\lim_{n\to\infty} \eta(n) = 0$.
*(iii)* *For* $\kappa \in \{1, 2\}$,

$$\lim_{n\to\infty} \frac{\log A_\kappa(n)}{\log \eta(n)} = -\frac{\kappa s + 1}{s}, \tag{14.13}$$

*in particular*

$$\lim_{n\to\infty} \frac{\log \eta(n)}{\log n} = -\frac{s}{s+1}, \tag{14.14}$$

*and for all integer* $\mu > 2$,

$$\limsup_{n\to\infty} \frac{\log |A_\mu(n)|}{-\log \eta(n)} \leq \frac{\mu s + 1}{s}. \tag{14.15}$$

*Proof.* (i) Formula $A_1(n) = n$ is proved by definition, and by construction $e^{-\eta(n)} \in ]0, 1[$ as a basic fact about the saddle-point equation. Notice that $g_2 = S_0 S_2 - S_1{}^2$ has positive coefficients. Indeed by an easy calculation,

$$g_2(x) = \sum_{0\leq h'\leq h\leq q-1} (h - h')^2 x^h + \sum_{0\leq h'\leq h\leq q-2} (h - h')^2 x^{2(q-1)-h}. \tag{14.16}$$

Hence $g_2(x) > 0$ if $x > 0$ and $A_2(n) > 0$ follows from (14.12).
   (ii) Using $S_0(x) \geq x^{q-1}$ and $S_1(x) \leq \frac{q(q-1)}{2} x^{q-2}$ for $x \geq 1$ one has

$$\frac{S_1(x)}{S_0(x)} \leq \frac{q(q-1)}{2x}.$$

Hence $n \leq \frac{q(q-1)}{2} \sum_{k=1}^{\infty} u_k e^{-\eta u_k}$ from (14.9). Consequently $\lim_{n\to\infty} \eta(n) = 0$.
   (iii) We consider $A_\mu$ as a function of $\eta$. By Condition (I), for all $\delta \in (0, 1)$ there exists $k_\delta$ such that for any $k \geq k_\delta$

$$(k + 1)^{s-\delta} \leq u_k \leq k^{s+\delta}.$$

Let $c = \sum_{k<k_\delta} u_k^\mu$. For $y \geq 1$, using $S_0(y) \leq qy^{q-1}$ and $g_2(y) \geq y^{2(q-1)-1}$ which is a consequence of (14.16) we obtain

$$\frac{1}{q} y^{-1} \leq \frac{g_1(y)}{S_0(y)} \leq \frac{q(q-1)}{2} y^{-1}$$

and

$$\frac{1}{q^2} y^{-1} \leq \frac{g_2(y)}{S_0(y)^2} \leq \frac{S_2(y)}{S_0(y)} \leq y^{-1} \sum_{h=1}^{q-1} h^2 = \frac{q(q-1)(2q-1)}{6} y^{-1}.$$

Therefore,

$$q^{-\kappa} \sum_{1}^{\infty} {}_U u^{\kappa} e^{-u\eta} \leq A_{\kappa}(n) \leq q^{\kappa+1} \sum_{1}^{\infty} {}_U u^{\kappa} e^{-u\eta} \quad (\kappa = 1, 2). \qquad (14.17)$$

It is not clear that for any $\mu > 2$ the value of $A_{\mu}(n)$ becomes positive for $n$ large enough. Let $K_{\mu}$ be the sum of the absolute values of the coefficients of the $g_{\mu}$, then $g_{\mu}(x) \leq K_{\mu} x^{(q-1)\mu-1}$ if $x \geq 1$ and

$$|A_{\mu}| \leq K_{\mu} \sum_{1}^{\infty} {}_U u^{\mu} e^{-\eta u}$$

$$\leq K_{\mu} \left( c + \sum_{v=1}^{\infty} v^{\mu(s+\delta)} e^{-\eta(v+1)^{s-\delta}} \right)$$

$$\leq K_{\mu} \left( c + \int_{1}^{\infty} x^{\mu(s+\delta)} e^{-\eta x^{s-\delta}} dx \right)$$

$$\leq K_{\mu} \left( c + \eta^{-(s-\delta)^{-1}(1+\mu(s+\delta))} \int_{0}^{\infty} y^{-1+(s-\delta)^{-1}(1+\mu(s+\delta))} e^{-y} dy \right).$$

The last integral is finite and positive, and does not depend on $\eta$. Since $\delta$ is arbitrarily in $(0, 1)$, the inequality

$$\limsup_{\eta \to 0} \frac{\log |A_{\mu}|}{-\log \eta} \leq s^{-1}(\mu s + 1)$$

holds.

To obtain the converse inequalities in cases $\mu = 1$ and $\mu = 2$, we use the first inequality in (14.17). For $\kappa \in \{1, 2\}$,

$$A_{\kappa} \geq q^{-\kappa} \sum_{1}^{\infty} {}_U u^{\kappa} e^{-\eta u}$$

$$\geq q^{-\kappa} \sum_{v=k_{\delta}}^{\infty} (v + 1)^{\kappa(s-\delta)} e^{-\eta v^{s+\delta}}$$

$$\geq q^{-\kappa} \int_{k_{\delta}}^{\infty} x^{\kappa(s-\delta)} e^{-\eta x^{s+\delta}} dx$$

$$\geq q^{-\kappa} \eta^{-(s+\delta)^{-1}(1+\kappa(s-\delta))} \int_{k_{\delta}^{3}}^{\infty} y^{-1+(s+\delta)^{-1}(1+\kappa(s-\delta))} e^{-y} dy.$$

The last integral is finite and positive, and does not depend on $\eta$, hence we have

$$\liminf_{\eta \to 0} \frac{\log A_{\kappa}}{-\log \eta} \geq s^{-1}(\kappa s + 1)$$

and finally (14.13) is proved. Since $A_1(n) = n$ and $\lim_{n\to\infty} \eta(n) = 0$ we have (14.14). The proof of Lemma 14.3 is complete. □

This lemma obviously implies the following two corollaries.

**Corollary 14.4**

$$\lim_{n\to\infty} \frac{\log A_2(n)}{\log n} = \frac{2s+1}{s+1}$$

and for $\mu > 2$,

$$\limsup_{n\to\infty} \frac{\log |A_\mu(n)|}{\log n} \le \frac{\mu s + 1}{s+1}.$$

**Corollary 14.5** *The terms $D_\rho$ in (14.8) verify asymptotically:*

$$\limsup_{n\to\infty} \frac{\log |D_\rho(n)|}{\log n} \le -\frac{\rho}{s+1}.$$

Now, Theorem 14.2 combined with the method of comparison between sums and integrals used in the proof of Lemma 14.3(iii), leads to the following.

**Corollary 14.6**

$$\lim_{n\to\infty} \frac{\log \left(\log M_{U,q}(n)\right)}{\log n} = \lim_{n\to\infty} \frac{\log \left[n\eta + \log(F_{U,q}(e^{-\eta}))\right]}{\log n} = \frac{1}{s+1}.$$

### 14.2.3 The monomial case

In this subsection we assume that $U = (k^s)_k$ ($s \in \mathbb{N}$) and derive better asymptotic behaviors for $\eta(n)$, $\log(F_{U,q}(e^{-\eta(n)}))$ and $A_\mu(n)$.

**Theorem 14.7** *In the monomial case $U = (k^s)_k$, set*

$$C_1(s,q) := \frac{1 - q^{-1/s}}{s} \Gamma\left(\frac{s+1}{s}\right) \zeta\left(\frac{s+1}{s}\right),$$

$$C_2(s,q) := \frac{1 - q^{-1/s}}{s} \Gamma\left(\frac{2s+1}{s}\right) \zeta\left(\frac{s+1}{s}\right).$$

*Then one has*

$$n\eta = \frac{C_1(s,q)}{\eta^{1/s}} + \mathcal{O}\left(\eta^{1/s}\right), \tag{14.18}$$

$$\log(F_{U,q}(e^{-\eta})) = -\frac{\log q}{2} + \frac{s C_1(s,q)}{\eta^{1/s}} + \mathcal{O}(\eta^{1/s}), \tag{14.19}$$

$$A_2 = \frac{C_2(s,q)}{\eta^{2+1/s}} + \mathcal{O}\left(\frac{1}{\eta^{2-1/s}}\right),\qquad(14.20)$$

$$A_\mu = \frac{1}{\eta^{\mu+1/s}}\frac{1}{s}\int_0^\infty y^{\mu-1+1/s}\frac{g_\mu(e^y)}{(S_0(e^y))^\mu}dy$$

$$+ \mathcal{O}\left(\frac{1}{\eta^{\mu-1/s}}\right),\quad \mu \geq 3,\qquad(14.21)$$

$$\eta = \frac{C_1(s,q)^{\frac{s}{s+1}}}{n^{\frac{s}{s+1}}}\left(1 + \mathcal{O}\left(\frac{1}{n^{\frac{2}{s+1}}}\right)\right).\qquad(14.22)$$

*Proof.* Let $f(\cdot)$ be the rational map defined by $f(x) = \frac{S_1(x)}{S_0(x)}$. Notice for the sequel that the equalities $S_1(x) = x^{q-2}S_0'(1/x)$ and $(x-1)S_1(x) = S_0(x) - q$ give respectively $f(x) = \frac{1}{x}\frac{S_0'(1/x)}{S_0(1/x)}$ and $f(x) = \frac{1}{x-1} - \frac{q}{x^q-1}$.

The Euler–Maclaurin formula at order 2 applied to (14.9) leads to

$$n = \sum_{k=0}^\infty k^s f(e^{\eta k^s})$$

$$= \int_0^\infty t^s f(e^{\eta t^s})dt + \frac{1}{12}\left[(t^s f(e^{\eta t^s}))'\right]_0^\infty - \frac{1}{2}\int_0^\infty \left(t^s f(e^{\eta t^s})\right)'' B_2(\{t\})dt$$

where $B_2(\cdot)$ is the Bernoulli polynomial of degree 2 and $\{t\} = t - \lfloor t \rfloor$. The change of variables $y = \eta t^s$ transforms the first integral into $s^{-1}\eta^{-1-1/s}\int_0^\infty y^{1/s}f(e^y)dy$ and by using a suitable expression for $f(\cdot)$ this integral becomes

$$\frac{1}{s\eta^{1+1/s}}(1 - q^{-1/s})\int_0^\infty y^{1/s}\frac{1}{e^y - 1}dy$$

with $\int_0^\infty y^{1/s}\frac{1}{e^y-1}dy = \Gamma(1 + 1/s)\zeta(1 + 1/s)$ (see [12, p. 353]). The second integral, after computing the second derivative, is bounded by

$$\frac{1}{12}\int_0^\infty \left|\left(t^s f(e^{\eta t^s})\right)''\right|dt \in \mathcal{O}\left(\frac{1}{\eta^{1-1/s}}\right).$$

In addition $\frac{1}{12}\left[(t^s f(e^{\eta t^s}))'\right]_0^\infty$ is equal to $\frac{f(1)}{12}$ for $s = 1$ and is null for $s \geq 2$. Collecting these results we get (14.18).

Now, from (14.7)

$$\log(F_{U,q}(e^{-\eta})) = \sum_{k=0}^\infty \log(S_0(e^{-\eta k^s})) - \log q.$$

$$\sum_{k=0}^{\infty} \log(S_0(e^{-\eta k^s})) = \int_0^{\infty} \log(S_0(e^{-\eta t^s}))dt + \frac{1}{2}\log q$$

$$+ \frac{1}{12}\left[(-\eta s t^{s-1} f(e^{\eta t^s}))\right]_0^{\infty}$$

$$+ \frac{\eta s}{2}\int_0^{\infty} (t^{s-1} f(e^{\eta t^s}))' B_2(\{t\})dt .$$

An integration by parts gives $\int_0^{\infty} \log(S_0(e^{-\eta t^s}))dt = \frac{1}{\eta^{1/s}}\int_0^{\infty} y^{1/s} f(e^y)dy$ and after a straightforward but cumbersome computation

$$\int_0^{\infty} |(t^{s-1} f(e^{\eta t^s}))'|dt \in \mathcal{O}\left(\frac{1}{\eta^{1-1/s}}\right).$$

With these results we arrive at (14.19).

In a similar way, still with the help of the Euler–Maclaurin formula, we obtain

$$A_2 = \frac{1}{\eta^2}\sum_{k=0}^{\infty} (\eta k^s)^2 h(e^{\eta k^s})$$

$$= \frac{1}{\eta^2}\int_0^{\infty} (\eta t^s)^2 h(e^{\eta t^s})dt - \frac{1}{2\eta^2}\int_0^{\infty} \left((\eta k^2)^s h(e^{\eta k^s})\right)'' B_2(\{t\})dt$$

where $h = g_2/S_0^2$. The computation of the first integral is based on the relation

$$g_2(x)/(S_0(x))^2 = x^{-1}/(1-x^{-1})^2 - q^2 x^{-q}/(1-x^{-q})^2 ,$$

as a consequence of $(x-1)S_2(x) = 2S_1(x) + S_0(x) - q^2$ and $(x-1)S_1(x) = S_0(x) - q$. The calculation introduces the integral $\int_0^{\infty} \frac{y^{1+1/s}e^y}{(e^y-1)^2}dy$ which is known to be equal to $\Gamma(2+1/s)\zeta(1+1/s)$ (see [12, p. 358]). The second integral is bounded by

$$\frac{1}{2\eta^2}\int_0^{\infty} \left|\left((\eta k^s)^2 h(e^{\eta k^s})\right)''\right|dt \in \mathcal{O}\left(\frac{1}{\eta^{2-1/s}}\right).$$

Therefore

$$A_2 = \frac{1}{\eta^{2+1/s}}\frac{1}{s}(1-q^{-1/s})\Gamma(2+1/s)\zeta(1+1/s) + \mathcal{O}\left(\frac{1}{\eta^{2-1/s}}\right)$$

as expected. Similarly, we get (14.21). Finally, formula (14.22) is an easy consequence of (14.18). □

Formula (14.22) allows us to express the asymptotic behaviors of the quantities $n\eta$, $\log(F_{U,q}(e^{-\eta}))$, $A_\mu$ ($\mu \geq 2$) with respect to $n$. As a consequence, we can easily derive the following estimate.

**Corollary 14.8** *In the monomial case* $U = (k^s)_k$, *with the notation of Theorem 14.7,*

$$M_{U,q}(n) = \left(2q\pi C_2 C_1^{-\frac{2s+1}{s+1}} n^{\frac{2s+1}{s+1}}\right)^{-1/2} e^{(1+s)C_1^{\frac{s}{s+1}} n^{\frac{1}{s+1}}} \left(1 + \mathcal{O}\left(n^{-\frac{1}{s+1}}\right)\right).$$

## 14.3 Proof of Theorem 14.2

Our proof takes into account the original proof of Roth and Szekeres, so that we only pay attention to the modifications which are needed to obtain Theorem 14.2. We first introduce the following function of the positive real number $u$, which is also a function of $\theta \in [-\pi, \pi]$:

$$L(u) = \log \frac{S_0(e^{-\eta u + i\theta u})}{S_0(e^{-\eta u})}.$$

Cauchy's formula

$$M_{U,q}(n) = \frac{1}{2i\pi} \int_C \frac{F_{U,q}(w)}{w^{n+1}} dw$$

can be developed as follows:

$$M_{U,q}(n) = \frac{1}{2\pi} \int_{-\pi}^{\pi} \left\{\prod_{1}^{\infty} S_0(e^{-\eta u + i\theta u})\right\} e^{n\eta - in\theta} d\theta$$

$$= \frac{1}{2\pi} \exp\left\{\eta n + \log(F_{U,q}(e^{-\eta}))\right\} \int_{-\pi}^{\pi} \exp\left\{-in\theta + \sum_{1}^{\infty} L(u)\right\} d\theta.$$

(14.23)

To evaluate the last integral above, we split it up into three parts

$$I_1 = \int_{-\theta_0}^{\theta_0}, \qquad I_2 = \int_{\theta_0}^{\pi}, \qquad I_3 = \int_{-\pi}^{-\theta_0}, \qquad (14.24)$$

with

$$\theta_0 = \eta^{1/5} \left(\frac{1}{2} A_2\right)^{-2/5}. \qquad (14.25)$$

The following lemma, independent of the context, will be used to expand $L(u)$. It simplifies some computations in the proof of Theorem 14.2, and takes into account the general case.

**Lemma 14.9** *(i) For any $t \in ]0, 1[$ the analytic function $f_t(\cdot)$ of $z$ defined by*

$$f_t(z) = \log \frac{S_0(te^z)}{S_0(t)}$$

*has a power series expansion* $f_t(z) = \sum_{v=1}^{\infty} a_v(t)z^v$ *in a disk of radius greater than* $r_t = \frac{1}{2(q-1)} \log \frac{1}{t}$.

(ii) *For* $v = 1, 2, 3, \ldots,$ *there exists a polynomial* $g_v(x)$ *whose highest-degree monomial is* $x^{(q-1)v-1}$ *and such that*

$$v!(S_0(x))^v a_v(1/x) = g_v(x). \tag{14.26}$$

*In particular,* $g_1 = S_1$, $g_2 = S_0 S_2 - S_1^2$.

(iii) *Let* $m \geq 1$ *be integer and let* $t_0 \in ]0, 1[$. *For any* $t \in ]0, t_0]$ *and* $|z| \leq r_t$, *one has*

$$f_t(z) = \sum_{v=1}^{m-1} a_v(t)z^v + \mathcal{O}(|z|^m \sqrt{t}).$$

*Proof.* (i) Write $f_t(z) = \log(1 + Z)$ where $Z$ stands for $Z_t(z)$ with

$$Z_t(z) = \frac{1}{S_0(t)} \sum_{h=1}^{q-1} t^h(e^{zh} - 1).$$

If $|z| \leq r_t = \frac{\log \frac{1}{t}}{2(q-1)}$ we have $|e^{zh} - 1| \leq e^{|z|h} \leq \frac{1}{\sqrt{t}}$, hence

$$|Z| \leq \frac{\sqrt{t}}{S_0(t)} \sum_{h=1}^{q-1} t^{h-1} \leq \sqrt{t}. \tag{14.27}$$

Since $\sqrt{t} < 1$, we can use the power series of the logarithm.

(ii) Using the variables $x = \frac{1}{t}$ and $T = \frac{z}{S_0(x)}$, the relation (14.26) means that the coefficients of the expansion of $f_t(z)$ in powers of $T$ are polynomial in $x$ and we have

$$Z = \frac{1}{S_0(t)} \sum_{h=1}^{q-1} t^h \sum_{k=1}^{\infty} \frac{(zh)^k}{k!} = \sum_{h=1}^{q-1} x^{q-1-h} \sum_{k=1}^{\infty} \frac{(Th)^k}{k!}(S_0(x))^{k-1}. \tag{14.28}$$

More simply

$$Z = \sum_{k=1}^{\infty} p_k(x)T^k \tag{14.29}$$

where $p_k(x)$ is a polynomial with highest degree term $\frac{1}{k!}x^{(q-1)k-1}$.

We now expand $f_t$ in power series of $Z$:

$$f_t(z) = \log(1 + Z) = \sum_{\ell=1}^{\infty}(-1)^{\ell-1}\frac{1}{\ell}Z^\ell. \tag{14.30}$$

From (14.29), the coefficient of $T^v$ in the expansion of $Z^\ell$ is

$$\sum_{k_1+\cdots+k_\ell=v} p_{k_1}(x) \ldots p_{k_\ell}(x),$$

which is a polynomial of degree $(q-1)v - \ell$. Hence the coefficient of $T^v$ from (14.30) is a polynomial whose highest degree term is obtained for $\ell = 1$; this term is $\frac{1}{v!}x^{(q-1)v-1}$.

(iii) From the first equality in (14.28) the coefficients of the expansion of $Z$, in power series of $z$, are positive. Using (14.30), the $a_v(t)$ are linear combinations of the coefficients of the expansion of $Z$, with positive or negative signs.

As above, we expand the function

$$h_t(z) = -\log(1-Z) = \sum_{\ell=1}^{\infty} \frac{1}{\ell} z^\ell$$

in powers of $z$. Therefore $h_t(z) = \sum_{v=1}^{\infty} b_v(t)z^v$ where the $b_v(t)$ are the same linear combinations of the coefficients of the expansion of $Z$, but with positive signs. Consequently $|a_v(t)| \le b_v(t)$ and

$$\left| \sum_{v=m}^{\infty} a_v(t)z^v \right| \le \sum_{v=m}^{\infty} |a_v(t)| \left(\frac{|z|}{r_t}\right)^m r_t^v \le \left(\frac{|z|}{r_t}\right)^m h_t(r_t). \qquad (14.31)$$

We bound $h_t(r_t) = -\log(1-Z)$, where $Z = Z_t(r_t)$ is real, by means of (14.27):

$$h_t(r_t) = -\log(1-Z) \le \frac{Z}{1-Z} \le \frac{\sqrt{t}}{1-\sqrt{t_0}}.$$

Moreover, we can bound $\frac{|z|}{r_t}$ in (14.31) by $\frac{|z|}{r_{t_0}}$ and conclude that

$$\sum_{v=m}^{\infty} a_v(t)z^v = \mathcal{O}(|z|^m \sqrt{t}).$$

$\square$

We apply Lemma 14.9 to estimate the integral $I_1$. Now we choose $t = e^{-\eta u}$, $z = i\theta u$ and replace $m$ by $2m$ in the lemma. The definition of $\theta_0$ in (14.25) implies $\theta_0 < \eta^\alpha < \frac{1}{2(q-1)}\eta$ for $n$ large enough, where $\alpha = 1 + \frac{2}{5s} - \delta$. So the condition $|z| \le r_t$ is satisfied by any $-\theta_0 \le \theta \le \theta_0$ and one has

$$I_1 = \int_{-\theta_0}^{\theta_0} \exp\left\{ -in\theta + \sum_{1}^{\infty} L(u) \right\} d\theta$$

$$= \int_{-\theta_0}^{\theta_0} \exp\left\{ -in\theta + \sum_{1}^{\infty} \left( \sum_{\mu=1}^{2m-1} a_\mu(e^{\eta u})(i\theta u)^\mu \right. \right.$$

$$\left. \left. + \mathcal{O}\left( (|\theta|u)^{2m} e^{-\frac{\eta u}{2}} \right) \right) \right\} d\theta.$$

Recall that $g_1 = S_1$. From (14.9), one simplifies $-in\theta$ with the first term of the sum $\sum_{\mu=1}^{2m-1}$. Then, using the definition of $A_\mu$ in (14.12),

$$I_1 = \int_{-\theta_0}^{\theta_0} \exp\left\{ \sum_{\mu=2}^{2m-1} \frac{1}{\mu!} A_\mu (i\theta)^\mu + \mathcal{O}\left( |\theta|^{2m} \sum_1^\infty u^{2m} e^{-\frac{\eta u}{2}} \right) \right\} d\theta. \quad (14.32)$$

Let

$$A'_{2m} = \sum_1^\infty u^{2m} e^{-\frac{\eta u}{2}}$$

and, for $2 \leq \mu \leq 2m-1$, set $A'_\mu = A_\mu$. Of course the estimation of $A_\mu$ given in Corollary 14.4 is also valid for $A'_{2m}$, the proof being similar. Since $\theta_0 = \mathcal{O}\left(n^{-\frac{s+2/5}{s+1}+\delta}\right)$ we get for any integer $\mu$ subject to $3 \leq \mu \leq 2m$,

$$A'_\mu \theta_0^\mu = \mathcal{O}\left(n^{-\frac{1}{5(s+1)}+\delta}\right).$$

We deduce that, for $|\theta| \leq \theta_0$,

$$\exp\left\{ \sum_{\mu=3}^{2m-1} \frac{1}{\mu!} A'_\mu (i\theta)^\mu + \mathcal{O}\left(A'_{2m}|\theta|^{2m}\right) \right\}$$

$$= 1 + \sum_{\nu=1}^{5m} \frac{1}{\nu!}\left\{ \sum_{\mu=3}^{2m-1} \frac{1}{\mu!} A'_\mu (i\theta)^\mu + \mathcal{O}\left(A'_{2m}|\theta|^{2m}\right) \right\}^\nu + \mathcal{O}\left(n^{-\frac{m}{s+1}+\delta}\right)$$

$$= 1 + E(\theta) + \mathcal{O}\{R(|\theta|)\} + \mathcal{O}\left(n^{-\frac{m}{s+1}+\delta}\right),$$

where

$$E(\theta) = \sum_{\nu=1}^{5m}\sum_{\mu_1=3}^{2m-1}\cdots\sum_{\mu_\nu=3}^{2m-1} \frac{1}{\nu!\mu_1!\dots\mu_\nu!} A'_{\mu_1}\dots A'_{\mu_\nu} (i\theta)^{\mu_1+\cdots+\mu_\nu}$$

$$R(|\theta|) = \sum_{\nu=1}^{5m}\sum_{\mu_1=3}^{2m}\cdots\sum_{\mu_\nu=3}^{2m} A'_{\mu_1}\dots A'_{\mu_\nu} |\theta|^{\mu_1+\cdots+\mu_\nu},$$

(14.33)

the last summation being subject to $\max(\mu_1,\dots,\mu_\nu) = 2m$.

Now, use (14.32) and the substitution $t = \left(\frac{1}{2}A_2\right)^{1/2}\theta$:

$$I_1 = \int_{-\theta_0}^{\theta_0} e^{-\frac{1}{2}A_2\theta^2}\left[1 + E(\theta) + \mathcal{O}\{R(|\theta|)\} + \mathcal{O}\left(n^{-\frac{m}{s+1}+\delta}\right)\right]d\theta$$

$$= \sqrt{\frac{2}{A_2}} \int_{-B^{1/10}}^{B^{1/10}} e^{-t^2}\left[1 + E\left(t(\tfrac{1}{2}A_2)^{-1/2}\right) + \mathcal{O}\left\{R(|t|A_2^{-1/2})\right\} + \mathcal{O}\left(n^{-\frac{m}{s+1}+\delta}\right)\right]dt,$$

where $B = \frac{1}{2}\eta^2 A_2$.

In (14.33) replace $\theta$ by $(\frac{1}{2}A_2)^{-1/2}t$ and group together the terms in the following way. Let

$$
C_{\rho,\nu} = \sum_{\mu_1=3}^{2m-1} \cdots \sum_{\mu_\nu=3}^{2m-1} \frac{1}{\nu!\mu_1!\ldots\mu_\nu!} A'_{\mu_1} \cdots A'_{\mu_\nu} \left(\frac{1}{2}A_2\right)^{-\frac{1}{2}(\rho+2\nu)}
$$

subject to $(\mu_1 - 2) + \cdots + (\mu_\nu - 2) = \rho$ and in the same way define $C'_{\rho,\sigma}$ by replacing in the summation the upper bounds $2m - 1$ by $2m$. Since each $\mu$ is between 3 and $2m$, one has $\nu \le \rho \le 10m^2$. Thus

$$
\begin{aligned}
E\left(t(\tfrac{1}{2}A_2)^{-1/2}\right) &= \sum_{\rho=1}^{10m^2}\sum_{\nu=1}^{\rho} C_{\rho,\nu}(\mathrm{i}t)^{\rho+2\nu}, \\
R(|t|A_2^{-1/2}) &= \sum_{\rho=2m-2}^{10m^2}\sum_{\nu=1}^{\rho} C'_{\rho,\nu}|t|^{\rho+2\nu}.
\end{aligned}
\tag{14.34}
$$

The estimations in Corollary 14.4 imply

$$
C_{\rho,\nu}, C'_{\rho,\nu} = \mathcal{O}\left(n^{-\frac{\rho}{2(s+1)}+\delta}\right).
$$

Thus, collecting together the above results,

$$
\begin{aligned}
I_1 = \sqrt{\frac{2}{A_2}} \int_{-B^{1/10}}^{B^{1/10}} e^{-t^2}\Bigg\{ &1 + \sum_{\rho=1}^{2m-3}\sum_{\nu=1}^{\rho} C_{\rho,\nu}(\mathrm{i}t)^{\rho+2\nu} \\
&+ \mathcal{O}\Big(n^{-\frac{m-1}{s+1}+\delta}\sum_{\sigma=0}^{30m^2}|t|^\sigma\Big)\Bigg\}\mathrm{d}t.
\end{aligned}
\tag{14.35}
$$

Since $B = \mathcal{O}\left(n^{\frac{1}{s+1}+\delta}\right)$, there exists a positive constant $c$ such that, for $\sigma \le 30m^2$,

$$
\int_{B^{1/10}}^{\infty} e^{-t^2}t^\sigma\,\mathrm{d}t = \mathcal{O}(e^{-n^c}).
$$

On the other side, for odd $\rho$ the function $e^{-t^2}t^{\rho+2\nu}$ is odd and has null integral, so we consider in (14.35) only the terms with index $\rho = 2\rho'$. This leads to

$$
\begin{aligned}
I_1 &= \sqrt{\frac{2}{A_2}}\Bigg[ \int_{-\infty}^{\infty} e^{-t^2}\Big\{ 1 + \sum_{\rho'=1}^{m-2}\sum_{\nu=1}^{2\rho'}(-1)^{\rho'+\nu}C_{2\rho',\nu}t^{2\rho'+2\nu}\Big\}\mathrm{d}t \\
&\qquad + \mathcal{O}(n^{-\frac{m-1}{s+1}+\delta})\Bigg] \\
&= \sqrt{\frac{2\pi}{A_2}}\Big\{ 1 + \sum_{\rho=1}^{m-2} D_\rho + \mathcal{O}(n^{-\frac{m-1}{s+1}+\delta})\Big\}\mathrm{d}t
\end{aligned}
\tag{14.36}
$$

where, for certain numerical constants $e_\nu$,

$$D_\rho = \sum_{\nu=1}^{2\rho} e_\nu C_{2\rho,\nu}.$$

We shall now show that the $D_\rho$ have the form (14.10) given in the statement of the theorem. We have

$$C_{2\rho,\nu} = 2^{\rho+\nu} A_2^{-6\rho} \sum_{\mu_1=3}^{2m-1} \cdots \sum_{\mu_\nu=3}^{2m-1} \frac{1}{\nu! \mu_1! \ldots \mu_\nu!} A_{\mu_1} \ldots A_{\mu_\nu} A_2^{5\rho-\nu}$$

subject to $(\mu_1 - 2) + \cdots + (\mu_\nu - 2) = 2\rho$, so we can write

$$C_{2\rho,\nu} = 2^{\rho+\nu} A_2^{-6\rho} \sum_{\mu_1=3}^{2m-1} \cdots \sum_{\mu_\nu=3}^{2m-1} \sum_{\mu_{\nu+1}=2}^{2} \cdots \sum_{\mu_{5\rho}=2}^{2} \frac{1}{\nu! \mu_1! \ldots \mu_\nu!} A_{\mu_1} \ldots A_{\mu_{5\rho}}$$

subject to $\mu_1 + \cdots + \mu_{5\rho} = 12\rho$. Thus $D_\rho$ has the form (14.10).

Taking into account the evaluation of $I_1$, it remains only to find suitable bounds for $I_2$ and $I_3$. The modulus of the integrand, say $G(\theta)$, is the exponential of

$$\frac{1}{2} \sum_{1}^{\infty} \Big|_U \log \left| \frac{S_0(te^{i\theta u})}{S_0(t)} \right|^2,$$

where $t = e^{-\eta u}$. Now we bound from above $\log \left| \frac{S_0(te^{i\theta u})}{S_0(t)} \right|^2$ by

$$\left| \frac{S_0(te^{i\theta u})}{S_0(t)} \right|^2 - 1 = -2 \cdot \frac{\displaystyle\sum_{h=1}^{q-1} t^h \sum_{1 \le j \le \min(h, 2q-2-h),\, h-j \text{ even}} (1 - \cos(j\theta u))}{(S_0(t))^2}.$$

Hence, this term is negative and we can bound

$$\sum_{1}^{\infty} \Big|_U \log \left| \frac{S_0(te^{i\theta u})}{S_0(t)} \right|^2$$

by the finite sum

$$\sum_{1}^{k} \Big|_U \log \left| \frac{S_0(te^{i\theta u})}{S_0(t)} \right|^2.$$

We choose the integer $k = k(n)$ such that

$$u_k \le \eta^{-1} < u_{k+1}.$$

Condition (I) says that $\log u_k \sim s \log k$ and by Lemma 14.3

$$\lim_{n \to \infty} \frac{\log k(n)}{\log n} = \frac{1}{s+1}.$$

On one side, for $1 \leq v \leq k$ we have $\frac{1}{e} \leq t = e^{-\eta u_v} \leq 1$, and clearly $\frac{t^h}{(S_0(t))^2}$ is bounded from 0 by a positive constant $c$ valid for any $1 \leq h \leq q - 1$. On the other side, setting $\theta = 2\pi\alpha$, the numbers $1 - \cos(j\theta u)$ are greater than $\|j u\alpha\|^2$. Finally we have, with another positive constant $c'$,

$$\log G(\theta) \leq -c' \sum_{1}^{k} {}_{U} \sum_{1 \leq j \leq q-1} \|j u\alpha\|^2. \tag{14.37}$$

Here, using Corollary 14.4 and (14.24), $\alpha = \frac{\theta}{2\pi}$ is at least equal to

$$\frac{\theta_0}{2\pi} \geq n^{-\frac{s+\frac{2}{5}}{s+1} - \delta} \geq k^{-s-\frac{2}{5} - 2\delta}.$$

By Condition (II) the lower bound of $(\log k)^{-1} \sum_{v=1}^{k} \|u_v\alpha\|^2$ for $\frac{1}{2u_k} < \alpha \leq \frac{1}{2}$ has infinite limit when $k$ tends to infinity. This remains true when we take the lower bound for $k^{-s-\frac{2}{5}-2\delta} \leq \alpha \leq \frac{1}{2u_k}$, because

$$\sum_{v=1}^{k} \|u_v\alpha\|^2 = \sum_{v=1}^{k} u_v^2 \alpha^2$$
$$\geq k^{2s+1-\delta-2s-\frac{4}{5}-4\delta}$$
$$\geq k^{\frac{1}{5}-5\delta}.$$

So we deduce from (14.37) that the lower bound of $(\log k)^{-1} \log G(\theta)$ for $\theta_0 \leq \theta \leq \pi$ or $-\pi \leq \theta \leq -\theta_0$ has infinite limit when $n$ tends to infinity, and for $n$ large enough

$$\log G(\theta) < -(m+1) \log n.$$

Consequently, $G(\theta)$ and $I_2, I_3$ are in $\mathcal{O}(n^{-m}) = \mathcal{O}\left(A_2^{-1/2} n^{-(m-1)}\right)$, proving (14.8). $\qquad\square$

## Acknowledgement

These authors are partially supported by Aix Marseille University, CNRS: UMR 7373, École Centrale de Marseille.

# References

[1] F. C. Auluck and C. B. Haselgrove, On Ingham's Tauberian theorem for partitions. *Proc. Cambridge Philos. Soc.* **48**, 566–570, 1952.

[2] W. De Azevedo Pribitkin, Revisiting Rademacher's formula for the partition function $p(n)$. *Ramanujan J.* **4**, 455–467, 2000.

[3] N. A. Brigham, A general asymptotic formula for partitions. *Proc. Am. Math. Soc.* **1**, 182–191, 1950.

[4] N. G. de Bruijn, On Mahler's partition problem. *Indag. Math.* **10**, 210–220, 1948.

[5] J.-M. Dumont, N. Sidorov and A. Thomas, Number of representations related to linear recurrent basis. *Acta Arith.* **88**(4), 371–396, 1999.

[6] M. Dutta, On new partition of numbers. *Rend. Sem. Mat. Univ. Padova* **25**, 138–143, 1956.

[7] P. Erdős, On an elementary proof of some asymptotic formulae in the theory of partitions. *Ann. Math. (2)* **43**, 437–450, 1942.

[8] P. Erdős and B. Richmond, Concerning periodicity in the asymptotic behavior of partition functions. *J. Aust. Math. Soc. Ser. A* **21**, 447–456, 1976.

[9] P. Erdős and B. Richmond, Partitions into summands of the form $[m\alpha]$. *Proceedings of the Seventh Manitoba Conference on Numerical Mathematics and Computing (Univ. Manitoba, Winnipeg, Man., 1977)*, pp. 371–377. Utilitas Math., Winnipeg, Man., 1978.

[10] D.-J. Feng, P. Liardet and A. Thomas, Partition functions in numeration systems with bounded multiplicity. *Unif. Distrib. Theory* **9**(1), 43–77, 2014.

[11] P. Flajolet and R. Sedgewick, *Analytic Combinatorics*. Cambridge University Press, Cambridge, 2009.

[12] I. S. Gradshteyn and I. M. Ryzhik, *Table of Integrals, Series, and Products*, seventh edition. Academic Press, San Diego, CA, 2007.

[13] P. Hagis, Jr., Partitions with a restriction on the multiplicity of the summands. *Trans. Am. Math. Soc.* **155**, 375–384, 1971.

[14] G. H. Hardy and S. Ramanujan, Asymptotic formulae in combinatory analysis. *Proc. London Math. Soc. Ser. 2* **17**, 75–115, 1918.

[15] L.-K. Hua, On the number of partitions of a number into unequal parts. *Trans. Am. Math. Soc.* **51**, 939–961, 1942.

[16] A. E. Ingham, A Tauberian theorem for partition. *Ann. Math.* **42**(5), 1075–1090, 1941.

[17] K. Mahler, On a special functional equation. *J. London Math. Soc.* **15**, 115–123, 1940.

[18] G. Menardus, Asymptotische Aussagen über Partitionen. *Math. Z.* **59**, 388–398, 1954.

[19] D. J. Newman, The evaluation of the constant in the formula for the number of partitions of $n$. *Am. J. Math.* **73**, 599–601, 1951.

[20] W. B. Pennington, On Mahler's partition problem. *Ann. Math.* **57**(3), 531–546, 1953.

[21] V. Yu Protasov, Asymptotic behavior of the partition function. *Sb. Math.* **191**, 381–414, 2000.

[22] H. Rademacher, On the expansion of the partition function in a series, *Ann. Math.* **44**, 416–422, 1943.

[23]  B. Reznick, Some binary partition functions. *Analytic Number Theory (AllertonPark, IL, 1989)*. Progress in Mathematics, volume 85, pp. 451–47. Birkhäuser, Boston, MA, 1990.

[24]  L. B. Richmond, Asymptotic relations for partitions. *J. Number Theory* **7**, 389–405, 1975.

[25]  B. Richmond, Mahler's partition problem. *Ars Combin.* **2**, 169–189, 1976.

[26]  K. F. Roth and G. Szekeres, Some asymptotic formulae in the theory of partitions. *Q. J. Math. Oxford* **5**(2), 241–259, 1954.

[27]  W. Schwartz, Einige Anwendungen Tauberscher Sätze in der Zahlentheorie. C: Mahler's Partitionsproblem. *J. Reine Angew. Math.* **228**, 182–188, 1967.

[28]  H. N. V. Temperly, Statistical mechanics and the partition of numbers I. The transition of liquid helium. *Proc. R. Soc. London* **109**, 361–375, 1949.

[29]  E. M. Wright, Asymptotic partition formulae, III. Partitions into $k^{th}$ powers. *Acta Math.* **63**, 143–191, 1934.

# 15

# A trigonometric approach for Chebyshev polynomials over finite fields

*Juliano B. Lima*
Federal University of Pernambuco, Recife

*Daniel Panario*
Carleton University, Ottawa

*Ricardo M. Campello de Souza*
Federal University of Pernambuco, Recife

*Dedicated to the 70th birthday of Harald Niederreiter.*

## Abstract

In this paper, we introduce trigonometric definitions for Chebyshev polynomials over finite fields $\mathbb{F}_q$, where $q = p^m$, $m$ is a positive integer and $p$ is an odd prime. From such definitions, we derive recurrence relations which are equivalent to those established for real valued Chebyshev polynomials and for Chebyshev polynomials of the first and second kinds over finite fields. Periodicity and symmetry properties of these polynomials are also studied. Such properties are then used to derive sufficient conditions for the Chebyshev polynomials of the second, third and fourth kinds over finite fields to be permutation polynomials.

## 15.1 Introduction

Since their introduction in the nineteenth century, Chebyshev polynomials have been widely studied. Usually, such polynomials are defined over the real numbers by means of ordinary trigonometric and inverse trigonometric functions cosine and sine. This produces four main kinds of polynomials, which have been applied, for instance, in problems related to orthogonal polynomials, polynomial approximation, numerical integration and spectral methods for partial differential equations [14].

Definitions for Chebyshev polynomials over finite fields also exist. However, in this scenario, such polynomials are better known as Dickson polynomials

255

and are evaluated by recurrence relations and a functional equation [11, 17]. Let $\mathbb{F}_q$ be a finite field of $q$ elements. For any integer $n \geq 2$ and a parameter $a \in \mathbb{F}_q$, the $n$th Dickson polynomial of the first kind $D_n(x, a) \in \mathbb{F}_q[x]$ is computed by

$$D_n(x, a) = x D_{n-1}(x, a) - a D_{n-2}(x, a), \qquad (15.1)$$

where $D_0(x, a) = 2$ and $D_1(x, a) = x$. For odd $q$, the corresponding $n$th Chebyshev polynomial of the first kind, $T_n(x)$, can be obtained from $T_n(x) = D_n(2x, 1)/2$. The $n$th Dickson polynomial of the second kind, $E_n(x, a) \in \mathbb{F}_q[x]$, can be obtained using the recurrence relation given in (15.1), with $E_0(x, a) = 1$ and $E_1(x, a) = x$. The corresponding $n$th Chebyshev polynomial of the second kind, $U_n(x)$, can be obtained from $U_n(x) = E_n(2x, 1)$. To the best of our knowledge, Chebyshev polynomials of the third and fourth kinds over finite fields have not been studied.

Concerning their applications, Chebyshev polynomials over finite fields have been used to design public-key cryptography schemes [6, 10, 13], where semigroup and chaotic properties of $T_n(x)$ are explored. Chebyshev polynomials over finite fields have also been used to describe maximal curves over finite fields [7] and to investigate reversibility of two-dimensional additive cellular automata on finite square grids [16]. Furthermore, permutation properties of these polynomials are particularly important in the construction of cryptographic systems for secure transmission of data [11].

In this paper, we define Chebyshev polynomials over finite fields using concepts of finite field trigonometry. Such concepts were first introduced by Campello de Souza *et al.*, as part of the requirements for defining the Hartley transform in a finite field [1]. Here, in Section 15.2, we extend these concepts by introducing hyperbolic trigonometric functions over finite fields and studying some basic properties. In Section 15.3 we give trigonometric definitions of the four kinds of Chebyshev polynomials over finite fields and develop the corresponding recurrence relations. Such relations coincide with those obtained for real valued Chebyshev polynomials and also with those obtained for Chebyshev polynomials derived from Dickson polynomials over finite fields for the first and second kinds. In Section 15.4 we study periodicity and symmetry properties of the introduced polynomials and, in Section 15.5, we discuss their permutation behavior. More specifically, we use the definitions proposed in Section 15.3 to obtain sufficient conditions for the Chebyshev polynomials of the second, third and fourth kinds to be permutation polynomials. The paper closes with some conclusions in Section 15.6.

## 15.2 Trigonometry in finite fields

In this section, we present some definitions which are slightly different from those given in previous works [1, 13]. Additionally, we introduce the concept of finite field hyperbolic trigonometric functions. Throughout this paper, we assume that we work in the finite field $\mathbb{F}_q$, where $q = p^m$, $m$ is a positive integer and $p$ is an odd prime. In what follows, we consider that $m \geq 2$ and the computations are carried out modulo an irreducible polynomial $f$ of degree $m$ over $\mathbb{F}_p$. If $m = 1$, we just have to substitute $q$ and (mod $f$) by $p$ and (mod $p$), respectively.

**Definition 15.1** *The set of Gaussian integers over $\mathbb{F}_q$ is the set $\mathbb{I}_q$ with elements in the form $\zeta = a + bj$, such that $a, b \in \mathbb{F}_q$ and $j^2 \equiv \kappa$ is a quadratic nonresidue over $\mathbb{F}_q$.*

An element $\zeta \in \mathbb{I}_q$ can be viewed as a "complex" element with "real" and "imaginary" parts given by $\Re\{\zeta\} = a$ and $\Im\{\zeta\} = b$, respectively. In this sense, $\zeta^* = a - bj$ denotes the finite field conjugate of $\zeta = a + bj$. Additionally, we observe that $\mathbb{I}_q$ is isomorphic to $\mathbb{F}_{q^2}$.

**Definition 15.2** *The unimodular set of $\mathbb{I}_q$ is the set $G_1$ of elements $\zeta = a + bj \in \mathbb{I}_q$, such that $\zeta \cdot \zeta^* = (a + bj) \cdot (a - bj) = a^2 - b^2 j^2 \equiv 1 \pmod{f}$.*

If $\zeta = a + bj$ is unimodular, $\zeta^* = \zeta^{-1} = a - bj$.

**Proposition 15.3** $\zeta^{q+1} \equiv a^2 - b^2 j^2 \pmod{f}$.

*Proof.* In $\mathbb{I}_q$, $\zeta^q = (a+bj)^q = a^q + b^q j^q$, since this field is isomorphic to $\mathbb{F}_{q^2}$, a field of characteristic $p$. Additionally, one has $j^q = \kappa^{\frac{q}{2}} = \kappa^{\frac{q-1}{2}} \kappa^{\frac{1}{2}} = -\kappa^{\frac{1}{2}} = -j$. Therefore, $\zeta^q = a - bj = \zeta^*$ and $\zeta^{q+1} = \zeta \cdot \zeta^* \equiv a^2 - b^2 j^2 \pmod{f}$. $\square$

**Proposition 15.4** *The structure $\langle G_1, \cdot \rangle$ is a cyclic group of order $q + 1$.*

*Proof.* Take a primitive element $\sigma$ of $\mathbb{I}_q$. Then we have $G_1 = \langle \sigma^{q-1} \rangle$ and the result follows. $\square$

Naturally, $\langle G_1, \cdot \rangle$ is isomorphic to the group formed by the $(q + 1)$th roots of unity under multiplication. This suggests that we can associate an angle to each unimodular element $\zeta \in G_1$. In this context, the following finite field trigonometric functions can be defined.

**Definition 15.5** *Let* $\zeta \in \mathbb{I}_q$ *be a unimodular element with multiplicative order denoted by* $\mathrm{ord}(\zeta)$. *The finite field cosine and sine of the angle related to* $\zeta^x$, *the xth power of* $\zeta$, *are defined respectively as*

$$\cos_\zeta(x) = \frac{\zeta^x + \zeta^{-x}}{2} \tag{15.2}$$

$$\sin_\zeta(x) = \frac{\zeta^x - \zeta^{-x}}{2j}, \tag{15.3}$$

*for* $x = 0, 1, \ldots, \mathrm{ord}(\zeta) - 1$.

Expressions (15.2) and (15.3) are evaluated modulo $f$. In Definition 15.5, an explicit reference to the angle related to $\zeta^x$ is not important; this allows us to view finite field cosines and sines as functions of $x$, given a unimodular element $\zeta$. The set of all possible values for finite field cosines with respect to $\zeta$ is denoted by $\mathbb{C}_\zeta$; the set of all possible values for finite field sines with respect to $\zeta$ is denoted by $\mathbb{S}_\zeta$. Such functions hold some properties similar to those of the real valued functions [1]. As an example, we consider the unit circle property, which can be written as

$$\cos^2_\zeta(x) - j^2 \sin^2_\zeta(x) \equiv 1 \pmod{f}.$$

This comes directly from Definition 15.2. Note that, if $q \equiv 3 \pmod{4}$ and we use $j = \sqrt{-1}$, the last expression becomes similar to the classical fundamental trigonometric identity. The formulas for the cosine and sine of the addition of two arcs, given respectively by

$$\begin{aligned}\cos_\zeta(x + y) &= \cos_\zeta(x) \cdot \cos_\zeta(y) + j^2 \sin_\zeta(x) \cdot \sin_\zeta(y) \\ \sin_\zeta(x + y) &= \sin_\zeta(x) \cdot \cos_\zeta(y) + \cos_\zeta(x) \cdot \sin_\zeta(y),\end{aligned} \tag{15.4}$$

can be derived from Definition 15.5.

Symmetry properties of the cosine and sine functions are preserved in the finite field context. The angle related to $\zeta^{-x}$ is the additive inverse of that related to $\zeta^x$. Therefore, one has $\cos_\zeta(-x) = \cos_\zeta(x)$ and $\sin_\zeta(-x) = -\sin_\zeta(x)$.

**Proposition 15.6** *Let* $\zeta \in \mathbb{I}_q$ *be a unimodular element. The finite field cosine and sine of the angle related to* $\zeta^x$, *the xth power of* $\zeta$, *can be computed respectively as*

$$\cos_\zeta(x) = \mathfrak{R}\{\zeta^x\}$$

$$\sin_\zeta(x) = \mathfrak{I}\{\zeta^x\},$$

*for* $x = 0, 1, \ldots, \mathrm{ord}(\zeta) - 1$.

Table 15.1 *Elements in the unimodular set of* $\mathbb{I}_7$ *and respective* $\cos(\cdot)$ *and* $\sin(\cdot)$*; the number* $\zeta = 2 + 2j \in \mathbb{I}_7$*, which has multiplicative order* $\text{ord}(\zeta) = 8$*, is considered*

| $x$ | $\zeta^x$ | $\cos_\zeta(x)$ | $\sin_\zeta(x)$ |
|-----|-----------|-----------------|-----------------|
| 0 | 1 | 1 | 0 |
| 1 | $2 + 2j$ | 2 | 2 |
| 2 | $j$ | 0 | 1 |
| 3 | $5 + 2j$ | 5 | 2 |
| 4 | 6 | 6 | 0 |
| 5 | $5 + 5j$ | 5 | 5 |
| 6 | $6j$ | 0 | 6 |
| 7 | $2 + 5j$ | 2 | 5 |

*Proof.* From Definition 15.5, we observe that

$$\zeta^x = \cos_\zeta(x) + j\sin_\zeta(x),$$

and the result follows. □

**Example 15.7** *Let us consider* $j = \sqrt{-1} \in \mathbb{I}_7$ *and the unimodular element* $\zeta = (2 + 2j) \in \mathbb{I}_7$*, which has multiplicative order* $\text{ord}(\zeta) = 8$*. The powers* $\zeta^x$*,* $x = 0, 1, \ldots, 7$*, of* $\zeta$*, i.e., the elements of* $G_1$*, as well as the respective finite field trigonometric functions are shown in Table 15.1.*

**Example 15.8** *Let* $\alpha$ *be a root of the primitive polynomial* $f(t) = t^2 + t + 2 \in \mathbb{F}_3[t]$ *and* $\mathbb{F}_9 = \mathbb{F}_3(\alpha)$*. Let us consider* $j = \sqrt{2\alpha} \in \mathbb{I}_9$ *and the unimodular element* $\zeta = (\alpha + 2j) \in \mathbb{I}_9$*, which has multiplicative order* $\text{ord}(\zeta) = 10$*. The powers* $\zeta^x$*,* $x = 0, 1, \ldots, 9$*, of* $\zeta$*, i.e., the elements of* $G_1$*, as well as the respective finite field trigonometric functions are shown in Table 15.2.*

From our previous discussion, we see that all possible different values for cosines of angles related to elements in $G_1$ can be obtained by considering a generator $\zeta_u$ of $G_1$. This is achieved by computing

$$\cos_{\zeta_u}(x) = \frac{\zeta_u^x + \zeta_u^{-x}}{2}, \quad x = 0, 1, \ldots, \frac{q+1}{2}. \tag{15.5}$$

Similarly, all possible different values for sines of angles related to elements in $G_1$ are obtained from

$$\sin_{\zeta_u}(x) = \frac{\zeta_u^x - \zeta_u^{-x}}{2j}, \quad x = -\left\lfloor \frac{q+1}{4} \right\rfloor, -\left\lfloor \frac{q+1}{4} \right\rfloor + 1, \ldots, \left\lfloor \frac{q+1}{4} \right\rfloor, \tag{15.6}$$

where $\lfloor \cdot \rfloor$ denotes the integer part of the argument.

Table 15.2 *Elements in the unimodular set of* $\mathbb{I}_9$
*and respective* $\cos(\cdot)$ *and* $\sin(\cdot)$; *the element*
$\zeta = \alpha + 2j \in \mathbb{I}_9$ *where* $\mathbb{F}_9 = \mathbb{F}_3(\alpha)$, *which has*
*multiplicative order* $\mathrm{ord}(\zeta) = 10$, *is considered*

| $x$ | $\zeta^x$ | $\cos_\zeta(x)$ | $\sin_\zeta(x)$ |
|---|---|---|---|
| 0 | 1 | 1 | 0 |
| 1 | $\alpha + 2j$ | $\alpha$ | 2 |
| 2 | $(\alpha + 1) + \alpha j$ | $\alpha + 1$ | $\alpha$ |
| 3 | $(2\alpha + 2) + \alpha j$ | $2\alpha + 2$ | $\alpha$ |
| 4 | $2\alpha + 2j$ | $2\alpha$ | 2 |
| 5 | 2 | 2 | 0 |
| 6 | $2\alpha + j$ | $2\alpha$ | 1 |
| 7 | $(2\alpha + 2) + 2\alpha j$ | $2\alpha + 2$ | $2\alpha$ |
| 8 | $(\alpha + 1) + 2\alpha j$ | $\alpha + 1$ | $2\alpha$ |
| 9 | $\alpha + j$ | $\alpha$ | 1 |

In fact, computing cosines over the range given in expression (15.5) produces exactly $(q + 3)/2$ different values. Analogously, computing sines over
the range given in expression (15.6) produces exactly $2\left\lfloor \frac{q+1}{4} \right\rfloor + 1$ different
values. Such results are supported by the following lemmas.

**Lemma 15.9** *Let* $\zeta_u \in \mathbb{I}_q$ *be a generator of* $\mathrm{G}_1$. *For* $0 \leq x, y \leq \frac{q+1}{2}$,

$$\cos_{\zeta_u}(x) = \cos_{\zeta_u}(y) \quad \textit{if and only if} \quad x = y.$$

*Proof.* If $x \equiv y \pmod{(q + 1)}$,

$$\cos_{\zeta_u}(x) = \frac{\zeta_u^x + \zeta_u^{-x}}{2} = \frac{\zeta_u^y + \zeta_u^{-y}}{2} = \cos_{\zeta_u}(y).$$

On the other hand, if $\cos_{\zeta_u}(x) = \cos_{\zeta_u}(y)$,

$$\zeta_u^y + \zeta_u^{-y} = 2\cos_{\zeta_u}(x) \quad \text{implies} \quad (\zeta_u^y)^2 - 2\cos_{\zeta_u}(x)\zeta_u^y + 1 = 0.$$

Solving the last equation, one has

$$\zeta_u^y = \cos_{\zeta_u}(x) \pm \sqrt{\cos_{\zeta_u}^2(x) - 1} = \cos_{\zeta_u}(x) \pm j \sin_{\zeta_u}(x) = \zeta_u^{\pm x},$$

which implies $y = x$, once the range $0 \leq x, y \leq \frac{q+1}{2}$ is considered. □

**Lemma 15.10** *Let* $\zeta_u \in \mathbb{I}_q$ *be a generator of* $\mathrm{G}_1$. *For* $-\left\lfloor \frac{q+1}{4} \right\rfloor \leq x, y \leq \left\lfloor \frac{q+1}{4} \right\rfloor$,

$$\sin_{\zeta_u}(x) = \sin_{\zeta_u}(y) \quad \textit{if and only if} \quad x = y.$$

*Proof.* If $x \equiv y \pmod{(q+1)}$,

$$\sin_{\zeta_u}(x) = \frac{\zeta_u^x - \zeta_u^{-x}}{2j} = \frac{\zeta_u^y - \zeta_u^{-y}}{2j} = \sin_{\zeta_u}(y).$$

On the other hand, if $\sin_{\zeta_u}(x) = \sin_{\zeta_u}(y)$,

$$\zeta_u^y - \zeta_u^{-y} = 2j\sin_{\zeta_u}(x) \quad \text{implies} \quad (\zeta_u^y)^2 - 2j\sin_{\zeta_u}(x)\zeta_u^y - 1 = 0.$$

Solving the last equation, one has

$$\zeta_u^y = j\sin_{\zeta_u}(x) \pm \sqrt{1 + j^2 \sin_{\zeta_u}^2(x)} = j\sin_{\zeta_u}(x) \pm \cos_{\zeta_u}(x) = \pm\zeta_u^{\pm x}.$$

Since $-1 = \zeta_u^{\frac{q+1}{2}}$, one has $-\zeta_u^{-x} = \zeta_u^{\frac{q+1}{2}-x}$. Due to the range considered, we are left with $y = x$ and the proof is complete. $\qquad\square$

Lemma 15.9 and the even symmetry of the finite field cosine indicate that the computation of such cosines does not produce all elements of $\mathbb{F}_q$. In other words, given an element $g \in \mathbb{F}_q$, there may not exist a number $x$, such that $\cos_{\zeta_u}(x) = g$. In Example 15.7, for instance, we see in the corresponding table that there are no cosines of the values $\pm 3 \pmod 7$. On the other hand, there may exist two different angles with the same cosine value. These facts, which are also verified for finite field sines, should be considered in the computation of inverse trigonometric functions, which are discussed in the sequel.

**Definition 15.11** *Let $C = \cos_\zeta(x_c)$ (respectively $S = \sin_\zeta(x_s)$) be the cosine (respectively sine) of the angle related to $\zeta^{x_c}$ (respectively $\zeta^{x_s}$), the $x_c$th (respectively $x_s$th) power of the unimodular element $\zeta \in \mathbb{I}_q$. The inverse cosine of $C$ and the inverse sine of $S$ are, respectively, the functions $\arccos_\zeta(C) = x_c$ and $\arcsin_\zeta(S) = x_s$, where the ranges of $x_c$ and $x_s$ are given in Equations (15.5) and (15.6).*

Considering Definition 15.11, we can derive a closed formula for the $\arccos(\cdot)$. From expression (15.2),

$$C = \frac{\zeta^{x_c} + \zeta^{-x_c}}{2},$$

$$(\zeta^{x_c})^2 - 2C\zeta^{x_c} + 1 = 0,$$

from which

$$\zeta^{x_c} = C \pm \sqrt{C^2 - 1}$$

so that

$$x_c = \log_\zeta(C \pm \sqrt{C^2 - 1}).$$

Analogously, from Equation (15.3), one obtains

$$x_s = \log_\zeta (jS \pm \sqrt{j^2 S^2 + 1}).$$

For the sake of simplicity, the inverse finite field trigonometric functions are written as

$$\text{arccos}_\zeta (x) = \log_\zeta (x \pm \sqrt{x^2 - 1}) \tag{15.7}$$

$$\text{arcsin}_\zeta (x) = \log_\zeta (jx \pm \sqrt{j^2 x^2 + 1}). \tag{15.8}$$

By substituting $\zeta$ by $\zeta_u$ and considering the ranges given in expressions (15.5) and (15.6), we can see that Equations (15.7) and (15.8) have at most one solution. As we previously mentioned, such equations may not have a solution for some values of $x$. As an illustrative comparison between such a possibility and that observed for real valued trigonometric relations, we could say that, in some sense, this is equivalent to the computation of ordinary inverse cosines and sines whose arguments are numbers outside the interval $[-1, 1]$. This has inspired the following definition.

**Definition 15.12** *Let* $\lambda \in \mathbb{F}_q$ *be an element with multiplicative order* $\text{ord}(\lambda)$. *The finite field hyperbolic cosine and sine functions, with respect to* $\lambda$, *are defined respectively as*

$$\cosh_\lambda (x) = \frac{\lambda^x + \lambda^{-x}}{2}$$

$$\sinh_\lambda (x) = \frac{\lambda^x - \lambda^{-x}}{2},$$

*where* $x = 0, 1, \ldots, \text{ord}(\lambda) - 1$.

The set of all possible values for finite field hyperbolic cosines with respect to $\lambda$ is denoted by $\mathbb{C}^h_\lambda$; the set of all possible values for finite field hyperbolic sines with respect to $\lambda$ is denoted by $\mathbb{S}^h_\lambda$. We verify that the hyperbolic cosine and the hyperbolic sine are, respectively, even and odd symmetric functions. For such functions, the relation

$$\cosh^2_\lambda (x) - \sinh^2_\lambda (x) = 1 \pmod{f}$$

holds. The formulas

$$\cosh_\lambda (x + y) = \cosh_\lambda (x) \cdot \cosh_\lambda (y) + \sinh_\lambda (x) \cdot \sinh_\lambda (y)$$

$$\sinh_\lambda (x + y) = \sinh_\lambda (x) \cdot \cosh_\lambda (y) + \cosh_\lambda (x) \cdot \sinh_\lambda (y)$$

are also valid. This can be demonstrated by using the formulas given in Definition 15.12. Additionally, if $\lambda_q \in \mathbb{F}_q$ is an element with multiplicative order $\mathrm{ord}(\lambda_q) = q - 1$, one has

$$\cosh_{\lambda_q}(x) = \cosh_{\lambda_q}(y) \quad \text{if and only if} \quad x = y, \quad 0 \le x, y \le \frac{q-1}{2},$$

and

$$\sinh_{\lambda_q}(x) = \sinh_{\lambda_q}(y) \quad \text{if and only if} \quad x = y,$$
$$-\left\lfloor \frac{q-1}{4} \right\rfloor \le x, y \le \left\lfloor \frac{q-1}{4} \right\rfloor.$$

The relations above can be demonstrated by using arguments analogous to those used in the proofs of Lemmas 15.9 and 15.10.

**Example 15.13** *Let us consider the element $\lambda = 3 \in \mathbb{F}_7$, which has multiplicative order $\mathrm{ord}(\lambda) = 6$. The powers $\lambda^x$, $x = 0, 1, \ldots, 5$, of $\lambda$, i.e., the nonzero elements of $\mathbb{F}_7$, as well as the respective hyperbolic trigonometric functions over $\mathbb{F}_7$ are shown in Table 15.3.*

**Example 15.14** *Let $\lambda = \alpha$ be a root of the primitive polynomial $f(t) = t^2 + t + 2 \in \mathbb{F}_3[t]$ and $\mathbb{F}_9 = \mathbb{F}_3(\alpha)$. The multiplicative order of $\lambda$ satisfies $\mathrm{ord}(\lambda) = 8$. The powers $\lambda^x$, $x = 0, \ldots, 7$, of $\lambda$, i.e., the nonzero elements of $\mathbb{F}_9$, as well as the respective hyperbolic trigonometric functions over $\mathbb{F}_9$ are shown in Table 15.4.*

**Theorem 15.15** *Let $\mathbb{C}_{\zeta_u}$ be the set of all values obtained from*

$$\cos_{\zeta_u}(x) = \frac{\zeta_u^x + \zeta_u^{-x}}{2}, \quad x = 0, 1, \ldots, \frac{q+1}{2},$$

Table 15.3 *Nonzero elements of $\mathbb{F}_7$ and respective $\cosh(\cdot)$ and $\sinh(\cdot)$; the number $\lambda = 3 \in \mathbb{F}_7$, which has multiplicative order $\mathrm{ord}(\lambda) = 6$, is considered*

| $x$ | $\lambda^x$ | $\cosh_\lambda(x)$ | $\sinh_\lambda(x)$ |
|---|---|---|---|
| 0 | 1 | 1 | 0 |
| 1 | 3 | 4 | 6 |
| 2 | 2 | 3 | 6 |
| 3 | 6 | 6 | 0 |
| 4 | 4 | 3 | 1 |
| 5 | 5 | 4 | 1 |

Table 15.4 *Nonzero elements of* $\mathbb{F}_9$ *and respective* $\cosh(\cdot)$ *and* $\sinh(\cdot)$; *the element* $\lambda = \alpha \in \mathbb{F}_9$, *which has multiplicative order* $\mathrm{ord}(\lambda) = 8$, *is considered*

| $x$ | $\lambda^x$ | $\cosh_\lambda(x)$ | $\sinh_\lambda(x)$ |
|-----|-------------|---------------------|---------------------|
| 0 | 1 | 1 | 0 |
| 1 | $\alpha$ | $\alpha + 2$ | 1 |
| 2 | $2\alpha + 1$ | 0 | $2\alpha + 1$ |
| 3 | $2\alpha + 2$ | $2\alpha + 1$ | 1 |
| 4 | 2 | 2 | 0 |
| 5 | $2\alpha$ | $2\alpha + 1$ | 2 |
| 6 | $\alpha + 2$ | 0 | $\alpha + 2$ |
| 7 | $\alpha + 1$ | $\alpha + 2$ | 2 |

*where* $\zeta_u \in \mathbb{I}_q$ *is a generator of* $\mathbf{G}_1$. *Let* $\mathbb{C}^h_{\lambda_q}$ *be the set of all values obtained from*

$$\cosh_{\lambda_q}(x) = \frac{\lambda_q^x + \lambda_q^{-x}}{2}, \quad x = 0, 1, \ldots, \frac{q-1}{2},$$

*where* $\lambda_q \in \mathbb{F}_q$ *is an element such that* $\mathrm{ord}(\lambda_q) = q - 1$. *Then* $\mathbb{C}_{\zeta_u} \cup \mathbb{C}^h_{\lambda_q} = \mathbb{F}_q$.

*Proof.* From Lemma 15.9, we know that the number of elements of $\mathbb{C}_{\zeta_u}$ is $\#\{\mathbb{C}_{\zeta_u}\} = \frac{q+1}{2} + 1 = \frac{q+3}{2}$. Similarly, the number of elements of $\mathbb{C}^h_{\lambda_q}$ is $\#\{\mathbb{C}^h_{\lambda_q}\} = \frac{q-1}{2} + 1 = \frac{q+1}{2}$. From Proposition 15.6, we know that any cosine corresponds to the real part of a unimodular element written as $c + dj = [\cos_{\zeta_u}(x)] + dj$. Therefore, any element in the intersection $\mathbb{C}_{\zeta_u} \cap \mathbb{C}^h_{\lambda_q}$ should satisfy $\cos_{\zeta_u}(x) = \cosh_{\lambda_q}(y)$ for some $x$ and $y$, and, consequently,

$$[\cosh_{\lambda_q}(y)]^2 - d^2 j^2 \equiv 1 \pmod{f}.$$

The above equation can be rewritten as

$$\left( \frac{\lambda_q^y + \lambda_q^{-y}}{2} \right)^2 - d^2 j^2 \equiv 1 \pmod{f},$$

so that

$$\lambda_q^y - \lambda_q^{-y} \equiv \pm 2dj \pmod{f}.$$

Since $\lambda_q \in \mathbb{F}_q$, the last equation holds only if $d = 0$. Consequently, the possible values for $y$ are 0 and $\frac{q-1}{2}$. Then, the number of elements of $\mathbb{C}_{\zeta_u} \cap \mathbb{C}^h_{\lambda_q}$ is $\#\{\mathbb{C}_{\zeta_u} \cap \mathbb{C}^h_{\lambda_q}\} = 2$ and

$$\#\{\mathbb{C}_{\zeta_u} \cup \mathbb{C}^h_{\lambda_q}\} = \#\{\mathbb{C}_{\zeta_u}\} + \#\{\mathbb{C}^h_{\lambda_q}\} - \#\{\mathbb{C}_{\zeta_u} \cap \mathbb{C}^h_{\lambda_q}\} = \frac{q+3}{2} + \frac{q+1}{2} - 2 = q.$$

$\square$

We remark that there is no theorem similar to Theorem 15.15 for $\mathbb{S}_{\zeta_u}$ and $\mathbb{S}^h_{\lambda_q}$. This can be seen, for example, by observing the columns related to the sines in Table 15.1 and Table 15.3, and also in Table 15.2 and Table 15.4; not all the elements of $\mathbb{F}_7$ and $\mathbb{F}_9$ appear in these columns.

## 15.3 Chebyshev polynomials over finite fields

In this section, trigonometric definitions for Chebyshev polynomials over finite fields are presented. These definitions, which are based on the trigonometric relations and functions discussed in Section 15.2, are in perfect analogy with the classical definitions of Chebyshev polynomials over the real numbers. This fact can be verified from Propositions 15.17, 15.19, 15.21 and 15.23, where recurrence equations for Chebyshev polynomials over finite fields of the first, second, third and fourth kinds, respectively, are deduced. These equations coincide with those of the real valued Chebyshev polynomials and also with those related to nontrigonometric approaches for Dickson polynomials of the first and second kinds over finite fields. In what follows, $\mathbb{C}_{\zeta_u}$ and $\mathbb{C}^h_{\lambda_q}$ are defined as in Theorem 15.15.

**Definition 15.16** *The Chebyshev polynomials of the first kind over $\mathbb{F}_q$ are defined as*

$$T_n(x) := \begin{cases} \cos_{\zeta_u}[n \arccos_{\zeta_u}(x)] & \text{if } x \in \mathbb{C}_{\zeta_u} \\ \cosh_{\lambda_q}[n \arccosh_{\lambda_q}(x)] & \text{if } x \in \mathbb{C}^h_{\lambda_q}. \end{cases}$$

**Proposition 15.17** *Let $T_n(x)$ be the Chebyshev polynomial of the first kind of degree $n$ over $\mathbb{F}_q$. The recurrence relation*

$$T_{n+2}(x) = 2x \cdot T_{n+1}(x) - T_n(x),$$

*where $T_0(x) = 1$ and $T_1(x) = x$, holds.*

*Proof.* Let us consider that $x \in \mathbb{C}_{\zeta_u}$ and $y = \arccos_{\zeta_u}(x)$. From Definition 15.16, we know that $T_n(x) = \cos_{\zeta_u}(ny)$ and, consequently, $T_0(x) = 1$ and $T_1(x) = x$. Using the formula for the cosine of the addition of two arcs, one has the expansion

$$\begin{aligned} T_{n+1}(x) &= \cos_{\zeta_u}(ny + y) = \cos_{\zeta_u}(ny) \cdot \cos_{\zeta_u}(y) + j^2 \sin_{\zeta_u}(ny) \cdot \sin_{\zeta_u}(y) \\ &= T_n(x) \cdot x + j^2 \cdot \sin_{\zeta_u}(ny) \cdot \sin_{\zeta_u}(y). \end{aligned} \tag{15.9}$$

We also use the expansion

$$T_{n+2}(x) = \cos_{\zeta_u}[(ny + y) + y]$$
$$= \cos_{\zeta_u}(ny + y) \cdot \cos_{\zeta_u}(y) + j^2 \sin_{\zeta_u}(ny + y) \cdot \sin_{\zeta_u}(y).$$

Together with the formula for the sine of the addition of two arcs, one has

$$T_{n+2}(x) = T_{n+1}(x) \cdot x + j^2[\sin_{\zeta_u}(ny) \cdot \cos_{\zeta_u}(y)$$
$$+ \sin_{\zeta_u}(y) \cdot \cos_{\zeta_u}(ny)] \cdot \sin_{\zeta_u}(y)$$
$$= T_{n+1}(x) \cdot x + j^2 \sin_{\zeta_u}(ny) \cdot \sin_{\zeta_u}(y) \cdot x + j^2 \sin^2_{\zeta_u}(y) \cdot T_n(x).$$

Using Equation (15.9) and the unit circle property in the last equation, one has

$$T_{n+2}(x) = T_{n+1}(x) \cdot x + T_{n+1}(x) \cdot x - T_n(x) \cdot x^2 + T_n(x) \cdot x^2 - T_n(x)$$
$$= 2x \cdot T_{n+1}(x) - T_n(x).$$

If $x \in \mathbb{C}^h_{\lambda_q}$, the last equation can also be obtained following the same steps as for $x \in \mathbb{C}_{\zeta_u}$. □

**Definition 15.18** *The Chebyshev polynomials of the second kind over $\mathbb{F}_q$ are defined as*

$$U_n(x) := \begin{cases} \dfrac{\sin_{\zeta_u}[(n+1)\arccos_{\zeta_u}(x)]}{\sin_{\zeta_u}[\arccos_{\zeta_u}(x)]} & \text{if } x \in \mathbb{C}_{\zeta_u} \quad \text{and } x \not\equiv \pm 1 \ (\text{mod } p) \\[2ex] \dfrac{\sinh_{\lambda_q}[(n+1)\operatorname{arccosh}_{\lambda_q}(x)]}{\sinh_{\lambda_q}[\operatorname{arccosh}_{\lambda_q}(x)]} & \text{if } x \in \mathbb{C}^h_{\lambda_q} \quad \text{and } x \not\equiv \pm 1 \ (\text{mod } p) \\[2ex] n + 1 & \text{if } x = 1 \\[1ex] left(-1)^n (n + 1) & \text{if } x = p - 1. \end{cases}$$

$$(15.10)$$

We remark that specific expressions of $U_n(x)$, $x \equiv \pm 1 \pmod{p}$, are necessary because, for such values of $x$, direct use of the first or the second line of (15.10) would lead to the indeterminate form $0/0$. A similar fact occurs in the definition of $V_n(x)$ and $W_n(x)$. The third and the fourth lines of (15.10) are in accordance with the following proposition.

**Proposition 15.19** *Let $U_n(x)$ be the Chebyshev polynomial of the second kind of degree $n$ over $\mathbb{F}_q$. The recurrence relation*

$$U_{n+2}(x) = 2x U_{n+1}(x) - U_n(x),$$

*where $U_0(x) = 1$ and $U_1(x) = 2x$, holds.*

Juliano B. Lima, Daniel Panario and Ricardo M. Campello de Souza 267

*Proof.* Let us consider that $x \in \mathbb{C}^h_{\lambda_q}$ and $y = \mathrm{arccosh}_{\lambda_q}(x)$. The proof for $x \in \mathbb{C}_{\zeta_u}$ is analogous. Using Definition 15.18, one has $U_0(x) = \frac{\sinh_{\lambda_q}(y)}{\sinh_{\lambda_q}(y)} = 1$ and $U_1(x) = \frac{\sinh_{\lambda_q}(2y)}{\sinh_{\lambda_q}(y)} = \frac{2\sinh_{\lambda_q}(y)\cosh_{\lambda_q}(y)}{\sinh_{\lambda_q}(y)} = 2x$. Also from Definition 15.18, one has

$$U_{n-1}(x) = \frac{\sinh_{\lambda_q}(ny)}{\sinh_{\lambda_q}(y)} \tag{15.11}$$

and, using the formula for the hyperbolic sine of the addition of two arcs, one has

$$U_n(x) = \frac{\sinh_{\lambda_q}(ny+y)}{\sinh_{\lambda_q}(y)} = \frac{\sinh_{\lambda_q}(ny)\cdot\cosh_{\lambda_q}(y)}{\sinh_{\lambda_q}(y)} + \frac{\sinh_{\lambda_q}(y)\cdot\cosh_{\lambda_q}(ny)}{\sinh_{\lambda_q}(y)}$$
$$= U_{n-1}(x)\cdot\cosh_{\lambda_q}(y) + \cosh_{\lambda_q}(ny). \tag{15.12}$$

The formula for the hyperbolic cosine of the addition of two arcs allows the expansion

$$U_{n+1}(x) = \frac{\sinh_{\lambda_q}[(n+1)y+y]}{\sinh_{\lambda_q}(y)}$$
$$= \frac{\sinh_{\lambda_q}[(n+1)y]\cdot\cosh_{\lambda_q}(y)}{\sinh_{\lambda_q}(y)} + \frac{\sinh_{\lambda_q}(y)\cdot\cosh_{\lambda_q}[(n+1)y]}{\sinh_{\lambda_q}(y)}$$
$$= U_n(x)\cdot x + \cosh_{\lambda_q}(ny)\cdot x + \sinh_{\lambda_q}(ny)\cdot\sinh_{\lambda_q}(y).$$

Finally, from the unit circle property and Equations (15.11) and (15.12), one obtains

$$U_{n+1}(x) = U_n(x)\cdot x + U_n(x)\cdot x - U_{n-1}(x)\cdot x^2 + U_{n-1}(x)\cdot x^2 - U_{n-1}(x)$$
$$= 2x\cdot U_n(x) - U_{n-1}(x).$$

The given recurrence relation can also be verified for $x = \pm 1 \pmod p$. □

**Definition 15.20** *The Chebyshev polynomials of the third kind over $\mathbb{F}_q$ are defined as*

$$V_n(x) := \begin{cases} \dfrac{\cos_{\zeta_u}\left[\left(n+\frac{1}{2}\right)\mathrm{arccos}_{\zeta_u}(x)\right]}{\cos_{\zeta_u}\left[\frac{1}{2}\mathrm{arccos}_{\zeta_u}(x)\right]} & \text{if } x \in \mathbb{C}_{\zeta_u} \text{ and } x \neq p-1 \\[3ex] \dfrac{\cosh_{\lambda_q}\left[\left(n+\frac{1}{2}\right)\mathrm{arccosh}_{\lambda_q}(x)\right]}{\cosh_{\lambda_q}\left[\frac{1}{2}\mathrm{arccosh}_{\lambda_q}(x)\right]} & \text{if } x \in \mathbb{C}^h_{\lambda_q} \text{ and } x \neq p-1 \\[3ex] (2n+1)(-1)^n & \text{if } x = p-1. \end{cases}$$

**Proposition 15.21** *Let $V_n(x)$ be the Chebyshev polynomial of the third kind of degree $n$ over $\mathbb{F}_q$. The recurrence relation*

$$V_{n+2}(x) = 2x\, V_{n+1}(x) - V_n(x),$$

*where $V_0(x) = 1$ and $V_1(x) = 2x - 1$, holds.*

*Proof.* Let us consider that $x \in \mathbb{C}_{\zeta_u}$ and $y = \arccos_{\zeta_u}(x)$. Using Definition 15.20, one has

$$V_0(x) = \frac{\cos_{\zeta_u}\left(\frac{1}{2}y\right)}{\cos_{\zeta_u}\left(\frac{1}{2}y\right)} = 1.$$

Also from Definition 15.20, one has

$$V_1(x) = \frac{\cos_{\zeta_u}\left(y + \frac{1}{2}y\right)}{\cos_{\zeta_u}\left(\frac{1}{2}y\right)},$$

which can be developed as

$$V_1(x) = \frac{\cos_{\zeta_u}(y) \cdot \cos_{\zeta_u}\left(\frac{1}{2}y\right)}{\cos_{\zeta_u}\left(\frac{1}{2}y\right)} + j^2 \frac{\sin_{\zeta_u}(y) \cdot \sin_{\zeta_u}\left(\frac{1}{2}y\right)}{\cos_{\zeta_u}\left(\frac{1}{2}y\right)}$$

$$= x + 2j^2 \sin_{\zeta_u}^2\left(\frac{1}{2}y\right) = x + 2\left[\cos_{\zeta_u}^2\left(\frac{1}{2}y\right) - 1\right]$$

$$= x + 2\left[\frac{1 + \cos_{\zeta_u}(y)}{2} - 1\right] = 2x - 1.$$

The proof can be concluded using similar steps to those in the proof of Proposition 15.17. This is also valid for $x \in \mathbb{C}_{\lambda_q}^h$ and for $x = p - 1$. $\qquad\square$

**Definition 15.22** *The Chebyshev polynomials of the fourth kind over $\mathbb{F}_q$ are defined as*

$$W_n(x) := \begin{cases} \dfrac{\sin_{\zeta_u}\left[\left(n+\frac{1}{2}\right)\arccos_{\zeta_u}(x)\right]}{\sin_{\zeta_u}\left(\frac{1}{2}\arccos_{\zeta_u}(x)\right)} & \text{if } x \in \mathbb{C}_{\zeta_u} \text{ and } x \neq 1 \\[4mm] \dfrac{\sinh_{\lambda_q}\left[\left(n+\frac{1}{2}\right)\operatorname{arccosh}_{\lambda_q}(x)\right]}{\sinh_{\lambda_q}\left(\frac{1}{2}\operatorname{arccosh}_{\lambda_q}(x)\right)} & \text{if } x \in \mathbb{C}_{\lambda_q}^h \text{ and } x \neq 1 \\[4mm] 2n + 1 & \text{if } x = 1. \end{cases}$$

**Proposition 15.23** *Let $W_n(x)$ be the Chebyshev polynomial of the fourth kind of degree $n$ over $\mathbb{F}_q$. The recurrence relation*

$$W_{n+2}(x) = 2x\,W_{n+1}(x) - W_n(x),$$

*where $W_0(x) = 1$ and $W_1(x) = 2x + 1$, holds.*

*Proof.* Let us consider that $x \in \mathbb{C}_{\zeta_u}$ and $y = \arccos_{\zeta_u}(x)$. Using Definition 15.22, one has

$$W_0(x) = \frac{\sin_{\zeta_u}\left(\frac{1}{2}y\right)}{\sin_{\zeta_u}\left(\frac{1}{2}y\right)} = 1.$$

Also from Definition 15.22, one has

$$W_1(x) = \frac{\sin_{\zeta_u}\left(y + \frac{1}{2}y\right)}{\sin_{\zeta_u}\left(\frac{1}{2}y\right)},$$

which can be developed as

$$W_1(x) = \frac{\sin_{\zeta_u}(y) \cdot \cos_{\zeta_u}\left(\frac{1}{2}y\right)}{\sin_{\zeta_u}\left(\frac{1}{2}y\right)} + \frac{\sin_{\zeta_u}\left(\frac{1}{2}y\right) \cdot \cos_{\zeta_u}(y)}{\sin_{\zeta_u}\left(\frac{1}{2}y\right)}$$

$$= x + 2\cos_{\zeta_u}^2\left(\frac{1}{2}y\right) = x + 2\left[\frac{1 + \cos_{\zeta_u}(y)}{2}\right] = 2x + 1.$$

The proof can be concluded using similar steps to those in the proof of Proposition 15.17. This is also valid for $x \in \mathbb{C}_{\lambda_q}^h$ and for $x = 1$. □

The recurrence relations given in Propositions 15.17, 15.19, 15.21 and 15.23 are rigorously equal to those for real valued Chebyshev polynomials. This means that the proposed trigonometric definitions for Chebyshev polynomials over finite fields correspond to the conventional Chebyshev polynomials evaluated in a finite field. The aforementioned recurrence relations are useful to compute the value of a Chebyshev polynomial of a specific kind, given $n$ and $x$. In addition, the trigonometric form of such polynomials is suitable in the establishment of some of the properties of Chebyshev polynomial. This can be observed, for example, in the investigation of periodicity properties of Chebyshev polynomials over finite fields, which are discussed in the following section.

## 15.4 Periodicity and symmetry properties of Chebyshev polynomials over finite fields

In what follows, periodicity and symmetry properties of Chebyshev polynomials over finite fields are established. For Propositions 15.24, 15.25, 15.26 and 15.27, we define

$$\eta(x) = \zeta_u^{\text{arccos}_{\zeta_u}(x)} = \lambda_q^{\text{arccosh}_{\lambda_q}(x)} = x \pm \sqrt{x^2 - 1},$$

such that $\text{ord}(\eta(x)) = N_x$. The validity of these propositions is demonstrated by using Definitions 15.16, 15.18, 15.20 and 15.22, respectively, and also Definitions 15.5 and 15.12. In this context, the key point is to observe that $[\eta(x)]^{N_x} = 1$ and $[\eta(x)]^{N_x/2} = -1$ ($N_x$ even). Moreover, it is also important to remark that the possible values for $\text{ord}(\eta(x)) = N_x$ are the divisors of $q + 1$ and $q - 1$. This is explained by the fact that $\eta(x)$ corresponds to powers of $\zeta_u$ or $\lambda_q$, which are, respectively, generators of $\langle G_1, \cdot \rangle$ (see Proposition 15.4) and the cyclic group $\mathbb{F}_q^*$.

**Proposition 15.24** *For* $x \in \mathbb{F}_q$,

- $T_{n'}(x) = T_n(x)$, *if* $n' \equiv \pm n \pmod{N_x}$,
- $T_{n'}(x) = -T_n(x)$, *if* $n' \equiv \frac{N_x}{2} \pm n \pmod{N_x}$ *(*$N_x$ *even).*

*Proof.* Let us consider that $x \in \mathbb{C}_{\zeta_u}$. If $n' = \pm n + kN_x$, for any integer $k$, one has

$$T_{n'}(x) = \cos_{\zeta_u}(n' \arccos_{\zeta_u}(x)) = \frac{\zeta_u^{n' \arccos_{\zeta_u}(x)} + \zeta_u^{-n' \arccos_{\zeta_u}(x)}}{2}$$

$$= \frac{[\eta(x)]^{n'} + [\eta(x)]^{-n'}}{2} = \frac{[\eta(x)]^{\pm n + kN_x} + [\eta(x)]^{-(\pm n + kN_x)}}{2}$$

$$= \frac{[\eta(x)]^{\pm n} + [\eta(x)]^{-(\pm n)}}{2} = T_n(x).$$

If $n' = \frac{N_x}{2} \pm n + kN_x$, one has

$$T_{n'}(x) = \frac{[\eta(x)]^{\frac{N_x}{2} \pm n + kN_x} + [\eta(x)]^{-(\frac{N_x}{2} \pm n + kN_x)}}{2}$$

$$= -\frac{[\eta(x)]^{\pm n} + [\eta(x)]^{-(\pm n)}}{2} = -T_n(x).$$

The proof for $x \in \mathbb{C}_{\lambda_q}^h$ is analogous. $\qquad\square$

In order to illustrate the period distribution of a polynomial $T_n(x)$ over a finite field, we can plot a histogram where the number of $x$ with a given period

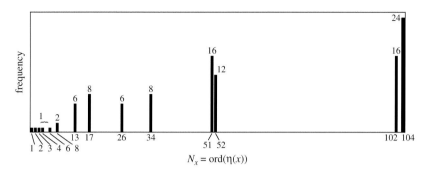

Figure 15.1  Period distribution of polynomials $T_n(x)$ over $\mathbb{F}_{103}$.

$\operatorname{ord}(\eta(x)) = N_x$ can be viewed. As an example, let us consider $q = 103$. The corresponding period distribution is shown in Figure 15.1.

**Proposition 15.25**  *For $x \in \mathbb{F}_q$ and $x \not\equiv \pm 1 \pmod{p}$,*

- $U_{n'}(x) = U_n(x)$, *if* $n' \equiv n \pmod{N_x}$ *or* $n' \equiv \frac{N_x}{2} - n - 2 \pmod{N_x}$ ($N_x$ *even*),
- $U_{n'}(x) = -U_n(x)$, *if* $n' \equiv -n - 2 \pmod{N_x}$ *or* $n' \equiv \frac{N_x}{2} + n \pmod{N_x}$ *($N_x$ even).*

*For $x = 1$,*

- $U_{n'}(x) = U_n(x)$, *if* $n' \equiv n \pmod{p}$,
- $U_{n'}(x) = -U_n(x)$, *if* $n' \equiv -n - 2 \pmod{p}$.

*For $x = p - 1$,*

- $U_{n'}(x) = U_n(x)$, *if* $n' \equiv n \pmod{2p}$ *or* $n' \equiv p - n - 2 \pmod{2p}$,
- $U_{n'}(x) = -U_n(x)$, *if* $n' \equiv -n - 2 \pmod{2p}$ *or* $n' \equiv p + n \pmod{2p}$.

*Proof.*  Let us consider that $x \in \mathbb{C}_{\zeta_u}$ and $x \not\equiv \pm 1 \pmod{p}$. If $n' = n + kN_x$, for any integer $k$, one has

$$
\begin{aligned}
U_{n'}(x) &= \left[ \frac{1}{\sin_{\zeta_u}(\arccos_{\zeta_u}(x))} \right] \cdot \sin_{\zeta_u}[(n' + 1)\arccos_{\zeta_u}(x)] \\
&= \left[ \frac{1}{\sin_{\zeta_u}(\arccos_{\zeta_u}(x))} \right] \cdot \frac{\zeta_u^{(n'+1)\arccos_{\zeta_u}(x)} - \zeta_u^{-(n'+1)\arccos_{\zeta_u}(x)}}{2j} \\
&= \left[ \frac{1}{\sin_{\zeta_u}(\arccos_{\zeta_u}(x))} \right] \cdot \frac{[\eta(x)]^{n'+1} - [\eta(x)]^{-(n'+1)}}{2j}
\end{aligned}
$$

$$= \left[ \frac{1}{\sin_{\zeta_u}(\arccos_{\zeta_u}(x))} \right] \cdot \frac{[\eta(x)]^{n+1+kN_x} - [\eta(x)]^{-(n+1+kN_x)}}{2j}$$

$$= \left[ \frac{1}{\sin_{\zeta_u}(\arccos_{\zeta_u}(x))} \right] \cdot \frac{[\eta(x)]^{n+1} - [\eta(x)]^{-(n+1)}}{2j} = U_n(x).$$

Analogous developments are obtained for all other cases where $x \not\equiv \pm 1 \pmod{p}$. If $x = 1$ and $n' = n+kp$, we use the third line of Equation (15.10) to write

$$U_{n'}(x) = n' + 1 = n + kp + 1 \equiv n + 1 \pmod{p} = U_n(x).$$

If $x = 1$ and $n' = -n - 2 + kp$, one has

$$U_{n'}(x) = n' + 1 = -n - 2 + kp + 1 \equiv -n - 1 \pmod{p} = -U_n(x).$$

If $x = p - 1$ and $n' = n + k2p$, we use the fourth line of Equation (15.10) to write

$$U_{n'}(x) = (-1)^{n'}(n' + 1) = (-1)^{n+k2p}(n + k2p + 1)$$
$$\equiv (-1)^n (n + 1) \pmod{2p} = U_n(x).$$

The same occurs for $n' = p - n - 2 + k2p$. If $x = p - 1$ and $n' = -n - 2 + k2p$, one has

$$U_{n'}(x) = (-1)^{n'}(n' + 1) = (-1)^{-n-2+k2p}(-n - 2 + k2p + 1)$$
$$\equiv (-1)^n(-n - 1) \pmod{2p} = -U_n(x).$$

The same occurs for $n' = p + n + k2p$.     □

The period distribution of a polynomial $U_n(x)$ over a finite field equals that of $T_n(x)$, except for $x \equiv \pm 1 \pmod{p}$. Once $U_n(1)$ and $U_n(-1) = U_n(p - 1)$ are computed from particular equations in Definition 15.18, their periods do not correspond to $\mathrm{ord}(\eta(x))$. According to Proposition 15.25, $U_n(1)$ and $U_n(p - 1)$ have periods $p$ and $2p$, respectively.

**Proposition 15.26** *For $x \in \mathbb{F}_q$,*

- $V_{n'}(x) = V_n(x)$, *if $n' \equiv n \pmod{N_x}$ or $n' \equiv -n - 1 \pmod{N_x}$,*
- $V_{n'}(x) = -V_n(x)$, *if $n' \equiv \frac{N_x}{2} - n - 1 \pmod{N_x}$ ($N_x$ even) or $n' \equiv \frac{N_x}{2} + n \pmod{N_x}$ ($N_x$ even).*

*For $x = p - 1$,*

- $V_{n'}(x) = V_n(x)$, *if $n' \equiv n \pmod{2p}$ or $n' \equiv -n - 1 \pmod{2p}$,*
- $V_{n'}(x) = -V_n(x)$, *if $n' \equiv p - n - 1 \pmod{2p}$ or $n' \equiv p + n \pmod{2p}$.*

**Proposition 15.27** *For $x \in \mathbb{F}_q$ and $x \neq 1$,*

- $W_{n'}(x) = W_n(x)$, if $n' \equiv n \pmod{N_x}$ or $n' \equiv \frac{N_x}{2} - n - 1 \pmod{N_x}$ ($N_x$ even),
- $W_{n'}(x) = -W_n(x)$, if $n' \equiv -n - 1 \pmod{N_x}$ or $n' \equiv \frac{N_x}{2} + n \pmod{N_x}$ ($N_x$ even).

*For $x = 1$,*

- $W_{n'}(x) = W_n(x)$, if $n' \equiv n \pmod{p}$,
- $W_{n'}(x) = -W_n(x)$, if $n' \equiv -n - 1 \pmod{p}$.

The proofs for Propositions 15.26 and 15.27 are omitted, since they are similar to those of Propositions 15.24 and 15.25. The period distributions of $V_n(x)$ and $W_n(x)$ have behavior analogous to that of $T_n(x)$ and $U_n(x)$.

## 15.5 Permutation properties of Chebyshev polynomials over finite fields

Permutation properties of Dickson and Chebyshev polynomials of the first and second kinds over finite fields have been widely investigated. It is a well-known fact that a necessary and sufficient condition for the Dickson polynomial of the first kind $T_n(x)$ to be a permutation polynomial over $\mathbb{F}_q$ is that $\gcd(n, q^2 - 1) = 1$ [11]. The first result related to Dickson polynomials of the second kind was obtained by Matthews [15]; it was shown that if $q$ is odd and $n$ satisfies the three congruences

$$\begin{cases} n + 1 \equiv \pm 2 \pmod{p}, \\ n + 1 \equiv \pm 2 \left(\mod \frac{q+1}{2}\right), \\ n + 1 \equiv \pm 2 \left(\mod \frac{q-1}{2}\right), \end{cases} \tag{15.13}$$

then $U_n(x)$ is a permutation polynomial over $\mathbb{F}_q$. If $q = p$ is an odd prime or $q = p^2$, $p \geq 7$ prime, conditions given in (15.13) are also necessary. This was shown by Cohen [4] and then by Cipu [2]; see also [3]. Related results about Dickson polynomials of the second kind are given in [5, 8]. The monograph [12] presents several results for Dickson polynomials and related polynomials [9].

In this section, we show that the symmetry and periodicity properties of Chebyshev polynomials, developed in Section 15.4 in the light of the proposed trigonometric definitions of such polynomials, can be used to derive results related to their permutation properties. Since the permutation behavior of Chebyshev polynomials of the first kind is already completely established, the developments related to $T_n(x)$ are not presented here; we focus on permutation properties of Chebyshev polynomials of the second kind, whose results have not been generalized. More specifically, we present a different way to obtain conditions (15.13). Furthermore, we conduct a preliminary discussion related to the permutation properties of $V_n(x)$ and $W_n(x)$.

**Lemma 15.28** *If* $\operatorname{ord}(\zeta)$ *is even,*

$$\operatorname{arccos}_\zeta(-x) = \frac{\operatorname{ord}(\zeta)}{2} + \operatorname{arccos}_\zeta(x).$$

*The same occurs, if we consider* $\operatorname{arccosh}_\lambda(-x)$.

*Proof.* From Equation (15.7), one has

$$\operatorname{arccos}_\zeta(x) = \log_\zeta(x \pm \sqrt{x^2 - 1})$$

$$\operatorname{arccos}_\zeta(-x) = \log_\zeta(-x \pm \sqrt{x^2 - 1}).$$

The last equation can be written as

$$\operatorname{arccos}_\zeta(-x) = \log_\zeta\left[(-1)(x \mp \sqrt{x^2 - 1})\right]$$
$$= \log_\zeta(-1) + \log_\zeta(x \mp \sqrt{x^2 - 1})$$
$$= \frac{\operatorname{ord}(\zeta)}{2} + \operatorname{arccos}_\zeta(x).$$

The proof is analogous for $\operatorname{arccosh}_\lambda(-x)$.                     □

**Proposition 15.29** *For* $x \in \mathbb{F}_q$, $U_n(-x) = (-1)^n U_n(x)$.

*Proof.* Let us consider that $x \in \mathbb{C}_{\zeta_u}$. Since $\operatorname{ord}(\zeta_u) = q + 1$, from Definition 15.18 and also considering Lemma 15.28, one has

$$U_n(-x) = \frac{\sin_{\zeta_u}\left[(n+1)\left(\frac{q+1}{2} + \operatorname{arccos}_{\zeta_u}(x)\right)\right]}{\sin_{\zeta_u}\left[\frac{q+1}{2} + \operatorname{arccos}_{\zeta_u}(x)\right]}.$$

Observing that

$$\sin_{\zeta_u}\left[(n+1)\left(\frac{q+1}{2}\right)\right] = 0, \quad \cos_{\zeta_u}\left[(n+1)\left(\frac{q+1}{2}\right)\right] = (-1)^{n+1},$$

and using Equation (15.4) to expand both numerator and denominator of the last equation, one has

$$U_n(-x) = \frac{(-1)^{n+1}\sin_{\zeta_u}[(n+1)\arccos_{\zeta_u}(x)]}{-\sin_{\zeta_u}[\arccos_{\zeta_u}(x)]} = (-1)^n U_n(x).$$

The proof is analogous for $x \in \mathbb{C}^h_{\lambda_q}$. $\qquad\qquad\qquad\qquad\qquad\square$

**Theorem 15.30** *Assume $U_n(x)$ is a permutation polynomial over $\mathbb{F}_q$. If $n'$ satisfies one of the following conditions for $N = q + 1$ and $N = q - 1$ (the satisfied condition may be different for $N = q + 1$ and $N = q - 1$):*

$$\begin{cases} n' \equiv n \pmod{N}, \\ n' \equiv \frac{N}{2} - n - 2 \pmod{N}, \\ n' \equiv -n - 2 \pmod{N}, \\ n' \equiv \frac{N}{2} + n \pmod{N}, \end{cases} \qquad (15.14)$$

*and if $n'$ satisfies*

$$n' \equiv n \pmod{2p} \quad or \quad n' \equiv -n - 2 \pmod{2p}, \qquad (15.15)$$

*then $U_{n'}(x)$ is a permutation polynomial over $\mathbb{F}_q$.*

*Proof.* From Proposition 15.29, we can conclude that $n$ must be odd. Otherwise, $U_n(x) = U_n(-x)$ and $U_n(x)$ is not a permutation polynomial. Our proof is based on the fact that, if $U_n(x)$ is a permutation polynomial, then a polynomial $U_{n'}(x)$ such that, for each value of $x$,

$$U_{n'}(x) = U_n(x) \quad \text{and, consequently,} \quad U_{n'}(-x) = U_n(-x) = -U_n(x)$$
$$(15.16)$$

or

$$U_{n'}(x) = -U_n(x) \quad \text{and, consequently,} \quad U_{n'}(-x) = -U_n(-x) = U_n(x),$$
$$(15.17)$$

is also a permutation polynomial. Conditions given in Equation (15.14) come from the first part of Proposition 15.25. If one of these conditions is satisfied for $N = q + 1$, (15.16) or (15.17) is satisfied for all $x$ such that $N_x | N$. The same occurs for $N = q - 1$. Conditions given in Equation (15.15) are related to the cases where $x \equiv \pm 1 \pmod{p}$ and come from the second and third parts of Proposition 15.25. For such values of $x$, (15.16) is satisfied if $n' \equiv n \pmod{p}$ and $n' \equiv n \pmod{2p}$. Such conditions can be summarized as $n' \equiv n \pmod{2p}$. The condition $n' \equiv p - n - 2 \pmod{2p}$, which is given in the third part of Proposition 15.25, is discarded because it leads to

$U_{n'}(1) = U_{n'}(-1)$; (15.17) is satisfied if $n' \equiv -n - 2 \pmod{p}$ and $n' \equiv -n - 2 \pmod{2p}$. Such conditions can be summarized as $n' \equiv -n -2 \pmod{2p}$. Analogously, the condition $n' \equiv p + n \pmod{2p}$ is discarded because it leads to $U_{n'}(1) = U_{n'}(-1)$.          □

We remark that, if $n = 1$, $U_1(x)$ is a permutation polynomial and the conditions given in Theorem 15.30 are equivalent to those derived in [15]. In other words, we can assert that our conditions are also necessary, if $q = p$ is an odd prime or $q = p^2$, $p \geq 7$ [4]. This can be seen by substituting $n$ by $n'$ and adding 1 on both sides of all congruences given in (15.14), which gives

$$\begin{cases} n' + 1 \equiv 2 \pmod{N}, \\ n' + 1 \equiv \frac{N}{2} - 2 \pmod{N}, \\ n' + 1 \equiv -2 \pmod{N}, \\ n' + 1 \equiv \frac{N}{2} + 2 \pmod{N}. \end{cases} \tag{15.18}$$

For $N = q + 1$ and $N = q - 1$, the first and fourth equations of (15.18) can be summarized as

$$n' + 1 \equiv 2 \left( \bmod \ \frac{N}{2} \right).$$

Analogously, the second and third equations of (15.18) can be summarized as

$$n' + 1 \equiv -2 \left( \bmod \ \frac{N}{2} \right).$$

The last two congruences correspond to the second and third congruences of (15.13), for $N = q + 1$ and $N = q - 1$. Conditions (15.15) with $n = 1$ correspond to $n' + 1 \equiv \pm 2 \pmod{2p}$. This is more restrictive than $n' + 1 \equiv \pm 2 \pmod{p}$, the first congruence of (15.13). However, no permutation polynomials are missed as a result of this change. If we write $n' + 1 \equiv \pm 2 \pmod{p}$ as $n' + 1 = \pm 2 + kp$, we see that $k$ must be an even integer. Otherwise, $n'$ is even and $U_{n'}(x)$ is not a permutation polynomial. Therefore, condition $n' + 1 \equiv \pm 2 \pmod{2p}$ may be used.

We can also derive conditions related to the permutation behavior of $V_n(x)$ and $W_n(x)$ from their periodicity and symmetry properties. However, in contrast to $U_n(x)$, such polynomials do not hold any property similar to that given in Proposition 15.29. Consequently, the sufficient conditions given below in Theorems 15.31 and 15.32 are more restrictive than those given in Theorem 15.30.

**Theorem 15.31** *Assume $V_n(x)$ is a permutation polynomial over $\mathbb{F}_q$. If $n'$ satisfies, for $N = q + 1$ and $N = q - 1$ (the satisfied condition may be different for $N = q + 1$ and $N = q - 1$),*

$$n' \equiv n \pmod{N} \quad or \quad n' \equiv -n - 1 \pmod{N}, \tag{15.19}$$

and if $n'$ satisfies

$$n' \equiv n \pmod{2p} \quad or \quad n' \equiv -n - 1 \pmod{2p}, \tag{15.20}$$

then $V_{n'}(x)$ is a permutation polynomial over $\mathbb{F}_q$.

*Proof.* From Proposition 15.26, we have that the conditions given in (15.19) and (15.20) imply that $V_{n'}(x) = V_n(x)$, for all $x \in \mathbb{F}_q$. Therefore, if $V_n(x)$ is a permutation polynomial, $V_{n'}(x)$ is also a permutation polynomial. $\qquad\square$

We remark that the conditions under which $V_{n'}(x) = -V_n(x)$ in Proposition 15.26 cannot be used to find a polynomial $V_{n'}(x)$ such that $V_{n'}(x) = -V_n(x)$, for all $x \in \mathbb{F}_q$, because $V_n(1) = 1$ for any integer $n$ (this follows from Proposition 15.21). In other words, if $V_n(x)$ is a permutation polynomial, there is no (permutation) polynomial $V_{n'}(x)$ such that $V_{n'}(x) = -V_n(x)$.

Combining the conditions (15.19) and (15.20), we conclude that they are equivalent to $n' \equiv n \pmod{\overline{N}}$ or $n' \equiv -n - 1 \pmod{\overline{N}}$, where $\overline{N} = \text{lcm}(q + 1, q - 1, 2p) = \frac{p(q+1)(q-1)}{2}$. In this manner, $\overline{N}$ corresponds to a kind of *general period* of $V_n(x)$, i.e., $V_n(x) = V_{n+k\overline{N}}(x) = V_{-n-1+k\overline{N}}(x)$, for any integer $k$. Naturally, if $V_n(x)$ is a permutation polynomial, then $V_{n+k\overline{N}}(x)$ and $V_{-n-1+k\overline{N}}(x)$ are also permutation polynomials.

The next theorem considers the analogous result for Chebyshev polynomials of the fourth kind.

**Theorem 15.32** *Assume $W_n(x)$ is a permutation polynomial over $\mathbb{F}_q$. If $n'$ satisfies, for $N = q + 1$ and $N = q - 1$,*

$$n' \equiv n \pmod{N}, \tag{15.21}$$

and if $n'$ satisfies

$$n' \equiv n \pmod{p}, \tag{15.22}$$

then $W_{n'}(x)$ is a permutation polynomial over $\mathbb{F}_q$. If $n'$ satisfies, for $N = q + 1$ and $N = q - 1$,

$$n' \equiv -n - 1 \pmod{N}, \tag{15.23}$$

and if $n'$ satisfies

$$n' \equiv -n - 1 \pmod{p}, \tag{15.24}$$

then $W_{n'}(x)$ is a permutation polynomial over $\mathbb{F}_q$.

The proof of Theorem 15.32 is analogous to that of Theorem 15.31. Combining the conditions (15.21) and (15.22), we conclude that they are equivalent to $n' \equiv n \pmod{\overline{N}}$, where $\overline{N} = \mathrm{lcm}(q+1, q-1, p) = \frac{p(q+1)(q-1)}{2}$. There-fore, one has $W_n(x) = W_{n+k\overline{N}}(x)$, for any integer $k$; if $W_n(x)$ is a permutation polynomial, then $W_{n+k\overline{N}}(x)$ is also a permutation polynomial. Analogously, combining the conditions (15.23) and (15.24), we conclude that they are equiv-alent to $n' \equiv -n - 1 \pmod{\overline{N}}$. Therefore, one has $W_n(x) = -W_{-n-1+k\overline{N}}(x)$, for any integer $k$; if $W_n(x)$ is a permutation polynomial, then $W_{-n-1+k\overline{N}}(x)$ is also a permutation polynomial.

## 15.6  Conclusions

In this paper, we introduced trigonometric definitions for Chebyshev poly-nomials over finite fields. These definitions are supported by a finite field trigonometry, which includes the notion of trigonometric and hyperbolic trigonometric functions over prime and extension fields. From the proposed definitions, periodicity and symmetry properties of Chebyshev polynomials were investigated. A preliminary study concerning the permutation properties of the mentioned polynomials was conducted by means of the proposed defi-nitions. Sufficient conditions for the Chebyshev polynomial of the second kind $U_n(x)$ to be a permutation polynomial were derived by an alternative proof. We also showed that the permutation behavior of Chebyshev polynomials of the third and fourth kinds does not follow that of $U_n(x)$. This suggests that the permutation properties of $V_n(x)$ and $W_n(x)$ are in a certain sense not too interesting. Indeed, if the conditions given in Theorems 15.31 and 15.32 are also necessary, only one and two different permutation polynomials $V_n(x)$ and $W_n(x)$, respectively, exist in a given field. The necessity of such conditions was confirmed by means of a computer search carried out for $V_n(x)$ and $W_n(x)$ over $\mathbb{I}_p$, $p \leq 31$. Due to the *general periodicity* of such polynomials, in the search, only polynomials of degree up to $\frac{p(q+1)(q-1)}{2}$ were considered.

We believe that such facts and other more general results with respect to permutation properties of Chebyshev polynomials may be studied using the concepts introduced in this paper. This possibility, as well as the notion of Chebyshev polynomials of the $(k+1)$th kind [18], is currently under investigation.

## References

[1]  R. M. Campello de Souza, H. M. de Oliveira, A. N. Kauffman and A. J. A. Paschoal, Trigonometry in finite fields and a new Hartley transform. In: *Proc. IEEE Int. Symp.—Information Theory (ISIT'98)*, p. 293. IEEE, New York, 1998.

[2] M. Cipu, Dickson polynomials that are permutations. *Serdica Math. J.* **30**, 177–194, 2004.

[3] M. Cipu and S. D. Cohen, Dickson polynomial permutations. In: G. L. Mullen, D. Panario, and I. E. Shparlinski (eds.), *Finite Fields and Applications*. Contemporary Mathematics, volume 461, pp. 79–90. AMS, Providence, RI, 2008.

[4] S. D. Cohen, Dickson polynomials of the second kind that are permutations. *Can. J. Math.* **46**, 225–238, 1994.

[5] R. S. Coulter and R. W. Matthews, On the permutation behaviour of Dickson polynomials of the second kind. *Finite Fields Appl.* **8**(4), 519–530, 2002.

[6] G. Fee and M. Monagan, Cryptography using Chebyshev polynomials. *Proc. 2004 Maple Summer Workshop.* Preprint available at www.cecm.sfu.ca/CAG/products2003.shtml, 2004.

[7] A. Garcia and H. Stichtenoth, On Chebyshev polynomials and maximal curves. *Acta Arith.* **90**, 301–311, 1999.

[8] M. Henderson and R. Matthews, Permutation properties of Chebyshev polynomials of the second kind over a finite field. *Finite Fields Appl.* **1**(1), 115–125, 1995.

[9] S. Kang, Remarks on finite fields III. *Bull. Korean Math. Soc.* **23**(2), 103–111, 1986.

[10] X. Liao, F. Chen and K.-W. Wong, On the security of public-key algorithms based on Chebyshev polynomials over the finite field $Z_N$. *IEEE Trans. Comput.* **59**(10), 1392–1401, 2010.

[11] R. Lidl and H. Niederreiter, *Finite Fields*, second edition. Cambridge University Press, Cambridge, 1997.

[12] R. Lidl, G. L. Mullen and G. Turnwald, *Dickson Polynomials*. Longman Scientific and Technical, Pitman Monographs and Surveys in Pure and Applied Mathematics, volume 65, Longman, 1993.

[13] J. B. Lima, D. Panario and R. M. Campello de Souza, Public-key encryption based on Chebyshev polynomials over GF($q$). *Inf. Process. Lett.* **111**(2), 51–56, 2010.

[14] J. C. Mason and D. C. Handscomb, *Chebyshev Polynomials*. Chapman & Hall/CRC, Boca Raton, FL, 2003.

[15] R. W. Matthews, Permutation polynomials in one and several variables. PhD Thesis, University of Tasmania, 1982.

[16] P. L. Montgomery, Chebyshev polynomials over finite fields and reversibility of $\sigma$-automata on square grids. *Theor. Comput. Scie.* **320**(2–3), 465–483, 2004.

[17] G. L. Mullen and D. Panario, *Handbook of Finite Fields*. Chapman & Hall/CRC, Boca Raton, FL, 2013.

[18] Q. Wang and J. L. Yucas, Dickson polynomials over finite fields. *Finite Fields Appl.* **18**(4), 814–831, 2012.

# 16

# Index bounds for value sets of polynomials over finite fields

*Gary L. Mullen*
Pennsylvania State University, University Park, PA

*Daqing Wan*
University of California Irvine, CA

*Qiang Wang*
Carleton University, Ottawa

*We dedicate this paper to the occasion of Harald Niederreiter's 70th birthday. His work on permutation polynomials over finite fields, and more generally, his work in so many areas of finite fields and their applications, has been a huge and lasting inspiration to all of us.*

## Abstract

We provide an upper bound for the cardinality of the value set of a univariate polynomial over a finite field in terms of the index of the polynomial. Moreover, we study when a polynomial vector map in $n$ variables is a permutation polynomial map, again using the index tuple of the map. This also provides an upper bound for the value set of a polynomial map in $n$ variables.

## 16.1 Introduction

Let $\mathbb{F}_q$ be the finite field of $q$ elements with characteristic $p$. The *value set* of a polynomial $g$ over $\mathbb{F}_q$ is the set $V_g$ of images when we view $g$ as a mapping from $\mathbb{F}_q$ to itself. Clearly $g$ is a *permutation polynomial (PP)* of $\mathbb{F}_q$ if and only if the cardinality $|V_g|$ of the value set $V_g$ of $g$ is $q$. Asymptotic formulas such as $|V_g| = \lambda(g)q + O(q^{1/2})$, where $\lambda(g)$ is a constant depending only on certain Galois groups associated to $g$, can be found in Birch and Swinnerton-Dyer [7] and Cohen [14]. Later, Williams [39] proved that almost all polynomials $f$ are polynomials satisfying $\lambda(g) = 1 - \frac{1}{2!} + \frac{1}{3!} + \cdots + (-1)^{d-1}\frac{1}{d!}$, where $d$ is the degree of the polynomial $g$. There are also several results on explicit bounds for the cardinality of value sets if $g$ is not a PP over $\mathbb{F}_q$; see for example

[33, 34]. Perhaps the most well-known result is due to Wan [34] who proved that if $g$ is not a PP then

$$|V_g| \leq q - \frac{q-1}{d}. \tag{16.1}$$

Using results from group theory, Guralnick and Wan [19] further proved that if $(d, q) = 1$ then $|V_g| \leq (47/63)q + O_d(\sqrt{q})$. Some progress on lower bounds of $|V_g|$ can be found in [15, 36]. The classification of *minimal value set polynomials* (polynomials satisfying $|V_g| = \lceil q/d \rceil$) can be found in [11, 18, 27], and in [8] for all the minimal value set polynomials in $\mathbb{F}_q[x]$ whose set of values is a subfield of $\mathbb{F}_q$. More recently, algorithms and complexity in computing $|V_g|$ have been studied [13]. All of these results relate $|V_g|$ to the degree $d$ of $g$.

In this paper, we take a different approach to study value sets. We note that any nonconstant polynomial $g \in \mathbb{F}_q[x]$ of degree $\leq q - 1$ can be written *uniquely* as $g(x) = a(x^r f(x^{(q-1)/\ell})) + b$ with index $\ell$ defined below. Namely, write

$$g(x) = a(x^n + a_{n-i_1}x^{n-i_1} + \cdots + a_{n-i_k}x^{n-i_k}) + b,$$

where $a$, $a_{n-i_j} \neq 0$, $j = 1, \ldots, k$. The case that $k = 0$ is trivial. Thus, we shall assume that $k \geq 1$. Write $n - i_k = r$, the vanishing order of $x$ at 0 (i.e., the lowest degree of $x$ in $g(x) - b$ is $r$). Then $g(x) = a\left(x^r f(x^{(q-1)/\ell})\right) + b$, where $f(x) = x^{e_0} + a_{n-i_1}x^{e_1} + \cdots + a_{n-i_{k-1}}x^{e_{k-1}} + a_r$,

$$\ell = \frac{q-1}{\gcd(n-r, n-r-i_1, \ldots, n-r-i_{k-1}, q-1)} := \frac{q-1}{s},$$

and $\gcd(e_0, e_1, \ldots, e_{k-1}, \ell) = 1$. The integer $\ell = \frac{q-1}{s}$ is called the *index* of $h(x)$. The concept of the index of any polynomial was first introduced in [5] and is closely related to the concept of the least index of a cyclotomic mapping polynomial [16, 32]. Clearly, the study of the value set of $g(x) = a(x^r f(x^{(q-1)/\ell})) + b$ over $\mathbb{F}_q$ is equivalent to studying the value set of $g(x) = x^r f(x^{(q-1)/\ell}) = x^r f(x^s)$ over $\mathbb{F}_q$. If $(r, (q-1)/\ell) = 1$, we say $g$ is in *reduced form*. Otherwise, if $(r, (q-1)/\ell) = t$, then $g(x) = g'(x^t)$ where $g'(x) = x^{r/t} f(x^{s/t})$ is in reduced form. In fact a permutation polynomial $g$ must be in reduced form. We note that permutation polynomials of the form $x^r f(x^s)$ were studied by Wan and Lidl [35] in 1991 and more recently by many others [1, 2, 3, 4, 5, 21, 37, 43, 44, 45]. For more background material on permutation polynomials we refer to Chapter 7 of [24]. For a detailed survey see [22, 23, 28, 30] and for recent results see [6, 9, 10, 12, 17, 20, 38, 40, 41, 42]. We refer to Section 8.1 of [29] for a detailed discussion of PPs and Section 8.3 of [29] for a discussion of value sets of polynomials over finite fields.

In Section 16.2, we study the value set problem in terms of the index of the polynomial $g$. In Theorem 16.3 we prove that if $g$ is not a PP then

$$|V_g| \leq q - \frac{q-1}{\ell}. \tag{16.2}$$

Our result improves Wan's result when the index $\ell$ of a polynomial is strictly smaller than the degree $d$. We note that the index $\ell$ of a polynomial is always smaller than the degree $d$ as long as $\ell \leq \sqrt{q} - 1$. For example, the index of any permutation binomial is always less than or equal to the degree. In fact, the index of polynomials is closely related to the concept of the least index of cyclotomic permutations. These permutations in terms of cyclotomic cosets were studied by Niederreiter and Winterhof [32] and Wang [37, 38]. Also in Section 16.2 a generic formula for $|V_g|$ in terms of the number of certain distinct cyclotomic cosets is given in Proposition 16.5.

Let $g : \mathbb{F}_q^n \to \mathbb{F}_q^n$ be a polynomial map in $n$ variables defined over $\mathbb{F}_q$, where $n$ is a positive integer. Denote by $|V_g|$ the number of distinct values taken by $g(x_1, \ldots, x_n)$ as $(x_1, \ldots, x_n)$ runs over $\mathbb{F}_q^n$. It is clear that $|V_f| \leq q^n$. If $|V_f| = q^n$, then $f$ is a *permutation polynomial vector*, see [24, Chapter 7]. Motivated by an open problem raised by Lipton [25] in his computer science blog, we extended Wan's result on upper bounds of value sets for univariate polynomials to polynomial maps in $n$ variables [31]. More specifically, we write $g$ as a polynomial vector:

$$g(x_1, \ldots, x_n) = (g_1(x_1, \ldots, x_n), \ldots, g_n(x_1, \ldots, x_n)), \tag{16.3}$$

where each $g_i$ ($1 \leq i \leq n$) is a polynomial in $n$ variables over $\mathbb{F}_q$. The polynomial vector $g$ induces a map from $\mathbb{F}_q^n$ to $\mathbb{F}_q^n$. By reducing the polynomial vector $g$ modulo the ideal $(x_1^q - x_1, \ldots, x_n^q - x_n)$, we may assume that the degree of $g_i$ in each variable is at most $q - 1$ and we may further assume that $g$ is a nonconstant map to avoid the trivial case. Let $d_i$ denote the total degree of $g_i$ in the $n$ variables $x_1, \ldots, x_n$ and let $d = \max\{d_1, \ldots, d_n\}$. Then $d$ satisfies $1 \leq d \leq n(q - 1)$. In particular, we proved the following.

**Theorem 16.1** [31] *If $|V_g| < q^n$, then $|V_g| \leq q^n - \min\left\{\frac{n(q-1)}{d}, q\right\}$.*

In Section 16.3, we extend the concept of index of a univariate polynomial to index tuples for multivariate polynomials $g_i(x_1, \ldots, x_n)$ for $1 \leq i \leq n$ and the polynomial vector map $g(x_1, \ldots, x_n)$ respectively. We remark that any multivariate polynomial $g_i(x_1, \ldots, x_n)$ behaves as a monomial in each subset of $\mathbb{F}_q \times \cdots \times \mathbb{F}_q$ that is partitioned by the cyclotomic cosets determined by the index tuple $(\ell_1^{(i)}, \ldots, \ell_n^{(i)})$. Similarly, each coordinate $g_i(x_1, \ldots, x_n)$ of any polynomial vector map $g(x_1, \ldots, x_n)$ behaves as a monomial when we view the vector map as a cyclotomic mapping. It turns out that the index tuple $(\ell_1, \ldots, \ell_n)$ of $g(x_1, \ldots, x_n)$ can be obtained from index tuples $(\ell_1^{(i)}, \ldots, \ell_n^{(i)})$ of $g_i(x_1, \ldots, x_n)$. Namely, $\ell_i = \mathrm{lcm}(\ell_i^{(1)}, \ldots, \ell_i^{(n)})$ for $1 \leq i \leq n$. Then we study the extreme cases for the value set problem for polynomial maps of $n$ variables. Namely, we describe when a polynomial map $g$ in $n$ variables is a permutation polynomial map. Essentially, each coordinate polynomial of a permutation vector map behaves as a monomial in terms of only one variable in each subset of $\mathbb{F}_q \times \cdots \times \mathbb{F}_q$ that is partitioned by the cyclotomic cosets determined by the index tuple $(\ell_1, \ldots, \ell_n)$, along with other explicit conditions as described in Theorem 16.10. In other words, each permutation vector map in $n$ variables consists of $n$ univariate cyclotomic monomial permutations together with another permutation on coordinate variables. As a corollary, we obtain the following result.

**Theorem 16.2** *Let $g$ be a polynomial vector map from $\mathbb{F}_q^n$ to $\mathbb{F}_q^n$ with the index tuple $(\ell_1, \ldots, \ell_n)$ and $\ell = \max\{\ell_1, \ldots, \ell_n\} > 1$. If $|V_g| < q^n$ then $|V_g| \leq q^n - \frac{q-1}{\ell}$.*

This also provides another answer to Lipton's problem on the existence of a Picard jump for polynomial maps (roughly, if $g$ misses one value in $\mathbb{F}_q^n$ then $g$ misses quite a few). An example meeting this upper bound is also provided. We also note that our new bound improves the bound in Theorem 16.1 when $\frac{d}{n} > \ell$.

## 16.2 Value sets of univariate polynomials

As explained in the introductory section, the value set problem for a univariate polynomial is equivalent to that for the polynomial $g(x) = x^r f(x^s)$. Here we want to emphasize the parameter index $\ell$ instead of the degree $d$. We note from [37] that $g(x) = x^r f(x^s)$ is a PP if and only if $\gcd(r, s) = 1$, and $f(\zeta^i) \neq 0$ for $0 \leq i \leq \ell - 1$ where $\zeta$ is a primitive $\ell$th root of unity, and $g(x)$ induces a permutation among all the $\ell$ cyclotomic cosets

$$\{C_0, C_1, \ldots, C_{\ell-1}\} = \mathbb{F}_q^* / (\mathbb{F}_q^*)^\ell.$$

In fact, we can always write $g(x)$ in terms of *cyclotomic mappings* $g(x) = c_i x^r$ if $x \in C_i$, where $c_i = f(\zeta^i)$; more details can be found in [16, 32, 37, 38]. In particular, a polynomial $g(x) \in \mathbb{F}_q[x]$ is called an orthomorphism if both $g(x)$ and $g(x) - x$ are permutation polynomials. Given a finite, nonempty set of positive integers $R$, a polynomial $g(x)$ is called an $R$-orthomorphism if $g^{(r)}(x)$ is an orthomorphism of $\mathbb{F}_q$ for all $r \in R$. (Here $g^{(r)}$ denotes the function $g$ composed with itself $r$ times.) We note that Niederreiter and Winterhof [32] proved several existence results for cyclotomic orthomorphisms and cyclotomic $R$-orthomorphisms of finite fields. Here we only concentrate on the permutation behavior of $g(x)$.

**Theorem 16.3** *Let $g(x) = ax^r f(x^s) + b$ $(a \neq 0)$ be a polynomial in reduced form (i.e., $\gcd(r, s) = 1$) over $\mathbb{F}_q$ with index $\ell$. Then $|V_g| > q - \frac{q-1}{\ell}$ if and only if $g$ is a PP of $\mathbb{F}_q$.*

*Proof.* Without loss of generality, we can assume $a = 1$ and $b = 0$. Hence $g(0) = 0$. The conditions $\gcd(r, s) = 1$ and $f(\zeta^i) \neq 0$ guarantee that all the images of elements in each $C_i$ are distinct nonzero elements. Because $|C_i| = s = \frac{q-1}{\ell}$, we conclude that $|V_g| > q - \frac{q-1}{\ell}$ if and only if there are more than $\ell - 1$ nonzero distinct cyclotomic cosets. Since there are exactly $\ell$ nonzero distinct image sets of cyclotomic cosets, we deduce that $|V_g| > q - \frac{q-1}{\ell}$ if and only if $g$ is a PP of $\mathbb{F}_q$. $\square$

This result improves Wan's result (i.e., Equation (16.1)) for arbitrary polynomials with index $\ell \leq \sqrt{q} - 1$. Indeed, if the index $\ell \leq \sqrt{q} - 1$, then $s \geq \sqrt{q} + 1$ and thus the degree $d \geq s + 1 > \ell$. Our result also works at least as well as Wan's result [34] if we want to verify that an arbitrary binomial over a prime field is a permutation using the contrapositive lower bound. Indeed, let $g(x) = x^d + ax^m$ with $d > m > 0$, be an arbitrary permutation binomial over a prime field $\mathbb{F}_p$. It was proved by Masuda and Zieve [26] that $\gcd(d - m, p - 1) \geq \sqrt{p - 3/4} - 1/2 (> \sqrt{p} - 1)$. Here $\gcd(d - m, p - 1)$ turns out to be equal to $s$. So the index $\ell = \frac{p-1}{s} \leq \sqrt{p - 3/4} + 1/2 (< \sqrt{p} + 1)$ and thus $s > \sqrt{p} - 1$. Then $d = m + es \geq 1 + s \geq \sqrt{p - 3/4} + 1/2 \geq \ell$. Hence $p - \frac{p-1}{d} \geq p - \frac{p-1}{\ell}$. However, we note that $d$ is strictly greater than $\ell$ for any $m > 1$ or $e > 1$ as above.

**Corollary 16.4** *Let $g(x) = ax^r f(x^{\frac{q-1}{\ell}}) + b$ $(a \neq 0)$ be any polynomial over $\mathbb{F}_q$ with index $\ell > 1$ and $s = \frac{q-1}{\ell}$. Assume $|V_g| < q$. Then*

(i) *if $\gcd(r, s) = 1$ then $|V_g| \leq q - \frac{q-1}{\ell}$,*

(ii) *if $\gcd(r, s) = t > 1$ then $|V_g| \leq \frac{q-1}{t} + 1$.*

*Therefore we always have $|V_g| \leq q - \frac{q-1}{\ell}$.*

*Proof.* Without loss of generality, we can assume $a = 1$ and $b = 0$. Hence $g(0) = 0$. The case of $\gcd(r, s) = 1$ follows from Theorem 16.3. If $\gcd(r, s) = t > 1$, then $g(x) = g_1(x^t)$ for some polynomial $g_1 \in \mathbb{F}_q[x]$. Thus, $|V_g| \leq |V_{x^t}| = \frac{q-1}{t} + 1$. We note that $\frac{q-1}{t} + 1 = q - \frac{(t-1)(q-1)}{t} = q - (q - 1 - \frac{q-1}{t})$. Because $t > 1$ and $\ell > 1$, we must have $q - 1 - \frac{q-1}{t} \geq q - 1 - \frac{q-1}{2} = \frac{q-1}{2} \geq \frac{q-1}{\ell}$. Thus we have $|V_g| \leq q - \frac{q-1}{\ell}$ in both cases.   □

In fact, we can obtain the following formula for the cardinality of the value set.

**Proposition 16.5** *Let* $g(x) = ax^r f(x^s) + b$ ($a \neq 0$) *be any polynomial over* $\mathbb{F}_q$ *with index* $\ell = \frac{q-1}{s}$ *and let* $\gcd(r, s) = t$. *Let* $\xi$ *be a fixed primitive element of* $\mathbb{F}_q$. *Then*

$$|V_g| = c\frac{s}{t} + 1, \quad or \quad |V_g| = (c - 1)\frac{s}{t} + 1,$$

*where* $c = |\{(\xi^{ir} f(\xi^{si}))^{s/t} \mid i = 0, \ldots, \ell - 1\}|$.

*Proof.* Without loss of generality, we can assume $a = 1$ and $b = 0$. Hence $g(0) = 0$. Let $C_0$ be the subgroup of $\mathbb{F}_q^*$ consisting of all the $\ell$th powers of $\mathbb{F}_q^*$ and let $D_0$ be the subgroup of $\mathbb{F}_q^*$ consisting of all the $\ell t$th powers. Let $C_i = \xi^i C_0$ for $i = 0, \ldots, \ell - 1$ be cyclotomic cosets of $\mathbb{F}_q^*$ induced by $C_0$. Note that $g(x) = c_i x^r$ when $x \in C_i$, where $c_i = f(\xi^{si})$ for $i = 0, \ldots, \ell - 1$. We also note that $x^r$ maps $C_0$ onto $D_0$ which contains $\frac{s}{t}$ distinct elements. So $x^r$ maps each coset $C_i = \xi^i C_0$ onto $\xi^{ir} D_0$. Therefore $g$ maps $C_i$ onto $\xi^{ir} f(\xi^{si}) D_0$, which could be either the set $\{0\}$ or one of the nonzero cyclotomic cosets of index $\ell t$. We observe that $c$ is the number of distinct image sets of the form $\xi^{ir} f(\xi^{si}) D_0$. Hence we have $|V_g| = c\frac{s}{t} + 1$ or $(c - 1)\frac{s}{t} + 1$, the latter happens when some of the $c_i$ in $g(x) = c_i x^r$ equal $g(0) = 0$.   □

From here it is straightforward to obtain a generic lower bound $\frac{s}{(r,s)} + 1$ for any nonzero polynomial. However, this lower bound can be improved depending on how much information we know about the coefficients of $g$ in order to say more about $\xi^{ir} f(\xi^{si})$. We also refer to [15] for a matrix method which can be used to obtain a lower bound for the cardinality $|V_g|$ of the value set of a univariate polynomial $g$ over $\mathbb{F}_q$.

## 16.3  Permutation polynomial vectors

Let us first consider a multivariate polynomial $g(x_1, \ldots, x_n)$ over $\mathbb{F}_q$. As in the univariate case, we can write $g(x_1, \ldots, x_n) = x_1^{r_1} \cdots x_n^{r_n} f(x_1^{s_1}, \ldots, x_n^{s_n}) + b$

where $g(0, \ldots, 0) = b$, and $r_1, \ldots, r_n$ are vanishing orders of $x_1, \ldots, x_n$ in $g(x_1, \ldots, x_n) - b$ at $0$ respectively (i.e., the lowest degree of $x_i$ in $g(x_1, \ldots, x_n) - b$ is $r_i$), and each $s_i$ is the greatest common divisor of all the exponents of $x_i$ from all monomial terms after factoring $x_i^{r_i}$, together with $q - 1$ for $1 \le i \le n$ (i.e., $s_i$ is the greatest common divisor of all the exponents of $x_i$ in $f(x_1^{s_1}, \ldots, x_n^{s_n})$ together with $q - 1$). We note that $r_i \ge 0$ in this case, instead of $r \ge 1$ for univariate polynomials. Let $\ell_i = \frac{q-1}{s_i}$ with $1 \le i \le n$. Then $(\ell_1, \ldots, \ell_n)$ is called the *index tuple* of the multivariate polynomial $g(x_1, \ldots, x_n)$.

**Example 16.6** *Let $g(x_1, x_2) = x_1^4 x_2^5 - x_1^2 x_2^5 + 3x_2^5$ over $\mathbb{F}_7$. So $(r_1, r_2) = (0, 5)$ is the pair of vanishing orders of $x_1$ and $x_2$ at $0$ respectively. We can write $g(x_1, x_2) = x_2^5(x_1^4 + x_1^2 + 3) = x_1^0 x_2^5 f(x_1^2, x_2^6)$ with $f(x_1, x_2) = x_1^2 + x_1 + 3$ because $s_1 = \gcd(4, 2, 0, 6) = 2$ and $s_2 = \gcd(0, 0, 0, 6) = 6$. Namely, $(\ell_1, \ell_2) = (3, 1)$ is the index tuple of $g$.*

*Similarly, $h(x_1, x_2) = 3x_1 x_2^3 - 2x_1$ over $\mathbb{F}_7$ can be written as $h(x_1, x_2) = x_1(3x_2^3 - 2) = x_1^1 x_2^0 f(x_1^6, x_2^3)$ where $f(x_1, x_2) = 3x_2 - 2$. Namely, $r_1 = 1$, $r_2 = 0$, $s_1 = 6$, $s_2 = 3$, $\ell_1 = 1$, and $\ell_2 = 2$.*

*Finally, $t(x_1, x_2) = 3x_1^2 x_2^3 - 2x_1^3 x_2 + 5$ over $\mathbb{F}_7$ can be written as $t(x_1, x_2) = x_1^2 x_2(3x_2^2 - 2x_1) + 5 = x_1 x_2 f(x_1, x_2^2) + 5$ where $f(x_1, x_2) = 3x_2 - 2x_1$. Namely, $r_1 = 1$, $r_2 = 1$, $s_1 = 1$, $s_2 = 2$, $\ell_1 = 6$, and $\ell_2 = 3$.*

**Definition 16.7** *The multivariate polynomial*

$$g(x_1, \ldots, x_n) = x_1^{r_1} \cdots x_n^{r_n} f(x_1^{(q-1)/\ell_1}, \ldots, x_n^{(q-1)/\ell_n}) + b$$

*is said to be in index form if $r_1, \ldots, r_n$ are vanishing orders of $x_1, \ldots, x_n$ at $0$ respectively, $g(0, \ldots, 0) = b$, and $(\ell_1, \ldots, \ell_n)$ is the index tuple of $g$.*

Without loss of generality, we assume $g(0, \ldots, 0) = 0$ and $s_i = \frac{q-1}{\ell_i}$ for $1 \le i \le n$. Hence

$$g(x_1, \cdots, x_n) = x_1^{r_1} \cdots x_n^{r_n} f(x_1^{s_1}, \ldots, x_n^{s_n}).$$

For each $1 \le i \le n$, let $C_{i,0}$ be the multiplicative subgroup of $\mathbb{F}_q^*$ containing all the $\ell_i$th powers and let $C_{i,j_i}$ be the $j_i$th coset of $C_{i,0}$ in $\mathbb{F}_q^*$ where $0 \le j_i \le \ell_i - 1$. Let $\xi$ be a fixed primitive element in $\mathbb{F}_q^*$ and $\zeta_i = \xi^{s_i}$ be a primitive $\ell_i$th root of unity where $1 \le i \le n$. Hence $C_{i,j_i} = \xi^{j_i} C_{i,0}$. Moreover, if $x \in C_{i,j_i}$ then $x^{s_i} = \zeta_i^{j_i}$. If $r_i > 0$, then $g(x_1, \ldots, x_{i-1}, 0, x_{i+1}, \ldots, x_n) = 0$. Otherwise, $g(x_1, \ldots, x_{i-1}, 0, x_{i+1}, \ldots, x_n)$ may not be zero. Hence, for a given index tuple $(\ell_1, \ldots, \ell_n)$, we can partition $\mathbb{F}_q \times \cdots \times \mathbb{F}_q$ as a union of $A_1 \times \cdots \times A_n$ where $A_i$ is either the set $\{0\}$ or one of the cosets $C_{i,j_i}$ determined by the index $\ell_i$. We define constants $a_1, \ldots, a_n$ over these sets $A_1, \ldots, A_n$ as follows:

$$a_i = \begin{cases} \zeta_i^{j_i} & \text{if } A_i = C_{i,j_i} \\ 0 & \text{if } A_i = \{0\}. \end{cases}$$

Then $g(x_1, \ldots, x_n) = x_1^{r_1} \cdots x_n^{r_n} f(x_1^{s_1}, \ldots, x_n^{s_n})$ can be written as a *cyclotomic mapping* as follows:

$$g(x_1, \ldots, x_n) = \begin{cases} 0 & \text{if } (x_1, \ldots, x_n) = (0, \ldots, 0) \\ f(a_1, \ldots, a_n)x_1^{r_1} \cdots x_n^{r_n} & \text{if } (x_1, \ldots, x_n) \in A_1 \times \cdots \times A_n. \end{cases}$$
(16.4)

Because $a_1, \ldots, a_n$ are constants over $A_1 \times \cdots \times A_n$, we remark that any multivariate polynomial $g$ behaves as a monomial $f(a_1, \ldots, a_n)x_1^{r_1} \cdots x_n^{r_n}$ in the subset $A_1 \times \cdots \times A_n$, determined by the partition of $\mathbb{F}_q \times \cdots \times \mathbb{F}_q$ according to the index tuple of $g$. We note that the similar concept for univariate polynomials can be found in [32, 37, 38].

**Example 16.8**  Consider $g(x_1, x_2) = x_1^2 x_2(4x_1^3 x_2^2 - 2x_2^4)$ over $\mathbb{F}_7$. We can write $g(x_1, x_2) = x_1^2 x_2 f(x_1^3, x_2^2)$ with the index tuple $(2, 3)$, where $f(x_1, x_2) = 4x_1 x_2 - 2x_2^2$. Let $\xi = 3$ be the fixed primitive element in $\mathbb{F}_7$. So $\zeta_1 = \xi^{6/2} = 6$ and $\zeta_2 = \xi^{6/3} = 2$. We can partition $\mathbb{F}_7^*$ into either $C_{1,0} = \{1, 2, 4\}$ and $C_{1,1} = \{3, 5, 6\}$ corresponding to $\ell_1 = 2$, or $C_{2,0} = \{1, 6\}$, $C_{2,1} = \{3, 4\}$, and $C_{2,2} = \{2, 5\}$ corresponding to $\ell_2 = 3$. Then $\mathbb{F}_7 \times \mathbb{F}_7$ can be partitioned into the union of all these $A_1 \times A_2$, where $A_1$ denotes any one of the sets $\{0\}$, $C_{1,0}$, and $C_{1,1}$, and $A_2$ denotes any one of the sets $\{0\}$, $C_{2,0}$, $C_{2,1}$ and $C_{2,2}$. Hence $g(x_1, x_2)$ can be represented by

$$g(x_1, x_2) = \begin{cases} 0 & \text{if } (x_1, x_2) = (0, 0) \\ 0 & \text{if } (x_1, x_2) \in \{0\} \times C_{2,0} \\ 0 & \text{if } (x_1, x_2) \in \{0\} \times C_{2,1} \\ 0 & \text{if } (x_1, x_2) \in \{0\} \times C_{2,2} \\ 0 & \text{if } (x_1, x_2) \in C_{1,0} \times \{0\} \\ 2x_1^2 x_2 & \text{if } (x_1, x_2) \in C_{1,0} \times C_{2,0} \\ 0 & \text{if } (x_1, x_2) \in C_{1,0} \times C_{2,1} \\ 5x_1^2 x_2 & \text{if } (x_1, x_2) \in C_{1,0} \times C_{2,2} \\ 0 & \text{if } (x_1, x_2) \in C_{1,1} \times \{0\} \\ x_1^2 x_2 & \text{if } (x_1, x_2) \in C_{1,1} \times C_{2,0} \\ 5x_1^2 x_2 & \text{if } (x_1, x_2) \in C_{1,1} \times C_{2,1} \\ x_1^2 x_2 & \text{if } (x_1, x_2) \in C_{1,1} \times C_{2,2}, \end{cases}$$
(16.5)

where the coefficients of $x_1^2 x_2$ in all these branches are computed using $f(a_1, a_2) = 4a_1 a_2 - 2a_2^2$ where $a_1 = 0, 1, -1$ and $a_2 = 0, 1, 2, 4$ respectively.

By an abuse of notation, let us now consider $g$ as a polynomial vector map in $n$ variables from $\mathbb{F}_q^n$ to $\mathbb{F}_q^n$:

$$g(x_1, \ldots, x_n) = (g_1(x_1, \ldots, x_n), \ldots, g_n(x_1, \ldots, x_n)), \qquad (16.6)$$

where each $g_i$ $(1 \leq i \leq n)$ is a polynomial in $n$ variables over $\mathbb{F}_q$. Using the previous definition of the index tuple and cyclotomic mappings, for each $1 \leq i \leq n$, we can write each $g_i$ in the index form. Namely,

$$g_i(x_1, \ldots, x_n) = x_1^{r_1^{(i)}} \cdots x_n^{r_n^{(i)}} f_i(x_1^{s_1^{(i)}}, \ldots, x_n^{s_n^{(i)}}) + b_i,$$

with the index tuple $(\ell_1^{(i)}, \ldots, \ell_n^{(i)})$ and $b_i \in \mathbb{F}_q$. Without loss of generality, we assume further that $b_i = 0$ for all $1 \leq i \leq n$.

Hence

$$g(x_1, \ldots, x_n)$$
$$= \left( x_1^{r_1^{(1)}} \cdots x_n^{r_n^{(1)}} f_1(x_1^{s_1^{(1)}}, \ldots, x_n^{s_n^{(1)}}), \ldots, x_1^{r_1^{(n)}} \cdots x_n^{r_n^{(n)}} f_n(x_1^{s_1^{(n)}}, \ldots, x_n^{s_n^{(n)}}) \right).$$

For each $1 \leq i \leq n$, we let

$$s_i = \gcd(s_i^{(1)}, \ldots, s_i^{(n)}) \quad \text{and} \quad \ell_i = \frac{q-1}{s_i}.$$

Then we call $(\ell_1, \ldots, \ell_n)$ the *index tuple* of the polynomial vector map $g$ in $n$ variables.

Let $\zeta_i = \xi^{s_i}$ be a primitive $\ell_i$th root of unity and $C_{i,j_i}$ be the $j_i$th coset of $C_{i,0}$ in $\mathbb{F}_q^*$ where $0 \leq j_i \leq \ell_i$. We note again that $\mathbb{F}_q \times \cdots \times \mathbb{F}_q$ can be partitioned as a union of $A_1 \times \cdots \times A_n$ where $A_i$ is either the set $\{0\}$ or one of the cosets $C_{i,j_i}$ determined by the index tuple $(\ell_1, \ldots, \ell_n)$. Again, as defined before, we let

$$a_i = \begin{cases} \zeta_i^{j_i} & \text{if } A_i = C_{i,j_i} \\ 0 & \text{if } A_i = \{0\}. \end{cases}$$

Hence, if $(x_1, \ldots, x_n) \in A_1 \times \cdots \times A_n$ then

$$g(x_1, \ldots, x_n)$$
$$= \left( x_1^{r_1^{(1)}} \cdots x_n^{r_n^{(1)}} f_1(a_1^{s_1^{(1)}/s_1}, \ldots, a_n^{s_n^{(1)}/s_n}), \ldots, x_1^{r_1^{(n)}} \cdots x_n^{r_n^{(n)}} f_n(a_1^{s_1^{(n)}/s_1}, \ldots, a_n^{s_n^{(n)}/s_n}) \right).$$

Let $c_i = f_i(a_1^{s_1^{(i)}/s_1}, \ldots, a_n^{s_n^{(i)}/s_n})$. Then $g(x_1, \ldots, x_n)$ maps $(A_1, \ldots, A_n)$ to

$$\left( c_1 A_1^{r_1^{(1)}} \cdots A_n^{r_n^{(1)}}, \ldots, c_n A_1^{r_1^{(n)}} \cdots A_n^{r_n^{(n)}} \right),$$

where we use the convention $0^0 = 1$ and $A^r = \{x^r \mid x \in A\}$.

**Example 16.9** *Let* $g(x_1, x_2) = (x_2(x_1^4 + 4x_1^2 + 4), x_1(3x_2^3 + 1))$ *be a map from* $\mathbb{F}_7 \times \mathbb{F}_7$ *to itself. Using the previous definitions, we obtain* $r_1^{(1)} = 0$, $r_2^{(1)} = 1$, $r_1^{(2)} = 1$, $r_2^{(2)} = 0$, $s_1^{(1)} = 2$, $s_2^{(1)} = 6$, $s_1^{(2)} = 6$, *and* $s_2^{(2)} = 3$. *Moreover,* $f_1(x_1^2, x_2^6) = x_1^4 + 4x_1^2 + 4$, $f_2(x_1^6, x_2^3) = 3x_2^3 + 1$. *Hence* $s_1 = \gcd(2, 6) = 2$, $s_2 = \gcd(6, 3) = 3$, $\ell_1 = 3$ *and* $\ell_2 = 2$. *Therefore we use cyclotomic cosets of orders* 3 *and* 2 *respectively in the partition of* $\mathbb{F}_7^*$. *Namely,* $C_{1,0} = \{1, 6\}$, $C_{1,1} = \{3, 4\}$ *and* $C_{1,2} = \{2, 5\}$ *are the cyclotomic cosets of order* 3, $C_{2,0} = \{1, 2, 4\}$, *and* $C_{2,1} = \{3, 5, 6\}$ *are cyclotomic cosets of order* 2. *Then* $\mathbb{F}_7 \times \mathbb{F}_7$ *is partitioned into* $\{0\} \times \{0\}$, $\{0\} \times C_{2,0}$, $\{0\} \times C_{2,1}$, $C_{1,0} \times \{0\}$, $C_{1,1} \times \{0\}$, $C_{1,2} \times \{0\}$, $C_{1,0} \times C_{2,0}$, $C_{1,0} \times C_{2,1}$, $C_{1,1} \times C_{2,0}$, $C_{1,1} \times C_{2,1}$, $C_{1,2} \times C_{2,0}$, $C_{1,2} \times C_{2,1}$. *Note that* $a_1 \in \{0, 1, 2, 4\}$ *and* $a_2 \in \{0, 1, 6\}$. *We can check that* $f_1(a_1, a_2^2) \neq 0$ *and* $f_2(a_1^3, a_2) \neq 0$. *For example,* $g(x_1, x_2) = (x_2(x_1^4 + 4x_1^2 + 4), x_1(3x_2^3 + 1))$ *maps* $C_{1,1} \times C_{2,1}$ *into* $C_{2,1} \times C_{1,0}$ *because* $g(x_1, x_2)$ *behaves as the map* $(2x_2, 5x_1)$ *over* $C_{1,1} \times C_{2,1}$. *Indeed,* $f_1(2, 6) = 2^2 + 8 + 4 = 2$ *and* $f_2(2, 6) = 18 + 1 = 5$.

**Theorem 16.10** *Let* $g$ *be a polynomial vector map from* $\mathbb{F}_q^n$ *to* $\mathbb{F}_q^n$ *defined by*

$$g(x_1, \ldots, x_n) = (g_1(x_1, \ldots, x_n) + b_1, \ldots, g_n(x_1, \ldots, x_n) + b_n),$$

*where* $b_1, \ldots, b_n \in \mathbb{F}_q$ *and* $g$ *has index tuple* $(\ell_1, \ldots, \ell_n)$ *such that for each* $1 \leq i \leq n$,

$$g_i(x_1, \ldots, x_n) = x_1^{r_1^{(i)}} \cdots x_n^{r_n^{(i)}} f_i(x_1^{s_1^{(i)}}, \ldots, x_n^{s_n^{(i)}})$$

*is a polynomial in* $n$ *variables over* $\mathbb{F}_q$ *in the index form with index tuple* $(\ell_1^{(i)}, \ldots, \ell_n^{(i)})$ *satisfying* $g_i(0, \ldots, 0) = 0$ *and* $s_j^{(i)} = \frac{q-1}{\ell_j^{(i)}}$ *for* $1 \leq j \leq n$. *Let* $s_i = \gcd(s_i^{(1)}, \ldots, s_i^{(n)})$ *and* $\ell_i = \frac{q-1}{s_i}$ *for* $1 \leq i \leq n$. *Then* $g$ *is a permutation of* $\mathbb{F}_q^n$ *if and only if the following holds.*

(i) *For all* $1 \leq i \leq n$, *we must have* $f_i(a_1^{s_1^{(i)}/s_1}, \ldots, a_n^{s_n^{(i)}/s_n}) \neq 0$ *as long as not all the* $a_i$ *are zero, where* $a_i = 0$ *or* $a_i = \xi^{s_i j_i}$ *with* $0 \leq j_i \leq \ell_i - 1$ *and* $\xi$ *is a fixed primitive element of* $\mathbb{F}_q$.

(ii) *The matrix*

$$R := \begin{bmatrix} r_1^{(1)} & r_2^{(1)} & \cdots & r_n^{(1)} \\ r_1^{(2)} & r_2^{(2)} & \cdots & r_n^{(2)} \\ \vdots & \vdots & \cdots & \vdots \\ r_1^{(n)} & r_2^{(n)} & \cdots & r_n^{(n)} \end{bmatrix}$$

*contains exactly one nonzero entry for each row and each column. Moreover, for each nonzero* $r_i^{(k)}$ *we must have* $\gcd(r_i^{(k)}, s_i^{(k)}) = 1$.

(iii)  $g$ induces a bijection between the set of all the parts $A_1 \times \cdots \times A_n$ of the partition of $\mathbb{F}_q \times \cdots \times \mathbb{F}_q$ corresponding to the index tuple $(\ell_1, \ldots, \ell_n)$, and the set of all the parts $A'_{i_1} \times \cdots \times A'_{i_n}$ of the partition of $\mathbb{F}_q \times \cdots \times \mathbb{F}_q$ corresponding to the index tuple $(\ell_{i_1}, \ldots, \ell_{i_n})$, where $(i_1, \ldots, i_n)^T = P(1, \ldots, n)^T$ and the permutation matrix $P$ is associated with $R$ defined by $p_{ij} = 1$ if $r_j^{(i)} \neq 0$.

*Proof.* Without loss of generality, we can assume that $b_1 = b_2 = \cdots = b_n = 0$. Assume that $g$ is a permutation polynomial vector from $\mathbb{F}_q^n$ to $\mathbb{F}_q^n$. It is easy to see that condition (i) holds. Otherwise, at least two elements in $\mathbb{F}_q^n$ are mapped into the tuple consisting of all 0s.

We now prove condition (ii). First, each row of $R$ must contain at least one nonzero entry. Otherwise, suppose the $i$th row is the zero row, all the tuples $(x_1, \ldots, x_n)$ satisfying that the $i$th entry $x_i \in A_i$ must be mapped into tuples with the same $i$th entry, contradicting that $g$ is a permutation. Moreover, each row contains exactly one nonzero entry. Indeed, without loss of generality, suppose the first row contains two nonzero entries $r_1^{(1)}$ and $r_2^{(1)}$. Then tuples of the form $\{0\} \times A_2 \times \cdots \times A_n$ and $A_1 \times \{0\} \times \cdots \times A_n$ are both mapped into $\{0\} \times A_2 \times \cdots \times A_n$, which is a contradiction.

Similarly, each column of $R$ should also contain exactly one nonzero entry. Indeed, if one column is a zero column, for example the first column, then $g(x_1, x_2, \ldots, x_n) = g(x'_1, x_2, \ldots, x_n)$ for any $x_1, x'_1 \in C_{1,j_1}$, contradicting that $g$ is a permutation map of $\mathbb{F}_q^n$. If one column contains at least two nonzero entries, for example $r_1^{(1)}, r_1^{(2)}$, then tuples of the form $\{0\} \times A_2 \times \cdots \times A_n$ are mapped to tuples of the form $\{0\} \times \{0\} \times \cdots \times A_n$, contradicting that $g$ is a permutation map. Moreover, if $r_i^{(k)} > 0$ then we consider two distinct tuples which differ only in the coordinate $x_i$ but both values in coordinate $x_i$ are in the same coset $C_{i,j_i}$. These tuples must be mapped to different images; this forces that $\gcd(r_i^{(k)}, s_i^{(k)}) = 1$.

Using condition (ii), we write

$$g(x_1, \ldots, x_n) = \left( x_{i_1}^{r_1^{(1)}} f_1(x_1^{s_1^{(1)}}, \ldots, x_n^{s_n^{(1)}}), \ldots, x_{i_n}^{r_n^{(n)}} f_n(x_1^{s_1^{(n)}}, \ldots, x_n^{s_n^{(n)}}) \right)$$

so that $i_1, \ldots, i_n$ is a permutation of $1, \ldots, n$ induced by the permutation matrix $P$. Let $A_1 \times \cdots \times A_n$ be a part in the partition of $\mathbb{F}_q \times \cdots \times \mathbb{F}_q$ determined by the index tuple $(\ell_1, \ldots, \ell_n)$. Recall that $c_i = f_i(a_1^{s_1^{(i)}/s_1}, \ldots, a_n^{s_n^{(i)}/s_n})$. Then $g(x_1, \ldots, x_n)$ maps $(A_1, \ldots, A_n)$ to $(c_1 A_{i_1}^{r_{i_1}^{(1)}}, \ldots, c_n A_{i_n}^{r_{i_n}^{(n)}})$. Because $\gcd(r_i^{(i)}, s_i^{(i)}) = 1$ for each $i$, the image becomes $(A'_{i_1}, \ldots, A'_{i_n})$, which gives one part $A'_{i_1} \times \cdots \times A'_{i_n}$ of the partition of $\mathbb{F}_q^n$ corresponding to the index tuple $(\ell_{i_1}, \ldots, \ell_{i_n})$. Hence condition (iii) holds. The converse also holds using the same arguments as above. □

We now consider the following examples to illustrate Theorem 16.10.

**Example 16.11** *Let* $g(x_1, x_2) = (x_2(x_1^4 + 4x_1^2 + 4), x_1(3x_1^3 + 1))$ *be a map from* $\mathbb{F}_7 \times \mathbb{F}_7$ *to itself as shown in Example 16.9. Obviously* $f_1(a_1, a_2^2) = a_1^2 + 4a_1 + 4 \neq 0$ *and* $f_2(a_1^3, a_2) = 3a_2 + 1 \neq 0$ *where* $a_1 \in \{0, 1, 2, 4\}$ *and* $a_2 \in \{0, 1, 6\}$. *Here,* $P = R = \begin{bmatrix} 0 & 1 \\ 1 & 0 \end{bmatrix}$ *is a permutation matrix. Moreover,* $\gcd(r_2^{(1)}, s_2^{(1)}) = 1$ *and* $\gcd(r_1^{(2)}, s_1^{(2)}) = 1$. *Furthermore,* $g$ *maps a part* $A_1 \times A_2$ *of the partition* $\mathbb{F}_7 \times \mathbb{F}_7$ *corresponding to the index tuple* $(3, 2)$ *into a part* $A'_2 \times A'_1$ *of the same size corresponding to the index tuple* $(2, 3)$. *Indeed,* $g$ *maps*

| | | |
|---|---|---|
| $\{0\} \times \{0\}$ | $\longmapsto$ | $\{0\} \times \{0\}$ |
| $\{0\} \times C_{2,0}$ | $\xrightarrow{(4x_2,0)}$ | $C_{2,0} \times \{0\}$ |
| $\{0\} \times C_{2,1}$ | $\xrightarrow{(4x_2,0)}$ | $C_{2,1} \times \{0\}$ |
| $C_{1,0} \times \{0\}$ | $\xrightarrow{(0,x_1)}$ | $\{0\} \times C_{1,0}$ |
| $C_{1,1} \times \{0\}$ | $\xrightarrow{(0,x_1)}$ | $\{0\} \times C_{1,1}$ |
| $C_{1,2} \times \{0\}$ | $\xrightarrow{(0,x_1)}$ | $\{0\} \times C_{1,2}$ |
| $C_{1,0} \times C_{2,0}$ | $\xrightarrow{(2x_2,4x_1)}$ | $C_{2,0} \times C_{1,1}$ |
| $C_{1,0} \times C_{2,1}$ | $\xrightarrow{(2x_2,5x_1)}$ | $C_{2,1} \times C_{1,2}$ |
| $C_{1,1} \times C_{2,0}$ | $\xrightarrow{(2x_2,4x_1)}$ | $C_{2,0} \times C_{1,2}$ |
| $C_{1,1} \times C_{2,1}$ | $\xrightarrow{(2x_2,5x_1)}$ | $C_{2,1} \times C_{1,0}$ |
| $C_{1,2} \times C_{2,0}$ | $\xrightarrow{(x_2,4x_1)}$ | $C_{2,0} \times C_{1,0}$ |
| $C_{1,2} \times C_{2,1}$ | $\xrightarrow{(x_2,5x_1)}$ | $C_{2,1} \times C_{1,1}$. |

*By Theorem 16.10,* $g$ *is a permutation of* $\mathbb{F}_7^2$.

**Example 16.12** *Let* $g(x_1, x_2) = (x_2, x_1(2 + x_2^3(x_1^4 - 2x_2^3)))$ *be a map from* $\mathbb{F}_7 \times \mathbb{F}_7$ *to itself. So* $s_1^{(1)} = 6$, $s_2^{(1)} = 6$, $s_1^{(2)} = 2$, *and* $s_2^{(2)} = 3$. *Hence* $s_1 = \gcd(6, 2) = 2$ *and* $s_2 = \gcd(6, 3) = 3$. *Thus* $\ell_1 = 3$ *and* $\ell_2 = 2$. *So* $C_{1,0} = \{1, 6\}$, $C_{1,1} = \{3, 4\}$, *and* $C_{1,2} = \{2, 5\}$ *are the cyclotomic cosets of order* 3, $C_{2,0} = \{1, 2, 4\}$, *and* $C_{2,1} = \{3, 5, 6\}$ *are cyclotomic cosets of order* 2. *Then* $\mathbb{F}_7 \times \mathbb{F}_7$ *is partitioned into* $\{0\} \times \{0\}$, $\{0\} \times C_{2,0}$, $\{0\} \times C_{2,1}$, $C_{1,0} \times \{0\}$, $C_{1,1} \times \{0\}$, $C_{1,2} \times \{0\}$, $C_{1,0} \times C_{2,0}$, $C_{1,0} \times C_{2,1}$, $C_{1,1} \times C_{2,0}$, $C_{1,1} \times C_{2,1}$, $C_{1,2} \times C_{2,0}$, $C_{1,2} \times C_{2,1}$. *Moreover,* $f_1(x_1^6, x_2^6) = 1$ *and* $f_2(x_1^2, x_2^3) = 2 + x_2^3(x_1^4 - 2x_2^3)$. *Note that* $a_1 \in \{0, 1, 2, 4\}$ *and* $a_2 \in \{0, 1, 2\}$. *We can easily* *check that $f_1(a_1^3, a_2^2) \neq 0$ and $f_2(a_1, a_2) \neq 0$ as long as one of $a_1$ and $a_2$ is nonzero. Furthermore, $R = \begin{bmatrix} 0 & 1 \\ 1 & 0 \end{bmatrix}$. In fact, g maps*

$$
\begin{array}{lcl}
\{0\} \times \{0\} & \longmapsto & \{0\} \times \{0\} \\
\{0\} \times C_{2,0} & \overset{(x_2,0)}{\longmapsto} & C_{2,0} \times \{0\} \\
\{0\} \times C_{2,1} & \overset{(x_2,0)}{\longmapsto} & C_{2,1} \times \{0\} \\
C_{1,0} \times \{0\} & \overset{(0,2x_1)}{\longmapsto} & \{0\} \times C_{1,2} \\
C_{1,1} \times \{0\} & \overset{(0,2x_1)}{\longmapsto} & \{0\} \times C_{1,0} \\
C_{1,2} \times \{0\} & \overset{(0,2x_1)}{\longmapsto} & \{0\} \times C_{1,1} \\
C_{1,0} \times C_{2,0} & \overset{(x_2,x_1)}{\longmapsto} & C_{2,0} \times C_{1,0} \\
C_{1,0} \times C_{2,1} & \overset{(x_2,6x_1)}{\longmapsto} & C_{2,1} \times C_{1,0} \\
C_{1,1} \times C_{2,0} & \overset{(x_2,4x_1)}{\longmapsto} & C_{2,0} \times C_{1,2} \\
C_{1,1} \times C_{2,1} & \overset{(x_2,3x_1)}{\longmapsto} & C_{2,1} \times C_{1,2} \\
C_{1,2} \times C_{2,0} & \overset{(x_2,2x_1)}{\longmapsto} & C_{2,0} \times C_{1,1} \\
C_{1,2} \times C_{2,1} & \overset{(x_2,5x_1)}{\longmapsto} & C_{2,1} \times C_{1,1}.
\end{array}
$$

*By Theorem 16.10, g is a permutation of $\mathbb{F}_7^2$.*

We remark that from the proof of Theorem 16.10, each coordinate polynomial of the permutation vector map is a multivariate cyclotomic mapping, which in turn behaves as a monomial (in one variable) on every individual coset. In other words, each permutation vector map in $n$ variables consists of $n$ univariate cyclotomic permutations together with another permutation on coordinate variables. This fact may help us to construct permutation maps in $n$ variables from these simpler coordinate polynomials.

Also from the proof of Theorem 16.10, it is easy to see that if one element in a part of the partition of $\mathbb{F}_q^n$ belongs to the value set then the whole part of the partition belongs to the value set. Hence we also obtain the following.

**Corollary 16.13** *Let g be a polynomial vector map from $\mathbb{F}_q^n$ to $\mathbb{F}_q^n$ with the index tuple $(\ell_1, \ldots, \ell_n)$ and $\ell = \max\{\ell_1, \ldots, \ell_n\} > 1$. If $|V_g| < q^n$ then $|V_g| \leq q^n - \frac{q-1}{\ell}$.*

*Proof.* Under the assumptions (i) and (ii) in Theorem 16.10, the cardinality of the value set of $g$ satisfies $|V_g| > q^n - \frac{q-1}{\ell}$ if and only if $g$ induces a permutation of $\mathbb{F}_q^n$.

Also from the proof of Theorem 16.10, if assumption (i) fails, then at least two cosets corresponding to one coordinate, say $i$, are collapsed into the same image set, therefore $|V_g| \leq q^n - \frac{q-1}{\ell_i} \leq q^n - \frac{q-1}{\ell}$. When assumption (ii) fails, the matrix $R$ contains a zero row or a zero column, or at least two entries in one row or column. Hence the same discussion shows that at least two cosets are collapsed into one image set. This implies again that $|V_g| \leq q^n - \frac{q-1}{\ell_j}$ for some $j$ and thus $|V_g| \leq q^n - \frac{q-1}{\ell}$. Finally, if the matrix $R$ contains exactly one entry in each row and column, but $(r_i^{(k)}, s_i^{(k)}) = t_i > 1$ for some $i$, we miss at least $\frac{(t_i-1)(q-1)}{t_i} = (q-1) - \frac{q-1}{t_i} \geq \frac{q-1}{2} \geq \frac{q-1}{\ell}$ values in the value set. Hence $|V_g| \leq q^n - \frac{q-1}{\ell}$. $\qquad\qquad\square$

As the next simple example shows, the upper bound for nonpermutations can be achieved.

**Example 16.14** *Let* $g(x_1, x_2) = (x_1, x_2(-x_2^3(x_1^3 + 2)^2 + 4)^2)$ *be a map from* $\mathbb{F}_7 \times \mathbb{F}_7$ *to itself. So* $s_1^{(1)} = 6$, $s_2^{(1)} = 6$, $s_1^{(2)} = 3$, *and* $s_2^{(2)} = 3$. *Hence* $s_1 = \gcd(6, 3) = 3$ *and* $s_2 = \gcd(6, 3) = 3$. *Thus* $\ell_1 = 2$ *and* $\ell_2 = 2$. *Here we use cyclotomic cosets of order 2,* $C_{1,0} = C_{2,0} = \{1, 2, 4\}$ *and* $C_{1,1} = C_{2,1} = \{3, 5, 6\}$, *to partition* $\mathbb{F}_7 \times \mathbb{F}_7$ *into* $\{0\} \times \{0\}$, $\{0\} \times C_{2,0}$, $\{0\} \times C_{2,1}$, $C_{1,0} \times \{0\}$, $C_{1,1} \times \{0\}$, $C_{1,0} \times C_{2,0}$, $C_{1,0} \times C_{2,1}$, $C_{1,1} \times C_{2,0}$, $C_{1,1} \times C_{2,1}$. *Moreover,* $f_1(x_1^6, x_2^6) = 1$ *and* $f_2(x_1^3, x_2^3) = (-x_2^3(x_1^3 + 2)^2 + 4)^2$. *Note that* $f_2(a_1, a_2) = (-a_2(a_1 + 2)^2 + 4)^2 \in \{0, 2, 4\}$ *where* $a_1 \in \{0, 1, 6\}$ *and* $a_2 \in \{0, 1, 6\}$. *In fact,* $g$ *maps*

$$
\begin{array}{lcl}
\{0\} \times \{0\} & \xmapsto{\quad} & \{0\} \times \{0\} \\
\{0\} \times C_{2,0} & \xmapsto{(0,0)} & \{0\} \times \{0\} \\
\{0\} \times C_{2,1} & \xmapsto{(0,x_2)} & \{0\} \times C_{2,1} \\
C_{1,0} \times \{0\} & \xmapsto{(x_1,2x_2)} & C_{1,0} \times \{0\} \\
C_{1,1} \times \{0\} & \xmapsto{(x_1,2x_2)} & C_{1,1} \times \{0\} \\
C_{1,0} \times C_{2,0} & \xmapsto{(x_1,4x_2)} & C_{1,0} \times C_{2,0} \\
C_{1,0} \times C_{2,1} & \xmapsto{(x_1,x_2)} & C_{1,0} \times C_{2,1} \\
C_{1,1} \times C_{2,0} & \xmapsto{(x_1,2x_2)} & C_{1,1} \times C_{2,0} \\
C_{1,1} \times C_{2,1} & \xmapsto{(x_1,4x_2)} & C_{1,1} \times C_{2,1}.
\end{array}
$$

*Here* $g$ *maps the* $\{0\} \times C_{2,1}$ *into* $\{0\} \times \{0\}$, *and all other parts of the partition corresponding to the index tuple* $(2, 2)$ *to distinct parts of the partition corresponding to the index tuple* $(2, 2)$. *Therefore* $|V_g| = 46 = q^2 - \frac{q-1}{2}$.

# Acknowledgements

We thank the referee and Arne Winterhof for their helpful suggestions.

The research of Daqing Wan and Qiang Wang was partially supported by NSF, NSERC of Canada, and National Natural Science Foundation of China (Grant No. 61170289).

# References

[1]   A. Akbary and Q. Wang, On some permutation polynomials. *Int. J. Math. Math. Sci.* **16**, 2631–2640, 2005.

[2]   A. Akbary and Q. Wang, A generalized Lucas sequence and permutation binomials. *Proc. Am. Math. Soc.* **134**(1), 15–22, 2006.

[3]   A. Akbary and Q. Wang, On polynomials of the form $x^r f(x^{(q-1)/l})$. *Int. J. Math. Math. Sci.* **2007**, 23408, 2007.

[4]   A. Akbary, S. Alaric and Q. Wang, On some classes of permutation polynomials. *Int. J. Number Theory* **4**(1), 121–133, 2008.

[5]   A. Akbary, D. Ghioca and Q. Wang, On permutation polynomials of prescribed shape. *Finite Fields Appl.* **15**, 195–206, 2009.

[6]   A. Akbary, D. Ghioca and Q. Wang, On constructing permutations of finite fields, *Finite Fields Appl.* **17**(1), 51–67, 2011.

[7]   B. J. Birch and H. P. F. Swinnerton-Dyer, Note on a problem of Chowla. *Acta Arith.* **5**, 417–423, 1959.

[8]   H. Borges and R. Conceicao, On the characterization of minimal value set polynomials. *J. Number Theory* **133**, 2021–2035, 2013.

[9]   X. Cao and L. Hu, New methods for generating permutation polynomials over finite fields. *Finite Fields Appl.* **17**, 493–503, 2011.

[10]  X. Cao, L. Hu and Z. Zha, Constructing permutation polynomials from piecewise permutations. *Finite Fields Appl.* **26**, 162–174, 2014.

[11]  L. Carlitz, D. J. Lewis, W. H. Mills and E. G. Straus, Polynomials over finite fields with minimal value sets, *Mathematika* **8**, 121–130, 1961.

[12]  P. Charpin and G. Kyureghyan, When does $F(x) + Tr(H(x))$ permute $\mathbb{F}_{p^n}$? *Finite Fields Appl.* **15**(5), 615–632, 2009.

[13]  Q. Cheng, J. Hill and D. Wan, Counting value sets: algorithms and complexity. *Tenth Algorithmic Number Theory Symposium ANTS-X*, University of California at San Deigo, 2012.

[14]  S. D. Cohen, The distribution of polynomials over finite fields. *Acta Arith.* **17**, 255–271, 1970.

[15]  P. Das and G. L. Mullen, Value sets of polynomials over finite fields. In: G. L. Mullen, H. Stichtenoth and H. Tapia-Recillas (eds.), *Finite Fields with Applications in Coding Theory, Cryptography and Related Areas*, pp. 80–85. Springer, 2002.

[16]  A. B. Evans, *Orthomorphism Graphs of Groups*, Lecture Notes in Mathematics, volume 1535. Springer, Berlin, 1992.

[17]  N. Fernando and X. Hou, A piecewise construction of permutation polynomial over finite fields. *Finite Fields Appl.* **18**, 1184–1194, 2012.

[18]  J. Gomez-Calderon and D. J. Madden, Polynomials with small value set over finite fields. *J. Number Theory* **28**(2), 167–188, 1988.

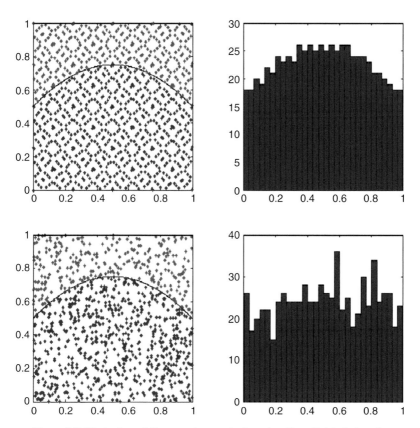

Figure 3.2 Illustration of the acceptance-rejection algorithm. Points below the curved line are accepted and then projected onto the $x$-axis. The top row shows the acceptance-rejection algorithm using a deterministic point set $P_{M,s+1}$, whereas the bottom row shows the acceptance-rejection sampler using random samples $P_{M,s+1}$. In both cases the number of points is $M = 2^7$. (See page 50.)

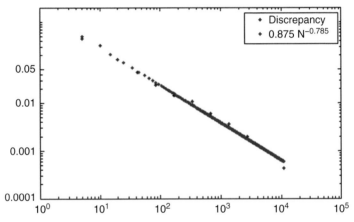

Figure 3.3 Numerical result of the acceptance-rejection algorithm using low-discrepancy point sets. The convergence rate is approximately of order $N^{-0.8}$, which is better than the rate one would expect when using random samples (which is $N^{-0.5}$). (See page 51.)

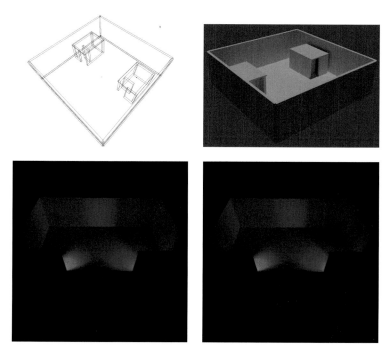

Figure 12.1 Convergence test using a more challenging test scene (top row) for light transport simulation. Inside each of the two symmetric rooms one polyhedral light source illuminates the inner court through the slit between the blockers in the door frame. The two images in the bottom row have been rendered using the same number of $2^{18}$ samples per pixel, where the left image used the Sobol' sequence, while the right image used the new rank-1 lattice construction to sample light transport paths. The images cannot be distinguished with respect to quality, however, sampling using the rank-1 lattice sequence algorithm is simpler and much more efficient. (See page 212.)

Figure 12.2 A visual comparison of the new methodology (top) and the Sobol'
sequence (bottom) in NVIDIA iray® . While the Sobol' sequence exposes the
typical structured artifacts in the form of rectangular patterns (for example on the
open book), the rank-1 lattice sequence exposes more noise (for example in the
highlight on the glass of the petrol lamp). Both images have been rendered using
only 64 path space samples per pixel. (See page 213.)

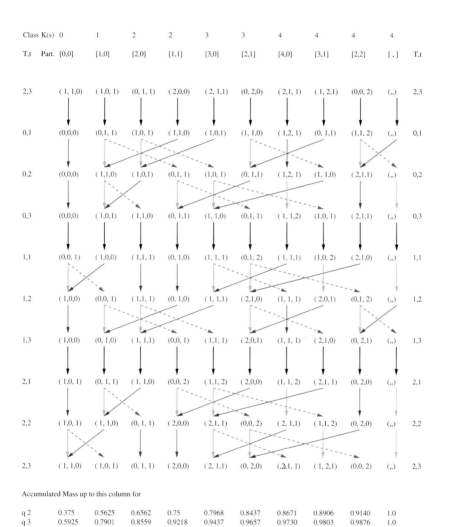

Figure 18.4  BDM states for $M = 2$ with class up to 4. (See page 323.)

[19] R. Guralnick and D. Wan, Bounds for fixed point free elements in a transitive group and applications to curves over finite fields. *Isr. J. Math.* **101**, 255–287, 1997.

[20] X. Hou, Two classes of permutation polynomials over finite fields. *J. Comb. Theory Ser. A* **118**(2), 448–454, 2011.

[21] Y. Laigle-Chapuy, Permutation polynomials and applications to coding theory. *Finite Fields Appl.* **13**, 58–70, 2007.

[22] R. Lidl and G. L. Mullen, When does a polynomial over a finite field permute the elements of the field? *Am. Math. Monthly* **95**, 243–246, 1988.

[23] R. Lidl and G. L. Mullen, When does a polynomial over a finite field permute the elements of the field? II. *Am. Math. Monthly* **100**, 71–74, 1993.

[24] R. Lidl and H. Niederreiter, *Finite Fields*, second edition. Cambridge University Press, Cambridge, 1997.

[25] R. Lipton, Claiming Picard's math may have gaps, http://rjlipton.wordpress. com/2011/09/26/claiming-picards-math-may-have-gaps/.

[26] A. Masuda and M. E. Zieve, Permutation binomials over finite fields. *Trans. Am. Math. Soc.* **361**(8), 4169–4180, 2009.

[27] W. H. Mills, Polynomials with minimal value sets. *Pacific J. Math* **14**, 225–241, 1964.

[28] G. L. Mullen, Permutation polynomials over finite fields. *Finite Fields, Coding Theory, and Advances in Communications and Computing.* Lecture Notes in Pure and Applied Mathematics, volume 141, pp. 131–151. Marcel Dekker, New York, 1992.

[29] G. L. Mullen and D. Panario, *Handbook of Finite Fields.* CRC Press, Boca Raton, FL, 2013.

[30] G. L. Mullen and Q. Wang, Permutation polynomials of one variable. In: G. L. Mullen and D. Panario (eds.), *Handbook of Finite Fields*, Section 8.1. CRC Press, Boca Raton, FL, 2013.

[31] G. L. Mullen, D. Wan and Q. Wang, Value sets of polynomial maps over finite fields, *Q. J. Math.* **64**(4), 1191–1196, 2013.

[32] H. Niederreiter and A. Winterhof, Cyclotomic $\mathcal{R}$-orthomorphisms of finite fields. *Discrete Math.* **295**, 161–171, 2005.

[33] G. Turnwald, A new criterion for permutation polynomials. *Finite Fields Appl.* **1**, 64–82, 1995.

[34] D. Wan, A $p$-adic lifting lemma and its applications to permutation polynomials. *Finite Fields, Coding Theory, and Advances in Communications and Computing.* Lecture Notes in Pure and Applied Mathematics, volume 141, pp. 209–216. Marcel Dekker, New York, 1992.

[35] D. Wan and R. Lidl, Permutation polynomials of the form $x^r f(x^{(q-1)/d})$ and their group structure. *Monatsh. Math.* **112**, 149–163, 1991.

[36] D. Wan, P. J. S. Shiue and C. S. Chen, Value sets of polynomials over finite fields. *Proc. Am. Math. Soc.* **119**, 711–717, 1993.

[37] Q. Wang, Cyclotomic mapping permutation polynomials over finite fields. *Sequences, Subsequences, and Consequences (International Workshop, SSC 2007, Los Angeles, CA, May 31–June 2, 2007)*, Lecture Notes in Computer Science, volume 4893, pp. 119–128. Springer, Berlin, 2007.

[38] Q. Wang, Cyclotomy and permutation polynomials of large indices. *Finite Fields Appl.* **22**, 57–69, 2013.

[39] K. S. Williams, On general polynomials. *Can. Math. Bull.* **10**(4), 579–583, 1967.

[40] P. Yuan and C. Ding, Permutation polynomials over finite fields from a powerful lemma. *Finite Fields Appl.* **17**(6), 560–574, 2011.

[41] P. Yuan and C. Ding, Further results on permutation polynomials over finite fields. *Finite Fields Appl.* **27**, 88–103, 2014.

[42] Z. Zha and L. Hu, Two classes of permutation polynomials over finite fields. *Finite Fields Appl.* **18**(4), 781–790, 2012.

[43] M. Zieve, Some families of permutation polynomials over finite fields. *Int. J. Number Theory* **4**, 851–857, 2008.

[44] M. Zieve, On some permutation polynomials over $\mathbb{F}_q$ of the form $x^r h(x^{(q-1)/d})$. *Proc. Am. Math. Soc.* **137**(7), 2209–2216, 2009.

[45] M. Zieve, Classes of permutation polynomials based on cyclotomy and an additive analogue. *Additive Number Theory*, pp. 355–361. Springer, New York, 2010.

# 17

# Rational points of the curve $y^{q^n} - y = \gamma x^{q^h+1} - \alpha$ over $\mathbb{F}_{q^m}$

*Ferruh Özbudak*
Middle East Technical University, Ankara

*Zülfükar Saygı*
TOBB, University of Economics and Technology, Ankara

*Dedicated to Harald Niederreiter on the occasion of his 70th birthday.*

## Abstract

Let $q$ be a power of an odd prime. For arbitrary positive integers $h, n, m$ with $n$ dividing $m$ and arbitrary $\gamma, \alpha \in \mathbb{F}_{q^m}$ with $\gamma \neq 0$ we determine the number of $\mathbb{F}_{q^m}$-rational points of the curve $y^{q^n} - y = \gamma x^{q^h+1} - \alpha$ in many cases.

## 17.1 Introduction

Let $p$ be an odd prime. For positive integers $e$ and $m$, let $q = p^e$ and let $\mathbb{F}_q$ and $\mathbb{F}_{q^m}$ denote the finite fields with $q$ and $q^m$ elements. Let $n$ be a positive integer dividing $m$. Let $\mathrm{Tr}_{\mathbb{F}_{q^m}/\mathbb{F}_{q^n}}$ denote the relative trace map.

Let $h$ be a nonnegative integer and $\alpha, \gamma \in \mathbb{F}_{q^m}$ with $\gamma \neq 0$. Let $N(m, n)$ denote the cardinality

$$N(m, n) = \left| \left\{ x \in \mathbb{F}_{q^m} \mid \mathrm{Tr}_{\mathbb{F}_{q^m}/\mathbb{F}_{q^n}} \left( \gamma x^{q^h+1} - \alpha \right) = 0 \right\} \right|.$$

Let $\chi$ be the Artin–Schreier type curve given by

$$\chi : \quad y^{q^n} - y = \gamma x^{q^h+1} - \alpha.$$

For the number $N(\chi)$ of its $\mathbb{F}_{q^m}$-rational points we have

$$N(\chi) = 1 + q^n N(m, n),$$

and hence determining $N(\chi)$ is the same as determining $N(m, n)$.

Algebraic curves over finite fields have many applications to various areas such as coding theory, cryptography and low-discrepancy point sets (see, for example [7, 8, 9, 10]). The number of rational points of algebraic curves is also important for these applications.

In this paper we are interested in determining $N(m, n)$. Note that the map

$$
\begin{aligned}
Q : \mathbb{F}_{q^m} &\rightarrow \mathbb{F}_{q^n} \\
x &\mapsto \mathrm{Tr}_{\mathbb{F}_{q^m}/\mathbb{F}_{q^n}}\left(\gamma x^{q^h+1}\right)
\end{aligned}
$$

is not necessarily a quadratic map as $n$ may not divide $h$. We denote $N(m, 1)$ by $N(m)$ in short.

Throughout the paper we choose and fix $q, m, n, \gamma$ and $\alpha$ as above. Moreover we define nonnegative integers $s, t$ and positive integers $r, m_1, h_1$ as follows:

$$
\begin{aligned}
m &= 2^s r m_1, \\
h &= 2^t r h_1,
\end{aligned}
$$

where $\gcd(m_1, h_1) = \gcd(2, r m_1 h_1) = 1$. Furthermore, let $u$ be the nonnegative integer and $\rho, n_1, m_2$ be positive integers so that

$$
n = 2^u \rho n_1 \quad \text{and} \quad m_1 = n_1 m_2
$$

such that $\gcd(2, \rho n_1) = 1$, $\rho | r$ and $n_1 | m_1$. Finally let

$$
A = \mathrm{Tr}_{\mathbb{F}_{q^m}/\mathbb{F}_{q^n}}(\alpha).
$$

Note that $u \leq s$ as $n | m$.

We determine $N(m, n)$ explicitly

- for $s \leq t$ in all cases in Theorem 17.1,
- for $s \geq t + 1$ and $u \leq t$ in all cases in Theorem 17.2,
- for $t + 1 \leq u \leq s$ and $A = 0$ in all cases in Theorem 17.3.

It remains to consider the case $t + 1 \leq u \leq s$ and $A \neq 0$.

In the literature $N(m, n)$ has been determined only for $n = 1$, $s \geq t + 1$ and $\alpha = 0$ [5]. This corresponds to a special subcase of Theorem 17.2.

For $n > 1$, it seems that determining $N(m, n)$ is more difficult than determining $N(m, 1)$ in general. For example Theorem 17.3 below explains this difficulty. Moreover it appears that the remaining final case of $t + 1 \leq u \leq s$ and $A \neq 0$ is rather complicated. We could not determine $N(m, n)$ explicitly for $t + 1 \leq u \leq s$ and $A \neq 0$ in all cases.

This paper is organized as follows. We state our result in the remainder of this section. We present some preliminaries in Section 17.2. We give the proof of the main result in Section 17.3.

Now we are ready to give our first result.

**Theorem 17.1** *Assume that $s \leq t$. Let $\eta$ and $\eta'$ denote the quadratic characters of $\mathbb{F}_q$ and $\mathbb{F}_{q^m}$, respectively. We have the following.*

- *If $m/n$ is even and $A = 0$, then*

$$N(m, n) = \begin{cases} q^{m-n} - (q^n - 1)\, q^{m/2-n} & \text{if } \eta\left((-1)^{m/2}\right) \eta'(\gamma) = 1 \\ q^{m-n} + (q^n - 1)\, q^{m/2-n} & \text{if } \eta\left((-1)^{m/2}\right) \eta'(\gamma) = -1. \end{cases}$$

- *If $m/n$ is even and $A \neq 0$, then*

$$N(m, n) = \begin{cases} q^{m-n} + q^{m/2-n} & \text{if } \eta\left((-1)^{m/2}\right) \eta'(\gamma) = 1 \\ q^{m-n} - q^{m/2-n} & \text{if } \eta\left((-1)^{m/2}\right) \eta'(\gamma) = -1. \end{cases}$$

- *If $m/n$ is odd and $A = 0$, then*

$$N(m, n) = q^{m-n}.$$

- *If $m/n$ is odd, $A \neq 0$ and $n$ is even, then*

$$N(m, n) = \begin{cases} q^{m-n} + q^{(m-n)/2} & \text{if } (u_1, u_2) \in \{(1, 1), (-1, -1)\} \\ q^{m-n} - q^{(m-n)/2} & \text{if } (u_1, u_2) \in \{(1, -1), (-1, 1)\}, \end{cases}$$

*where $u_1$ and $u_2$ are the integers in the set $\{-1, 1\}$ given by*

$$u_1 = \eta\left((-1)^{m/2}\right)\eta'(\gamma) \quad and \quad u_2 = \eta\left((-1)^{n/2}\right)\eta'(A).$$

- *If $m/n$ is odd, $A \neq 0$ and $n$ is odd, then*

$$N(m, n) = \begin{cases} q^{m-n} + q^{(m-n)/2} & \text{if } (u_1, u_2) \in \{(1, 1), (-1, -1)\} \\ q^{m-n} - q^{(m-n)/2} & \text{if } (u_1, u_2) \in \{(1, -1), (-1, 1)\}, \end{cases}$$

*where $u_1$ and $u_2$ are the integers in the set $\{-1, 1\}$ given by*

$$u_1 = \eta\left((-1)^{(m-1)/2}\right)\eta'(\gamma) \quad and \quad u_2 = \eta\left((-1)^{(n-1)/2}\right)\eta'(A).$$

Next we present our second result.

**Theorem 17.2** *Assume that $s \geq t + 1$ and $u \leq t$. Let $\omega$ be a generator of the multiplicative group $\mathbb{F}_{q^m} \setminus \{0\}$ and let $a$ be the integer with $0 \leq a < q^m - 1$ such that $\gamma = \omega^a$.*

*We have the following cases.*

- *Case $s = t + 1$: put $q_1 = q^{2^t r}$.*
  *If $a \not\equiv m_1 \frac{q_1+1}{2} \mod (q_1 + 1)$, then*

$$N(m, n) = \begin{cases} q^{m-n} + q^{m/2-n} & \text{if } A \neq 0 \\ q^{m-n} - (q^n - 1)q^{m/2-n} & \text{if } A = 0. \end{cases}$$

*If* $a \equiv m_1 \frac{q_1+1}{2}$ *mod* $(q_1 + 1)$, *then for* $k = 2^{t+1}r$ *we have that*

$$N(m, n) = \begin{cases} q^{m-n} - q^{(m+k)/2-n} & \text{if } A \neq 0 \\ q^{m-n} + (q^n - 1)q^{(m+k)/2-n} & \text{if } A = 0. \end{cases}$$

- *Case* $s \geq t + 2$: *put* $q_1 = q^{2^t r}$.
  *If* $a \not\equiv 0$ *mod* $(q_1 + 1)$, *then*

$$N(m, n) = \begin{cases} q^{m-n} - q^{m/2-n} & \text{if } A \neq 0 \\ q^{m-n} + (q^n - 1)q^{m/2-n} & \text{if } A = 0. \end{cases}$$

*If* $a \equiv 0$ *mod* $(q_1 + 1)$, *then for* $k = 2^{t+1}r$ *we have that*

$$N(m, n) = \begin{cases} q^{m-n} + q^{(m+k)/2-n} & \text{if } A \neq 0 \\ q^{m-n} - (q^n - 1)q^{(m+k)/2-n} & \text{if } A = 0. \end{cases}$$

Note that Theorem 17.3 comes into the picture only when $n > 1$. Finally we present Theorem 17.3, which can be considered to be the main theorem of this paper.

**Theorem 17.3** *Assume that* $t + 1 \leq u \leq s$ *and* $A = 0$. *Let* $\omega$ *be a generator of the multiplicative group* $\mathbb{F}_{q^m} \setminus \{0\}$ *and let a be the integer with* $0 \leq a < q^m - 1$ *such that* $\gamma = \omega^a$.
*We have the following cases.*

- *Case* $s = t + 1$: *put* $B_1 = \gcd\left(m_2, q^{2^t \rho} + 1\right)$.
  *If* $a \equiv n_1 m_2 \frac{q^{2^t r}+1}{2}$ *mod* $\left(\frac{q^{2^t r}+1}{q^{2^t \rho}+1} B_1\right)$, *then*

$$N(m, n) = q^{m-n} - (q^n - 1)q^{m/2-n} + B_1 \frac{q^n - 1}{q^{2^t \rho}}$$
$$+ 1\left(q^{m/2+2^t r-n} + q^{m/2-n}\right).$$

  *If* $a \not\equiv n_1 m_2 \frac{q^{2^t r}+1}{2}$ *mod* $\left(\frac{q^{2^t r}+1}{q^{2^t \rho}+1} B_1\right)$, *then*

$$N(m, n) = q^{m-n} - (q^n - 1)q^{m/2-n}.$$

- *Case* $s \geq t + 2$: *put* $B_1 = \gcd\left(2^{s-u} m_2, q^{2^t \rho} + 1\right)$.
  *If* $a \equiv 0$ *mod* $\left(\frac{q^{2^t r}+1}{q^{2^t \rho}+1} B_1\right)$, *then*

$$N(m, n) = q^{m-n} + (q^n - 1)q^{m/2-n}$$
$$- B_1 \frac{q^n - 1}{q^{2^t \rho} + 1}\left(q^{m/2+2^t r-n} + q^{m/2-n}\right).$$

If $a \not\equiv 0 \mod \left( \frac{q^{2^t r}+1}{q^{2^t \rho}+1} B_1 \right)$, then

$$N(m,n) = q^{m-n} + (q^n - 1)q^{m/2-n}.$$

## 17.2 Preliminaries

In this section we give some preliminaries that we use in Section 17.3. The following is the special subcase of Theorem 17.2 corresponding to $n = 1$. This result was obtained for $\alpha = 0$ by Klapper [5]. Using some facts on quadratic forms [6, Section 6.2], we extend these results to the case $\alpha \neq 0$ in the following proposition.

**Proposition 17.4** *Assume that $s \geq t + 1$ and $n = 1$. Then we determine $N(m) = N(m, 1)$ as follows: put $q_1 = q^{2^t r}$ and $k = 2^{t+1} r$.*

- *Case $s = t + 1$.*
  *If $a \not\equiv m_1 \frac{q_1+1}{2} \mod (q_1 + 1)$, then*

$$N(m) = \begin{cases} q^{m-1} + q^{m/2-1} & \text{if } A \neq 0 \\ q^{m-1} - (q-1)q^{m/2-1} & \text{if } A = 0. \end{cases}$$

  *If $a \equiv m_1 \frac{q_1+1}{2} \mod (q_1 + 1)$, then*

$$N(m) = \begin{cases} q^{m-1} - q^{(m+k)/2-1} & \text{if } A \neq 0 \\ q^{m-1} + (q-1)q^{(m+k)/2-1} & \text{if } A = 0. \end{cases}$$

- *Case $s \geq t + 2$.*
  *If $a \not\equiv 0 \mod (q_1 + 1)$, then*

$$N(m) = \begin{cases} q^{m-1} - q^{m/2-1} & \text{if } A \neq 0 \\ q^{m-1} + (q-1)q^{m/2-1} & \text{if } A = 0. \end{cases}$$

  *If $a \equiv 0 \mod (q_1 + 1)$, then*

$$N(m) = \begin{cases} q^{m-1} + q^{(m+k)/2-1} & \text{if } A \neq 0 \\ q^{m-1} - (q-1)q^{(m+k)/2-1} & \text{if } A = 0. \end{cases}$$

The following lemma is an important tool that we use in obtaining results for $N(m, n)$ with $n > 1$.

**Lemma 17.5** *Let $\omega$ be a generator of the multiplicative group $\mathbb{F}_{q^m} \setminus \{0\}$ and put $S = 1 + q^n + q^{2n} + \cdots + q^{(m/n-1)n}$. For $0 \leq i \leq \frac{q^n-1}{q-1} - 1$, let $\theta_i = \omega^{Si}$ and*

$$N_i = \left| \left\{ x \in \mathbb{F}_{q^m} : Tr_{n \mathbb{F}_{q^m}/\mathbb{F}_q} \left( \theta_i \left( \gamma x^{q^h+1} - \alpha \right) \right) = 0 \right\} \right|.$$

*Moreover put $N_i = q^{m-1} + R_i$. Then*

$$N(m, n) = q^{m-n} + R,$$

*where*

$$R = \frac{1}{q^{n-1}} \sum_{i=0}^{\frac{q^n-1}{q-1}-1} R_i.$$

*Proof.* Let $F = \mathbb{F}_{q^m}(x, y)$ be the function field

$$y^{q^n} - y = \gamma x^{q^h+1} - \alpha.$$

For $0 \le i \le \frac{q^n-1}{q-1} - 1$, let $F_i = \mathbb{F}_{q^m}(x, y_i)$ be the function field

$$y_i{}^q - y_i = \theta_i \left( \gamma x^{q^h+1} - \alpha \right).$$

Let $L_F(t)$ be the $L$-polynomial of $F$ (see [9, Definition 5.1.14]) and $L_{F_i}(t)$ be the $L$-polynomial of $F_i$. Let $N(F)$ and $N(F_i)$ denote the number of rational places of degree one of $F$ and $F_i$. We have

$$L_F(t) = \prod_{i=0}^{\frac{q^n-1}{q-1}-1} L_{F_i}(t),$$

which implies that

$$N(F) - (q^m + 1) = \sum_{i=0}^{\frac{q^n-1}{q-1}-1} \left( N(F_i) - (q^m + 1) \right).$$

This method of computing $N(F)$ using $N(F_i)$ was used, for example in [2, 3, 4]. By Hilbert's Theorem 90 we have

$$N(F) = 1 + q^n N(m, n) \quad \text{and} \quad N(F_i) = 1 + q N_i(m).$$

Combining these results we complete the proof. $\qquad\qquad\square$

## 17.3 Proof of the main theorem

The proof of Theorem 17.1 can be obtained either using [1] directly or alternatively using some facts from quadratic forms and Gauss sums in [6]. In this paper we give a detailed proof of Theorem 17.3. The proof of Theorem 17.2

is similar to the proof of Theorem 17.3 and it is simpler than the proof of Theorem 17.3.

*Proof of Theorem 17.3.* We first consider the case $s = t + 1$. Put $q_1 = q^{2^t r}$ and $k = 2^{t+1} r$. Moreover let

$$q_2 = q^{2^t \rho}, \quad \bar{r} = r/\rho, \quad \bar{m} = 2\bar{r}n_1 m_2, \quad \bar{h} = \bar{r}h_1, \quad \bar{n} = 2n_1, \quad \bar{k} = 2\bar{r}.$$

We have

$$q_1 = q_2^{\bar{r}}, \quad q^m = q_2^{\bar{m}}, \quad q^n = q_2^{\bar{n}}, \quad q^k = q_2^{\bar{k}}.$$

Let $S = 1 + q_2^{\bar{n}} + q_2^{2\bar{n}} + \cdots + q_2^{(\bar{r}m_2-1)\bar{n}}$. Note that

$$S \equiv m_2 \frac{q_2^{\bar{r}} + 1}{q_2 + 1} \mod (q_2^{\bar{r}} + 1).$$

For $0 \le i \le \frac{q_2^{\bar{n}}-1}{q_2-1} - 1$, let $\theta_i = \omega^{iS}$. We have

$$\theta_i \gamma = \omega^{a+iS} \quad \text{and} \quad \mathrm{Tr}_{n\mathbb{F}_{q^m}/\mathbb{F}_{q_2}}(\theta_i \alpha) = \mathrm{Tr}_{n\mathbb{F}_{q^n}/\mathbb{F}_{q_2}}(\theta_i A) = 0.$$

We need to determine the number of integers $i$ such that $0 \le i \le \frac{q_2^{\bar{n}}-1}{q_2-1} - 1$ and

$$a + iS \equiv n_1 m_2 \frac{q_2^{\bar{r}} + 1}{2} \mod (q_2^{\bar{r}} + 1),$$

or equivalently

$$a + im_2 \frac{q_2^{\bar{r}} + 1}{q_2 + 1} \equiv n_1 m_2 \frac{q_2^{\bar{r}} + 1}{2} \mod (q_2^{\bar{r}} + 1). \tag{17.1}$$

Note that

$$\gcd\left(m_2 \frac{q_2^{\bar{r}} + 1}{q_2 + 1}, q_2^{\bar{r}} + 1\right) = \frac{q_2^{\bar{r}} + 1}{q_2 + 1} B_1.$$

Hence if

$$a \equiv n_1 m_2 \frac{q_2^{\bar{r}} + 1}{2} \mod \left(\frac{q_2^{\bar{r}} + 1}{q_2 + 1} B_1\right) \tag{17.2}$$

does not hold, then there is no integer $i$ satisfying (17.1).

Assume that (17.2) holds. Put

$$B_2 = \frac{n_1 m_2 \frac{q_2^{\bar{r}}+1}{q_2+1} - a}{\frac{q_2^{\bar{r}}+1}{q_2+1} B_1}.$$

Then (17.1) is equivalent to

$$i\frac{m_2}{B_1} \equiv B_2 \quad \mathrm{mod}\ \left(\frac{q_2+1}{B_1}\right). \tag{17.3}$$

Note that $i$ is uniquely determined modulo $\frac{q_2+1}{B_1}$ by (17.3). As $\frac{q_2+1}{B_1}$ divides $\frac{q_2^{\bar{n}}-1}{q_2-1}$ we conclude that the number of integers $i$ such that $0 \le i \le \frac{q_2^{\bar{n}}-1}{q_2-1} - 1$ and (17.1) is satisfied is

$$N = \frac{q_2^{\bar{n}}-1}{q_2^2-1} B_1.$$

For $0 \le i \le \frac{q_2^{\bar{n}}-1}{q_2-1} - 1$, let (see Lemma 17.5)

$$N_i = q_2^{\bar{m}-1} + R_i = \left|\left\{x \in \mathbb{F}_{q^m} : \mathrm{Tr}_{\mathbb{F}_{q^m}/\mathbb{F}_{q_2}}\left(\theta_i\left(\gamma x^{q^h+1}-\alpha\right)\right)=0\right\}\right|.$$

If (17.2) does not hold, then using Proposition 17.4 we obtain that

$$R_i = -(q_2-1)q_2^{\bar{m}/2-1} \quad \text{for all}\quad 0 \le i \le \frac{q_2^{\bar{n}}-1}{q_2-1} - 1.$$

If (17.2) holds, then using Proposition 17.4 we have that the number of integers $i$ such that $0 \le i \le \frac{q_2^{\bar{n}}-1}{q_2-1} - 1$ and

$$R_i = (q_2-1)q_2^{(\bar{m}+\bar{k})/2-1} \quad \text{is}\quad N.$$

Again if (17.2) holds, then using Proposition 17.4 we have that the number of integers $i$ such that $0 \le i \le \frac{q_2^{\bar{n}}-1}{q_2-1} - 1$ and

$$R_i = -(q_2-1)q_2^{\bar{m}/2-1} \quad \text{is}\quad \frac{q_2^{\bar{n}}-1}{q_2-1} - N.$$

By Lemma 17.5 we have

$$N(m,n) = q_2^{\bar{m}-\bar{n}} + R \quad \text{and}\quad R = \frac{1}{q_2^{\bar{n}-1}}\sum_{i=0}^{\frac{q_2^{\bar{n}}-1}{q_2-1}-1} R_i.$$

Putting in the values above, we complete the proof for the case $s = t+1$.

Next we consider the case $s \ge t+2$. Put $q_1 = q^{2^t r}$, $k = 2^{t+1}r$. Moreover let $q_2 = q^{2^t \rho}$, $\bar{r} = r/\rho$, $\bar{m} = 2^{s-t}\bar{r}n_1 m_2$, $\bar{h} = \bar{r}h_1$, $\bar{n} = 2^{u-t}n_1$, $\bar{k} = 2\bar{r}$.

We have

$$q_1 = q_2^{\bar{r}}, \quad q^m = q_2^{\bar{m}}, \quad q^n = q_2^{\bar{n}}, \quad q^k = q_2^{\bar{k}}.$$

Let $S = 1 + q_2^{\bar{n}} + q_2^{2\bar{n}} + \cdots + q_2^{(2^{s-u}\bar{r}m_2 - 1)\bar{n}}$. Note that

$$S \equiv 2^{s-u}m_2 \frac{q_2^{\bar{r}} + 1}{q_2 + 1} \mod (q_2^{\bar{r}} + 1).$$

As in the first case above, we need to determine the number of integers $i$ such that $0 \le i \le \frac{q_2^{\bar{n}} - 1}{q_2 - 1} - 1$ and

$$a + iS \equiv 0 \mod (q_2^{\bar{r}} + 1),$$

or equivalently

$$a + i2^{s-u}m_2 \frac{q_2^{\bar{r}} + 1}{q_2 + 1} \equiv 0 \mod (q_2^{\bar{r}} + 1). \tag{17.4}$$

Note that

$$\gcd\left(2^{s-u}m_2 \frac{q_2^{\bar{r}} + 1}{q_2 + 1}, q_2^{\bar{r}} + 1\right) = \frac{q_2^{\bar{r}} + 1}{q_2 + 1} B_1,$$

where $B_1$ is defined differently in this case than the first case. Hence if

$$a \equiv 0 \mod \left(\frac{q_2^{\bar{r}} + 1}{q_2 + 1} B_1\right) \tag{17.5}$$

does not hold, then there is no integer $i$ satisfying (17.4).

Assume that (17.5) holds. Put

$$B_2 = \frac{-a}{\frac{q_2^{\bar{r}} + 1}{q_2 + 1} B_1}.$$

Then (17.4) is equivalent to

$$i \frac{2^{s-u}m_2}{B_1} \equiv B_2 \mod \left(\frac{q_2 + 1}{B_1}\right). \tag{17.6}$$

Therefore if (17.5) holds, then the number of integers $i$ such that $0 \le i \le \frac{q_2^{\bar{n}} - 1}{q_2 - 1} - 1$ and (17.4) is satisfied is

$$N = \frac{q_2^{\bar{n}} - 1}{q_2^2 - 1} B_1.$$

For $0 \le i \le \frac{q_2^{\bar{n}} - 1}{q_2 - 1} - 1$, let $R_i$ be defined as in the first case above. If (17.5) does not hold, then

$$R_i = (q_2 - 1)q_2^{\bar{m}/2 - 1} \quad \text{for all} \quad 0 \le i \le \frac{q_2^{\bar{n}} - 1}{q_2 - 1} - 1.$$

If (17.5) holds, then by Proposition 17.4 the number of integers $i$ such that $0 \leq i \leq \frac{q_2^{\overline{n}}-1}{q_2-1} - 1$ and

$$R_i = -(q_2 - 1)q_2^{(\overline{m}+\overline{k})/2-1} \quad \text{is} \quad N.$$

If (17.5) holds, then by Proposition 17.4 the number of integers $i$ such that $0 \leq i \leq \frac{q_2^{\overline{n}}-1}{q_2-1} - 1$ and

$$R_i = (q_2 - 1)q_2^{\overline{m}/2-1} \quad \text{is} \quad \frac{q_2^{\overline{n}} - 1}{q_2 - 1} - N.$$

We complete the proof of this case using Lemma 17.5 as in the proof of the first case above.                                                                 □

## Acknowledgments

F. Özbudak is partially supported by TÜBİTAK under Grant No. TBAG–112T011.

## References

[1]  R. S. Coulter, Explicit evaluations of some Weil sums. *Acta Arith.* **83**, 241–251, 1998.

[2]  I. Duursma, H. Stichtenoth and C. Voss, Generalized Hamming weights for duals of BCH, and maximal algebraic function fields. In: R. Pellikaan, M. Perret and S. G. Vladut (eds.), *Arithmetic Geometry and Coding Theory*, pp. 53–65. de Gruyter, Berlin, 1996.

[3]  C. Güneri and F. Özbudak, Improvements on generalized Hamming weights of some trace codes. *Des. Codes Cryptogr.* **39**(2), 215–231, 2006.

[4]  C. Güneri and F. Özbudak, Weil–Serre type bounds for cyclic codes. *IEEE Trans. Inf. Theory* **54**(1), 5381–5395, 2008.

[5]  A. Klapper, Cross-correlations of quadratic form sequences in odd characteristic. *Des. Codes Cryptogr.* **11**(3), 289–305, 1997.

[6]  R. Lidl and H. Niederreiter, *Finite Fields*. Cambridge University Press, Cambridge, 1997.

[7]  H. Niederreiter and C. Xing, *Rational Points on Curves over Finite Fields: Theory and Applications*. Cambridge University Press, Cambridge, 2001.

[8]  H. Niederreiter and C. Xing, *Algebraic Geometry in Coding Theory and Cryptography*. Princeton University Press, Princeton, NJ, 2009.

[9]  H. Stichtenoth, *Algebraic Function Fields and Codes*. Springer-Verlag, Berlin, 2009.

[10]  M. A. Tsfasman, S. G. Vladut and D. Nogin, *Algebraic Geometric Codes: Basic Notions*. American Mathematical Society, Providence, RI, 2007.

# 18

# On the linear complexity of multisequences, bijections between $\mathbb{Z}$ahlen and $\mathbb{N}$umber tuples, and partitions

*Michael Vielhaber*

Hochschule Bremerhaven, Bremerhaven

*Dedicated to Harald Niederreiter on the occasion of his 70th birthday.*

## Abstract

Stream ciphers employ pseudorandom sequences as a replacement for one-time pads. Such a cipher is deterministic, but should be indistinguishable from true randomness. A means of assessing the quality of such a pseudorandom symbol stream is its linear complexity profile.

This article assembles the known facts about linear complexity of single streams (over $\mathbb{F}_q$) and of multisequences, streams over $\mathbb{F}_q^M$, with $M$ typically a power of two, the wordlength.

We show how the approaches of Niederreiter and Wang and of Canales and the present author, the BDM, are equivalent.

In the course of modeling the linear complexity of multisequences, we will touch on partitions and stochastic infinite state machines. The bijection between both gives rise to a family of bijections between $\mathbb{N}_0^M$ and $\mathbb{Z}^M$.

## 18.1 Introduction and notation

This paper first collects known facts about the linear complexity of (single) sequences over any finite field $\mathbb{F}_q$ in Section 18.2 and then generalizes to multisequences in Section 18.3.

We are interested in the number of sequences of width $M$, i.e., $M$ symbol streams in parallel, of length $n$, with symbols from $\mathbb{F}_q$, having a linear complexity of $L$. How does this number behave, as a function $f(q, M, n, L)$?

Also, what is the average behavior of $L$, for how many sequences and how far does $L$ deviate from the expected value $E(L)$, again dependent on $M$ and $q$?

The final Section 18.4 covers interesting connections to partitions and bijections $\mathbb{N}_0^M \to \mathbb{Z}^M$. We close by stating some open problems.

### 18.1.1 The setting

Given a single sequence $a$ of symbols over the finite field $\mathbb{F}_q$, $q$ a prime power, we have three related ways to represent this sequence, or its prefixes.

**Linear feedback shift register (LFSR)** We can generate a given sequence $a = (a_1, a_2, \ldots)$ by an LFSR. The linear complexity of $a$, $L = L(n) = L(n, a)$ at length $n \in \mathbb{N}$, is the smallest number such that the $n$-symbol prefix of the sequence $a$ can be generated by an LFSR of length $L(n)$, with initial contents $(a_1, a_2, \ldots, a_{L(n)})$.

The connection polynomial of the LFSR represents the coefficients of the feedback taps. Its reciprocal is the minimal polynomial of the sequence prefix $(a_1, \ldots, a_n)$, the recursion of smallest degree satisfied by these $(a_i)$.

The sequence $(L(n, a), n \in \mathbb{N})$ is called the linear complexity profile (l.c.p.) of $a$. We set $L(0, a) = 0$, $\forall a$.

**Generating function I: Diophantine approximation** An equivalent description in terms of Diophantine approximation uses the generating function of $a$ and its rational approximation

$$G(a) = \sum_{k=1}^{\infty} a_k x^{-k} = \frac{u_n(x)}{v_n(x)} + o(x^{-n-1}).$$

The denominator $v_n(x)$ has the same degree as the minimal polynomial of the LFSR, the LFSR length $L(n, a)$.

**Generating function II: continued fraction expansion** The continued fraction expansion of $G(a)$ is

$$G(a) = \sum_{k=1}^{\infty} a_k x^{-k} = \cfrac{1}{p_1(x) + \cfrac{1}{p_2(x) + \cfrac{1}{p_3(x) + \frac{1}{\cdots}}}}$$

$$= \frac{1}{|p_1(x)|} + \frac{1}{|p_2(x)|} + \frac{1}{|p_3(x)|} + \cdots,$$

where the polynomials $p_j(x) \in \mathbb{F}_q[x] \backslash \mathbb{F}_q$ are called partial denominators. The sequence $\sum_{j=1}^{k} \deg(p_j(x))$, $k \in \mathbb{N}_0$ equals the sequence of different values in the linear complexity profile, starting with 0 for the empty sum.

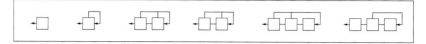

Figure 18.1 Six short LFSRs.

**Example 18.1** *The sequence* $a = 110101$ *has generating function* $G(a) = x^{-1} + x^{-2} + x^{-4} + x^{-6} = \frac{1|}{|x+1} + \frac{1|}{|x} + \frac{1|}{|x+1} = \frac{x^2+x+1}{x^3+x}$ *and linear complexity profile* $(1, 1, 2, 2, 3, 3)$. *The discrepancy sequence is* $(111011)$, *encoding the three partial denominators* $x + 1, x, x + 1 \equiv 11|10|11$ *(see [27]).*

*The six LFSRs generating the respective prefixes are given in Figure 18.1 (numbers 3 and 4 are the same, since* $\delta(4) = 0$). *The initial contents are 1, 11, 110, respectively (the prefix of the prefix). The full sequence has the connection polynomial* $x^2 + 1$ *and the minimal polynomial* $v(x) = x^3 + x$ *(the denominator).*

## 18.2 Single sequences

Harald Niederreiter gave a series of four talks at EUROCRYPT [9, 10, 11, 13], starting a quarter of a century ago, with the programmatic titles:

Sequences with almost perfect linear complexity Profile (1987)
The probabilistic theory of linear complexity (1988)
Keystream sequences with a good linear complexity profile for every starting point (1989)
The linear complexity profile and the jump complexity of keystream sequences (1990).

Prior to 1987, there were important results, two papers by Gustavson [7] and by Wang and Massey [26], and the encyclopedic work of Rueppel [12].

### 18.2.1 Just counting

**1976, Gustavson [7]** One starts with counting. We denote the number of sequence prefixes $a$ of length $n$ over $\mathbb{F}_q$ with linear complexity $L = L(n, a)$ by $N(n, L) = N^{(q)}(n, L)$, where $q$ may be dropped when apparent from the context.

In 1976, Gustavson derived the counting formula

$$N^{(q)}(n, L) = \begin{cases} 1 & \text{if } L = 0 \\ q^{2L-1}(q - 1) & \text{if } L = 1, 2, \ldots, \lfloor n/2 \rfloor \\ q^{2(n-L)}(q - 1) & \text{if } L = \lfloor n/2 \rfloor + 1, \lfloor n/2 \rfloor + 2, \ldots, n. \end{cases}$$

**1986, Rueppel [24]** In 1986, Rueppel published *Analysis and Design of Stream Ciphers* [24] which, for binary sequences, gives the expected value for $L(n)$ as

$$E(L(n)) = \left\lceil \frac{n}{2} \right\rceil + \frac{2}{9} \cdot (-1)^n + 2^{-n} \left( \frac{n}{3} + \frac{2}{9} \right),$$

and the variance of $L(n)$ as

$$\mathrm{Var}(L(n)) = E(L^2) - E(L)^2 = 86/81 + O(n2^{-n}).$$

He also conjectured that the sequence $1101000100000001(0)^{15}1(0)^{31}1\ldots$ satisfies $L(n) = \lceil \frac{n}{2} \rceil, \forall n$.

The formula given by Gustavson can be verified via the following recursion Figure 18.2 [24, Figure 4.2]. The update of $L$ from $L(n-1)$ to

$$
L(n) =
\begin{cases}
L(n-1) & 
\begin{cases}
\text{if } \delta(n) = 0 \wedge L(n-1) \geq n/2 & 0_{\mathrm{I}} \\
\text{if } \delta(n) = 0 \wedge L(n-1) < n/2 & 0_{\mathrm{II}} \\
\text{if } \delta(n) \neq 0 \wedge L(n-1) \geq n/2 & \mathrm{I} \\
\end{cases}
\\
n/2 - L(n-1) & \text{if } \delta(n) \neq 0 \wedge L(n-1) < n/2 \quad \mathrm{II}
\end{cases}
$$

depends on $\delta$ and $L(n-1)$. We call the four cases $0_{\mathrm{I}}$, $0_{\mathrm{II}}$, I, and II, respectively.

Given a shortest linear recurrence relation of length $L(n-1, a)$ that can generate a sequence $s_0, s_1, \ldots$ with $s_j = a_j$ for $j < n$, we define the discrepancy as $\delta = a_n - s_n$. Therefore, $\delta$ is zero, if the new symbol $a_n$ is correctly predicted/produced by the current approximation/LFSR. Hence, $\delta = 0$ corresponds to one symbol from $\mathbb{F}_q$. Otherwise $\delta \neq 0$, for the remaining $q-1$ elements of $\mathbb{F}_q$.

Also, a jump in the l.c.p. can only take place, if $L(n-1, a) < n/2$. The $n$th linear complexity is then $L(n, a) = n - L(n-1, a)$. Otherwise, a $\delta \neq 0$ will (necessarily) change the LFSR taps (the approximating function), but not the LFSR length or the polynomial degree. In terms of the Diophantine approximation of the generating function of $a$,

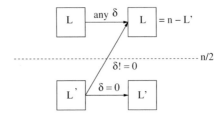

Figure 18.2 Dynamics of $L$ according to Rueppel [24, Figure 4.2].

$$G(a) := \sum_{k=1}^{\infty} a_k x^{-k} = \frac{u_n(x)}{v_n(x)} + O(x^{-n-1}),$$

we have $u_n = u_{n-1}$, $v_n = v_{n-1}$ in cases $0_I$, $0_{II}$, $\deg(v_n) = \deg(v_{n-1})$ in case I, and $\deg(v_n) = n - \deg(v_{n-1})$ in case II.

For the initial example $(a) = (110101)$, $(\delta) = (111011)$, only the 4th symbol maintains the recursion/LFSR (case $0_I$). The 1st, 3rd, and 5th symbols increase the LFSR length (case II), the 2nd and 6th symbols rearrange the LFSR taps without increasing the length (case I).

**1986, Wang and Massey [34]** Wang and Massey coined the notion of a "perfect linear complexity profile" (perfect l.c.p.), if $L(n) = \lceil n/2 \rceil$ for all $n \in \mathbb{N}$, and they showed for $q = 2$, i.e., for binary sequences, that exactly the sequences defined by $s_1 = 1$, $s_{2n}$ arbitrary, and $s_{2n+1} = s_{2n} + s_n$ have a perfect l.c.p., thus proving Rueppel's conjecture about $11010001\ldots$.
We now enter the cycle of Niederreiter's EUROCRYPT lectures.

**1987, Niederreiter [9]** Niederreiter generalized the notion of a perfect l.c.p. to "almost perfect l.c.p." if there exists a bound $K \in \mathbb{N}$ with $|L(n) - \lceil n/2 \rceil| \leq K$ for all $n$.

Let us define the linear complexity deviation (l.c.dev.) as $d(n) := L(n) - \lceil n/2 \rceil \in \mathbb{Z}$. Then for an almost perfect l.c.p., we require $|d(n)| \leq K$.

The continued fraction expansion (c.f.e.) of the generating function $G(a)$ with partial denominators $p_j(x)$ is correlated with the "perfectness" of the sequence $(a_k)$ as follows. Exactly the linear complexities $\sum_{i=1}^{k} \deg(p_i)$, $k \in \mathbb{N}_0$ appear in the l.c.p., the jumps in the l.c.p. are given by $\deg(p_i)$, and the deviations are maximal at the jumps, namely from $d(n-1) = -\lfloor \deg(p_i)/2 \rfloor$ to $d(n) = \lfloor \deg(p_i)/2 \rfloor$.

Niederreiter applied the theory of continued fractions to the notion of almost perfect l.c.p. and obtained that the degrees of all partial denominators of the c.f.e. have to be $\leq K$. Given that the variance of $L(n)$ is about 86/81, modest values for $K$ like 2 or 3 are sufficient to generate many sequences, whose single bits no longer satisfy recursions as in the case $K = 1$, perfect l.c.p.

### 18.2.2 Probability and dynamical systems

**1988, Niederreiter [10]** The probabilistic theory of linear complexity connects the representation of the sequence as a continued fraction of its generating function with dynamical systems theory. Niederreiter showed that the dynamical system defined by $H: G(a) \rightarrow \mathrm{Fr}(G(a)^{-1})$ (for $G(a) \in$

$\mathbb{F}_q[[x^{-1}]]\setminus\{0\}$), Fr being the fractional part, is isomorphic to the Bernoulli shift $\sigma : (p_1(x), p_2(x), \ldots) \mapsto (p_2(x), p_3(x), \ldots)$ on the set of infinite sequences of nonconstant polynomials $(\mathbb{F}_q[x]\setminus\mathbb{F}_q)^\infty$ with measure $\mu(p(x)) = q^{-2\deg(p)}$. This gives the *law of the iterated logarithm for continued fractions* [10, Corollary 3]:

$$\lim_{k\to\infty} \left\{ \begin{matrix} \sup \\ \inf \end{matrix} \right\} \frac{q-1}{\sqrt{2qk\log\log k}} \left( \sum_{j=1}^{k} \deg(p_j) - \frac{q}{q-1} \cdot k \right) = \left\{ \begin{matrix} +1 \\ -1. \end{matrix} \right.$$

We also have the following result.

**Theorem 18.2** [10, Theorem 7] *Let $h$ be the uniform Haar measure $h(a_1 a_2 \cdots a_k) = q^{-k}$ on the cylinder sets defined by sequence prefixes. Then*

$$\lim_{n\to\infty} \frac{L(n,a)}{n} = \frac{1}{2} \quad h\text{-a.e.}$$

Furthermore, the bound for obtaining "almost perfect" l.c.p.s can be drawn quite exactly. Let $K$ be a nonnegative nondecreasing function on $\mathbb{N}$. We distinguish two cases by (i) $\sum_{n=1}^{\infty} q^{-K(n)} < \infty$, and (ii) $\sum_{n=1}^{\infty} q^{-K(n)} = \infty$.

Then in case (i) $h$-a.e., $\left| L(n) - \frac{n}{2} \right| \leq \frac{1}{2} K(n)$ for all sufficiently large $n$ ($h$-almost all sequences have their l.c.dev. bounded by $K(n)/2$), while in case (ii), $L(n) - \frac{n}{2} > \frac{1}{2} K(n)$ and also $L(n) - \frac{n}{2} < -\frac{1}{2} K(n)$ for infinitely many $n$ ($h$-almost no l.c.dev. is bounded) [10, Theorems 8 and 9]. Typical functions on the border would be $K_{(i)}(n) = 1.001 \log_q(n)$ for case (i), and $K_{(ii)}(n) = \log_q(n)$ for case (ii).

The behavior of the l.c.p. can be described still more precisely by the following theorem.

**Theorem 18.3** [10, Theorem 10] *Law of the logarithm for the linear complexity:*

$$\lim_{n\to\infty} \left\{ \begin{matrix} \sup \\ \inf \end{matrix} \right\} \frac{L(n,a) - n/2}{\log n} = \left\{ \begin{matrix} +\frac{1}{2\log q} \\ -\frac{1}{2\log q} \end{matrix} \right. \quad h\text{-a.e.}$$

This implies that the linear complexity profile of a pseudorandom sequence should not be distinguishable from that of a random sequence as given by this "law of the logarithm."

**1989, Niederreiter [11]** Following a suggestion by Piper [21], keystream sequences should be "good," for example almost perfect, not only from the beginning, but for every starting point, omitting some leading prefix.

Niederreiter [11] defined the notion of a "good l.c.p." A sequence $a$ has a *good linear complexity profile*, if $L(n, a) \leq C \cdot \log(n), \forall n$, for some fixed constant $C > 0$ (in view of the last paragraph, any $C > 1$ will yield a good l.c.p. for $h$-almost all sequences). A sequence $a$ has a *uniformly good l.c.p.*, if there is a joint $C$ such that $L(n, \sigma^s(a)) \leq C \cdot \log(n), \forall n$, where $\sigma^k(a) = (a_{k+1}, a_{k+2}, a_{k+3}, \dots)$ is the sequence shifted by $k$ symbols. Niederreiter [11, Theorem 6] showed that the set of uniformly good sequences over a finite field has measure 0. It is in fact still an open question whether there exists even one such sequence.

**1990, Niederreiter [13]** For a sequence $a$, let $P(n) := P(n, a) = \#\{k \mid 1 \leq k \leq n, L(k, a) > L(k - 1, a)\}$ be the number of jumps in the l.c.p. of $a$ up to length $n$. Niederreiter showed that the expected value and the variance depend on $q$ and $n$ as follows.

**Theorem 18.4** [13, Theorem 3]

$$E(P(n)) = \frac{q - 1}{q} \cdot \frac{n}{2} + \frac{1 + (-1)^{n+1}}{2} + O\left(\frac{1}{q}\right)$$

*(the $O(1/q)$ term is given exactly in [13]).*

**Theorem 18.5** [13, Theorem 4]

$$\mathrm{Var}(P(n)) = \frac{q - 1}{q^2} \cdot \frac{n}{2} + O\left(\frac{1}{q}\right),$$

*where the $O(1/q)$ term is given exactly in [13].*

We can speak of a "good jump complexity profile" in the following sense.

**Theorem 18.6** [13, Theorem 5] *If $K(n)$ is a bound for the jump complexity profile with $\sum_{n=1}^{\infty} n \cdot e^{-K(n)^2/n} < \infty$, then with probability 1*

$$\left| P(n, a) - \frac{q - 1}{q} \cdot \frac{n}{2} \right| \leq K(n)$$

*for sufficiently large $n$.*

As a corollary, we also have [13, Corollary 2] $\lim_{n \to \infty} \frac{P(n,a)}{n} = \frac{q-1}{2q}$ $h$-a.e. All proofs use only combinatorial counting arguments.

**1990, Niederreiter [12]** Here, Niederreiter used combinatorial arguments (counting instead of the heavy machinery of dynamical systems and probability theory as in [10]) to obtain again

$$\left| L(n, a) - \frac{n}{2} \right| \leq \frac{1}{2} K(n)$$

with probability 1 for all functions $K(n)$ with $\sum_{n=1}^{\infty} q^{-K(n)} < \infty$.

**1997, Vielhaber [25]** The central result of this thesis, supervised by Harald Niederreiter, is an isometry $\mathbf{K}$ on the set $\mathbb{F}_q^{\infty}$, the domain being the sequences $a = (a_k)$, and the range an encoding $(\pi(p_i))$ of the partial denominators of the c.f.e. of $G(a)$, namely $\pi(p_j) = \pi(\sum_{i=0}^{d} b_i x^i) = 0^{d-1} b_d \cdots b_1 b_0$ (as turned out later, Dornstetter [3] had a very similar idea).

Also, $\mathbf{K}(a)$ directly is the discrepancy sequence of $a$.

Let $\mathbf{K_D}$ denote the concatenation of the first halves of the encodings, $\mathbf{K_D}(a) := 0^{d_1-1} b_{d_1} | 0^{d_2-1} b_{d_2} | 0^{d_3-1} b_{d_3} | \ldots$, where $d_j := \deg(p_j)$.

Based on $\mathbf{K}$, we can now apply precise theorems on coin tossing (see below [26]), and derive the shift commutator principle [16, 17]. We also see immediately why indeed $E(L(n))$ should be approximately $n/2$ (the length of the $\mathbf{K_D}$ part), and why $E(P(n))$ should be around $(q-1)/q \cdot n/2$ (the number of nonzeros in the $\mathbf{K_D}$ part).

### 18.2.3 Hausdorff dimensions and fractal sets

**1997, Niederreiter and Vielhaber [15]** $d$-perfect sequences satisfy $|d(n)| \leq d$ for all $n$. Let $\mathcal{A}(q, d)$ be the set of all $d$-perfect sequences over $\mathbb{F}_q$. This infinite set has measure 0 (see [10, Theorem 9]), and the next refinement in characterizing these sequences is the Hausdorff dimension of the set (after a suitable mapping $\iota$ onto the real unit interval). We have

$$D_H(\iota(\mathcal{A}(q, d))) = \frac{1 + \log_q \varphi(d, q)}{2},$$

where $\varphi(d, q)$ is the largest real root of the polynomial $x^d - (q-1) \sum_{i=0}^{d-1} x^i$, for example, $\varphi(2, 1) = 1$, $\varphi(2, 2) = \varphi = 1.618\ldots$.

The value for $D_H(\iota(\mathcal{A}(2, 1))) = \frac{1}{2}$, for instance, reflects the fact that (see Wang and Massey [34]) half of the sequence coefficients can be chosen freely.

A further section covers growing bounds $K(n)$ for the linear complexity deviation and we bound the measure $\mu(A)$ of those sequences whose l.c.p. stays within the bound $K(n)$,

$$A = \{a \in \mathbb{F}_q^{\infty} \mid |2 \cdot L(n, a) - n| \leq K(n), \forall n \in \mathbb{N}\}.$$

For example, for $K(n) = 1 + 2\lfloor \log_2(n) \rfloor$, the set of sequences satisfying $|2 \cdot L(n) - n| \leq 1 + 2\lfloor \log_2(n) \rfloor$, $\forall n \in \mathbb{N}$ has measure at least $e^{-2/21} \approx 0.909 > 0$.

### 18.2.4 Algorithms

**1998, Niederreiter and Vielhaber [16]** The l.c.p. can be computed in $O(n^2)$ time by for example the Berlekamp–Massey algorithm. To compute the l.c.p. for shifted versions $\sigma^k(S)$ of the sequence, for example testing for uniformly good complexity, straightforwardly requires $O(n^3)$ steps. This paper explains an algorithm that, for binary sequences, gives all linear complexities of all shifted versions still in $O(n^2)$ time, equal to the output complexity. The algorithm makes use of the shift commutator of the continued fraction operator **K** (see [25, 27]). The actual transduction from $a$ into the $n$ shifted l.c.p.s takes place via a transducer with state set $\{0, \ldots, 7\} \times \mathbb{Z}$.

**1999, Niederreiter and Vielhaber [17]** This paper generalizes the previous entry to arbitrary finite fields, using a total of $7\binom{n+1}{2}$ $\mathbb{F}_q$-operations to obtain the $2\binom{n+1}{2}$ $\mathbb{F}_q$-symbols of the continued fraction expansions of all shifted sequences, from where the linear complexity profile and the linear complexity deviation can easily be deduced (for more information about the "shift commutator principle" employed here, see [1, 30]).

### 18.2.5 Coin tossing: the ultimate complexity model

**2005, Vielhaber [26]** How useful is the determination of the linear and other complexities? This paper gives a unified view of complexity measures, compression schemes, and predictors as isometries on the input space $A^\infty \to \{1, \ldots, |A|\}^\infty$, for example $A = \mathbb{F}_q$. After mapping the sequence $(a_k)$ isometrically onto the sequence of ranks for example $(a_k) \to (\delta(k))$, we can apply the very precise bounds for *coin tossing*, for example checking the count of # 1 rankings against coin tossing with bias $1/|A|$.

The central advantage lies in the uniform "coin tossing" model, counting in $\mathbb{Z}$, compared to general complexities (e.g. the 2-adic complexity may even assume irrational values from $\mathbb{R}\backslash\mathbb{Q}$).

**2007, Vielhaber [27]** The law of the iterated logarithm for single sequences is defined here via Lévy classes of functions $f : \mathbb{F}_q^{\mathbb{N}} \times \mathbb{N}_0 \ni (a, n) \mapsto f(a, n) \in \mathbb{R}$ and bounds $\alpha : \mathbb{N} \ni n \mapsto \alpha(n) \in \mathbb{R}$ (ULC for upper lower class etc.):

$$
\begin{aligned}
\text{UUC}(f) &= \{\alpha \in \mathbb{R}^{\mathbb{N}} \mid \forall_\mu a, \exists n_0 \in \mathbb{N}, \forall n > n_0 : \quad f(a, n) \quad < \alpha(n)\} \\
\text{ULC}(f) &= \{\alpha \in \mathbb{R}^{\mathbb{N}} \mid \forall_\mu a, \forall n_0 \in \mathbb{N}, \exists n > n_0 : \quad f(a, n) \quad \geq \alpha(n)\} \\
\text{LUC}(f) &= \{\alpha \in \mathbb{R}^{\mathbb{N}} \mid \forall_\mu a, \forall n_0 \in \mathbb{N}, \exists n > n_0 : \quad f(a, n) \quad \leq \alpha(n)\} \\
\text{LLC}(f) &= \{\alpha \in \mathbb{R}^{\mathbb{N}} \mid \forall_\mu a, \exists n_0 \in \mathbb{N}, \forall n > n_0 : \quad f(a, n) \quad > \alpha(n)\}.
\end{aligned}
$$

$\forall_\mu a$ stands for "for $\mu$-almost all sequences $a$," where $\mu$ is the uniform product Haar measure on $\mathbb{F}_q^\infty$. UUC, LLC are of the "almost everywhere" type, ULC, LUC are of the "infinitely often" type. The idea is that for $h$-almost all sequences $a$, $f(a)$ stays within [LLC,UUC] from some $n_0$ on, while they all leave repeatedly the inner interval [LUC,ULC] of unavoidable oscillations.

**Theorem 18.7** [27, Theorem 17] *Let* $d^+(n) = \max_{k \leq n} |2 \cdot d(k)|$, *then*

$$\alpha \in \left\{ \begin{array}{c} \mathrm{UUC}(d^+(n)) \\ \mathrm{ULC}(d^+(n)) \end{array} \right. \iff \sum_{n=1}^\infty 2^{-\alpha(n)} \left\{ \begin{array}{l} < \infty \\ = \infty, \end{array} \right.$$

$1+\lfloor \log_2(n/2)-\log_2\log_2\log_2(n/2)+\log_2\log_2(\mathrm{e})-2-\varepsilon \rfloor \in \mathrm{LLC}(d^+(n)), \forall\varepsilon,$
$1+\lfloor \log_2(n/2)-\log_2\log_2\log_2(n/2)+\log_2\log_2(\mathrm{e})-1+\varepsilon \rfloor \in \mathrm{LUC}(d^+(n)), \forall\varepsilon.$

This is a difference of at most 2 between unavoidable fluctuations and strict containment! For UUC, ULC, this is the same result as previously given in [10], while the determination of LUC, LLC is new. The proof uses ideas of Révész [22, 23] as well as Erdős and Révesz [4] on coin tossing.

As a corollary [27, Corollary 18], $h$-a.a. sequences have a good l.c.p., and $h$-almost no sequence has a perfect l.c.p.

**Theorem 18.8** [27, Theorem 13] *The jump complexity* $P(n)$ *of binary sequences* (*see* [13]) *is bounded as follows:*

$$n/4 + 1 + \sqrt{n} \cdot \sqrt{\log_2\log_2(n)/4 + (3/8 + \varepsilon)\log_2\log_2\log_2(n)}$$
$$\in \mathrm{UUC}(P(n))$$
$$n/4 + \sqrt{n} \cdot \sqrt{\log_2\log_2(n)/4 + (1/8)\log_2\log_2\log_2(n)}$$
$$\in \mathrm{ULC}(P(n))$$
$$n/4 + 1 - \sqrt{n} \cdot \sqrt{\log_2\log_2(n)/4 + (1/8)\log_2\log_2\log_2(n)}$$
$$\in \mathrm{LUC}(P(n))$$
$$n/4 - \sqrt{n} \cdot \sqrt{\log_2\log_2(n)/4 + (3/8 + \varepsilon)\log_2\log_2\log_2(n)}$$
$$\in \mathrm{LLC}(P(n))$$

**2007, Vielhaber and Canales [32]** This paper deals with a normalized version of $L$, $\overline{L}_n := L_n/n \in [0, 1] \subset \mathbb{R}$ and derives the admissible pairs (see Figure 18.3)

$$(I, S) = (\liminf_{n \to \infty} \overline{L}_n, \limsup_{n \to \infty} \overline{L}_n) \in [0, 1]^2.$$

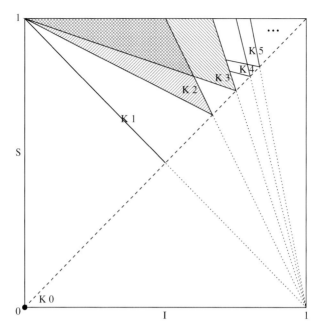

Figure 18.3  Allowed pairs $(I, S)$.

Furthermore, for each admissible pair $(I, S)$, we construct a sequence with these limits, we bound the Hausdorff dimension $D_H$ of the set of all sequences with these limits from both sides, and we show that $D_H$ is positive at least for $S < 1$.

## 18.3  Multilinear complexity

We now turn to several $\mathbb{F}_q$-sequences in parallel, as used for example for multimedia streams. Again, we are interested in the behavior of the linear complexity.

We denote the sequences under consideration $(a_m) = (a_{m,k}) \in (\mathbb{F}_q^M)^\infty$, $1 \leq m \leq M$, $k \in \mathbb{N}$, where $M$ is the degree of parallelism.

The linear complexity of such a bundle of sequences, $L(n) := L_M(n)$, $n \in \mathbb{N}$ is the smallest number such that the $n$-symbol prefixes can simultaneously be generated by the *same* LFSR of length $L(n)$, with initial contents $(a_{m,1}, \ldots, a_{m,L(n)})$, $\forall 1 \leq m \leq M$.

Equivalently, in terms of simultaneous Diophantine approximation, $L(n)$ is the smallest degree of a *common* denominator $v_n(x)$, such that for suitable polynomials $u_{m,n}(x)$, $G(a_m) = u_{m,n}(x)/v_n(x) + O(x^{-n-1})$.

The linear complexity deviation now is defined as $d(n) := L(n, a) - \lceil n \cdot \frac{M}{M+1} \rceil \in \mathbb{Z}$. Defining $d(n)$ thus instead of, for example, $d(n) = L(n, a) - n \cdot \frac{M}{M+1} \in \mathbb{Q}$, is justified since first this describes the l.c.p. for any sequence with $\delta(n) \neq 0, \forall n$ (where $\delta$ is again the discrepancy), or equivalently for $q \to \infty$, any $\delta$. Secondly, this definition leads to an average of $\bar{d} = 0$, when summing over the asymptotic measure for all $d \in \mathbb{Z}$, averaging over all timesteps mod $(M + 1)$ (see [29, Theorem 16]).

### 18.3.1 Counting and recursions over partitions

**2005, Wang and Niederreiter [35]** Let $K$ be the field of Laurent series $K = \mathbb{F}_q((x^{-1})) = \left\{ \sum_{i=i_0}^{\infty} a_i x^{-1}, a_i \in \mathbb{F}_q \right\}$ with valuation $V(0) = \infty$ and otherwise $V(\sum_{i=i_0}^{\infty} a_i x^{-i}) = i_0$ for $a_{i_0} \neq 0$.

A subset $\Lambda \subset K^{M+1}$ is an $\mathbb{F}_q$-lattice, if $\Lambda = \sum_{i=1}^{M+1} \mathbb{F}_q[x] \cdot \omega_i$ for some basis $(\omega_1, \ldots, \omega_{M+1})$ of $K^{M+1}$.

The refined lattice basis reduction algorithm for multisequences (refined LBRMS) of Wang et al. [36] computes a sequence of bases $(\omega_1, \ldots, \omega_{M+1})^{(n)}$, such that the generated lattices are equal to $\{(C(x) + D_1(x) \cdot a_1(x), \ldots, C(x) + D_M(x) \cdot a_M(x), C(x)x^{-L-1})\}$ for suitable polynomials $C(x), D_m(x) \in \mathbb{F}_q[x], 1 \leq m \leq M$.

The initial basis is

$$(\omega_1, \ldots, \omega_{M+1}) = \begin{pmatrix} & & & G(a_1) \\ & I_M & & \vdots \\ & & & G(a_M) \\ 0 & \ldots & 0 & x^{-n-1} \end{pmatrix}.$$

The polynomials $C^{(n)}$ are obtained recursively as [35, p. 616 f.]

$$C^{(n)} = \begin{cases} C^{(n-1)} & \text{cases } 0_{\mathrm{I}}, 0_{\mathrm{II}} \\ \quad \text{if } \Delta := \Delta^{(n-1)}(C^{(n-1)}) = 0 \\ C^{(n-1)} - \sum_{m=1}^{M} a_m x^{V(\omega_m) - (n - L(n-1))} \eta(\omega_m) & \text{case I} \\ \quad \text{if } \Delta \neq 0 \wedge n - L(n-1) \leq V(\omega_k), \\ x^{n - L(n-1) - V(\omega_k)} \cdot C^{(n-1)} - \sum_{m=1}^{M} a_m x^{V(\omega_m) - V(\omega_k)} \eta(\omega_m) & \text{case II} \\ \quad \text{if } \Delta \neq 0 \wedge n - L(n-1) > V(\omega_k), \end{cases}$$

where $\Delta = (\delta_1, \ldots, \delta_M)$ is the discrepancy vector, $\eta(\sum_{m=1}^{M} D_m(x) \cdot \varepsilon_m + C(x) \cdot (a_1, \ldots, a_M, x^{-n-1})) = C(x)$ is the coefficient of the $(M + 1)$-st standard basis vector in the representation of its argument, and $\omega_k$ is a current basis vector with smallest valuation, $V(\omega_k) \leq V(\omega_m), 1 \leq k \leq M$.

By [35, Lemma 2], $\sum_{m=1}^{M} V(\omega_m^{(n)}) = L(n)$. Wang and Niederreiter now extracted the crucial information of the refined LBRMS, namely the valuations

$i_m := V(\omega_m)$, in nondecreasing order as $\tilde{i}_1 \geq \tilde{i}_2 \geq \cdots \geq \tilde{i}_M$. The tuple
$\mathbf{I} = (\tilde{i}_1, \ldots, \tilde{i}_M)$ is used as the condensed information about the state of
approximation. Since $\sum_m i_m = L(n)$, $\mathbf{I}$ is a partition of the current linear
complexity $L$ into $M$ parts (or into at most $M$ nonzero parts).

Let $N_M(n, L)_{\mathbf{I}}$ be the number of $M$-multisequence prefixes of length $n$ with
linear complexity $L$, where $L$ is partitioned into parts according to $\mathbf{I}$. Furthermore, let $N_M(n, L) = \sum_{\mathbf{I}} N_M(n, L)_{\mathbf{I}}$ be the number of all $M$-multisequence
prefixes of length $n$ with linear complexity $L$.

Theorem 1 in [35] gives a recursion for $N_M(n, L)|_{\mathbf{I}}$ in terms of $N_M(n - 1, L)|_{\mathbf{I}_0}$ and $N_M(n-1, L)|_{\mathbf{I}_t}$, where $\mathbf{I}_0 = \mathbf{I}$ is the unchanged partition (cases $0_I$, $0_{II}$, $I$), while in $\mathbf{I}_t$ the $t$th element (not counting multiples) has been replaced
from $N - L(n - 1)$ to the current value, $1 \leq t \leq \mu(\mathbf{I})$, $\mu(\mathbf{I})$ being the number
of different positive entries in $\mathbf{I}$.

Cases $0_I$ and $0_{II}$ correspond to Theorem 1, case 1a ($\delta = 0$, $a_j = 0$ in (5)
$\forall j$), with a single occurrence, $c_0 = 1$.

Case I corresponds to Theorem 1, cases 2a–6a ($\mathbf{I} = \mathbf{I}_0$, $\Delta \neq 0$), with count
$c_0 = q^k$ for some $k$ (all symbols in all $k \leq M$ places permitted).

Case II corresponds to Theorem 1, cases 3b–6b ($\mathbf{I} = \mathbf{I}_t$, $\Delta \neq 0$), with count
$c_0 = (q^k - 1)q^l$ for some $k, l$.

The initial values for $n = 1$ are $N_M(1, 0)|_{(0,\ldots,0)} = 1$ ($\Delta = (0)$ for
$a_1 = (0, \ldots, 0)$), and $N_M(1, 1)|_{(1,0,\ldots,0)} = q^M - 1$ ($\Delta \neq (0)$ for $a_1 \in \mathbb{F}_q^M \setminus \{(0, \ldots, 0)\}$).

Observe that a step includes the whole $n$th column $a_{1,n} \ldots a_{M,n}$.

Summing up the counts $c_0$ over all predecessor partitions for a certain $\mathbf{I}$, we
obtain the number $N_M(n, L)|_{\mathbf{I}}$.

**Theorem 18.9** [35, Theorem 2]

$$N_M(n, L)|_{\mathbf{I}} = \frac{c(\mathbf{I})}{d(\mathbf{I})} q^{b(\mathbf{I}, n-L)},$$

*where*

$$c(\mathbf{I}) = \prod_{i=1}^{\lambda(\mathbf{I})} (q^{m+1-i} - 1)\frac{q^i - 1}{q - 1}, \qquad d(\mathbf{I}) = \prod_{j=1}^{\mu(\mathbf{I})} \prod_{i=1}^{s_{\mathbf{I},j}} \frac{q^i - 1}{q - 1}$$

$s_{\mathbf{I}, j}$ *is the count of the $j$th symbol in $\mathbf{I}$ ($j = 1$ the largest) and*

$$b(\mathbf{I}, n - L) = \begin{cases} e_{\lambda(\mathbf{I})} \circ [\mathbf{I}, n - L] - \frac{\lambda(\mathbf{I})(\lambda(\mathbf{I})-1)}{2} \\ \quad \text{if } 0 \leq n - L < i_{\lambda(\mathbf{I})} \\ e_{\lambda(\mathbf{I})} \circ [\mathbf{I}, n - L] - \left(\frac{\lambda(\mathbf{I})(\lambda(\mathbf{I})+1)-\sum_{k=1}^{w} s_{\mathbf{I},k}}{2}\right) \\ \quad \text{if } \exists w, 1 \leq w \leq \mu(\mathbf{I}) - 1 : \sum_{k=1}^{w+1} s_{\mathbf{I},k} \leq n - L < \sum_{k=1}^{w} s_{\mathbf{I},k} \\ e_{\lambda(\mathbf{I})} \circ [\mathbf{I}, n - L] - \frac{\lambda(\mathbf{I})(\lambda(\mathbf{I})+1)}{2} \\ \quad \text{if } n - L \geq i_1. \end{cases}$$

Adding over all partitions $\mathbf{I}$ with $\sum_m i_m = L$, we then obtain the following closed form for the number of sequences over $\mathbb{F}_q$ of length $n$, width $M = 2$, and with linear complexity $L$.

**Theorem 18.10** [35, Theorem 3]

$$N_2(n, L) =$$

$$
\begin{cases}
1 & L = 0 \\
(q^2 - 1)q^{3L-2} & 1 \le L \le \frac{1}{2}n \\
\frac{q^2-1}{q^2+1}q^{2(n-L)} + (q^2 - 1)q^{3L-2} - \frac{(q^2-1)(q^2-q+1)}{q^2+1}q^{6L-2n-3} & \frac{1}{2}n < L \le \frac{2}{3}n \\
\frac{q^2-1}{q^2+1}q^{2(n-L)} + (q^2 - 1)q^{4n-3L} - \frac{(q^2-1)(q^2-q+1)}{q^2+1}q^{6(n-L)+1} & \frac{2}{3}n < L \le n.
\end{cases}
$$

**Theorem 18.11** [35, Theorem 4] *The expected value* $E_2(L(n)) = q^{-2n} \sum_{L=0}^{n} L \cdot N_2(n, L)$ *is then* $E_2(L(n)) =$

$$
\frac{2}{3}n + \frac{3(q^5 + q^4 - q^3 + q^2 + q) + R_3(n)(q^6 - 2q^5 - q^4 + 4q^3 - q^2 - 2q + 1)}{3(q^2 + q + 1)(q^4 + q^2 + 1)}
$$
$$
+ O\left(\frac{n}{q^n}\right),
$$

where the $O(nq^{-n})$ function is given exactly in the paper, and $R_3(n)$ is the rest $n \mod 3$.

**2005, Niederreiter and Wang [18]** Niederreiter and Wang built upon their technique of what could be called the "recursive counting over partitions" machinery developed in [35] and they showed for any prime power $q$ and any integer $M \ge 1$ that the following holds.

**Theorem 18.12** [18, Theorem 1]

$$\lim_{n \to \infty} \frac{L_M(n, a)}{n} = \frac{M}{M + 1} \text{ with probability } 1.$$

**Theorem 18.13** [18, Theorem 2] *For all* $M \ge 1$,

$$E_M(L(n)) = \frac{M}{M + 1} + o(n).$$

The proof involves the formula for $N_M(n, L)|_{\mathbf{I}} = f(c(\mathbf{I}), d(\mathbf{I}), b(\mathbf{I}))$ from [35]. Sharper bounds, with $o(n)$ replaced by $O(1)$, were known then only for the cases $M = 1$ (Rueppel [24]), $M = 2$ and $q = 2$ (Feng and Dai [5]), and $M = 2$, any $q$ (Wang and Niederreiter [35]), but see [20] for improvements.

**2006, Vielhaber and Canales [28]** As in [15], the $d$-perfect multisequences form an infinite set of measure zero. For higher $d$ and $M \geq 2$, two different notions of $d$-perfect are possible. Either $|d_M| \leq d$, or all partial denominators (jumps) are of degree (height) deg $\leq d$. The former notion is more restrictive, since the latter allows for up to $M$ jumps of degree up to $d$, exceeding the bound $|d_M| \leq d$.

The paper gives numerical results for both notions in the cases $d = 1, \ldots, 5$ and $M = 1, \ldots, 5, 8, 16$. Furthermore, the Hausdorff dimension for $d = 1$, any $M$, under the restriction $|d_M| \leq 1$, is

$$D_H(\iota(\mathcal{A}(1; M, q))) = 1 + \frac{\log_q \left( \sqrt[M+1]{\prod_{k=1}^{M} 1 - \frac{1}{q^k}} \right)}{M} < 1.$$

### 18.3.2 Counting and stochastic infinite state machines

**2006, Vielhaber and Canales [29]** This approach starts with the multi-strict continued fraction algorithm (mSCFA) of Dai *et al.* [2] and then develops the "battery-discharge model" (BDM). The digest of the mSCFA (ignoring the actual approximation, keeping only the development of degrees) is as follows.

**Algorithm 18.14** mSCFA
deg := 0; $w_m$ := 0, $1 \leq m \leq M$
FOR $n := 1, 2, \ldots$
    FOR $m := 1, \ldots, M$
        compute $\delta(n, m)$ // discrepancy
        IF $\delta(n, m) \neq 0$ AND $n - \deg - w_m > 0$: // [2, Theorem 2 (2b)], Case II
            deg_copy := deg, deg := $n - w_m$, $w_m$ := $n - $ deg_copy
    ENDFOR
ENDFOR

We now focus on the l.c.dev. (instead of on the l.c.p.), and denote the deviation as $d := \deg - \lceil n \cdot M/(M + 1) \rceil \in \mathbb{Z}$ (l.c.deviation or "drain"), where deg is the degree of the current approximation denominator, and $b_m := \lfloor n/(M + 1) \rfloor - w_m \in \mathbb{Z}$ (the $M$ "batteries").

Case II, i.e., $\delta \neq 0$ while $L(n - 1, a) < \lceil n \cdot M/(M + 1) \rceil$, translates from the mSCFA to the BDM setting as follows: $b_m$ and $d$ interchange their values, if and only if $\delta \neq 0$ and $b_m > d$ (that is $n - \deg - w_m > 0$).

We have the following actions for the BDM:

$n_=, n_<$: Do nothing, since $b_m = d$ or $b_m < d$, cases $0_I$, I.
D: Discharge, swap $(b_m, d)$, if and only if $\delta \neq 0$ (modeled as prob $= (q-1)/q$) when $b_m > d$, case II.
I: Inhibition, do nothing in case $0_{II}$ ($\delta = 0$, modeled as prob $= 1/q$), although $b_m > d$.

We run the time cyclically through $(T, t) \in \{0, \ldots, M\} \times \{1, \ldots, M+1\}$, starting at $(0, M+1)$. We consider input symbol $a_{m,n}$ at timestep $(T, t) = (n \bmod (M+1), m)$, querying battery $b_m$.

For $t = M + 1$, we do not read the input, but either decrease $d$ (at $0 \le T < M$), action $d_-$, or increase all $M$ batteries at $(T, t) = (M, M+1)$, action $b_+$, thus adjusting for the jump in $\lceil n \cdot M/(M+1) \rceil$, then returning to $(0, 1)$.

The state set of the BDM is

$$S = \left\{ s = (b_1, \ldots, b_M, d; T, t) \colon b_m \in \mathbb{Z}, 1 \le m \le M, d \in \mathbb{Z}, \right.$$

$$\left. 0 \le T \le M, 1 \le t \le M+1, \text{ with } d + T + \sum_{m=1}^{M} b_m = 0 \right\}$$

with initial state $s_0 = (0, \ldots, 0, 0; 0, M+1)$.

Most mass lies on the states with low l.c.dev., and the initial state is recurrent. The distribution of l.c.dev.s repeats itself each time, when $s_0$ is reached again. In this respect, the distribution of linear complexity deviations is self-similar.

Every state has one or two allowed transitions $s = (b_1, \ldots, b_M, d; T, t) \xrightarrow{\alpha} s^+$, depending on the action $\alpha$, with successor state $s^+$ according to Table 18.1. Figure 18.4 gives all states with class (see below) up to 4 with $M = 2$.

Table 18.1 *BDM state transitions*

| $\alpha$ | Condition | $s^+$ | Probability |
|---|---|---|---|
| D | $b_t > d, t \le M$ | $(b_1, \ldots, b_{t-1}, d, b_{t+1}, \ldots, b_M, b_t; T, t+1)$ | $(q-1)/q$ |
| I | $b_t > d, t \le M$ | $(b_1, \ldots, b_M, d; T, t+1)$ | $1/q$ |
| $N_=$ | $b_t = d, t \le M$ | $(b_1, \ldots, b_M, d; T, t+1)$ | $1$ |
| $N_<$ | $b_t < d, t \le M$ | $(b_1, \ldots, b_M, d; T, t+1)$ | $1$ |
| $d_-$ | $T < M, t = M+1$ | $(b_1, \ldots, b_M, d-1; T+1, 1)$ | $1$ |
| $b_+$ | $T = M, t = M+1$ | $(b_1+1, b_2+1, \ldots, b_M+1, d; 0, 1)$ | $1$ |

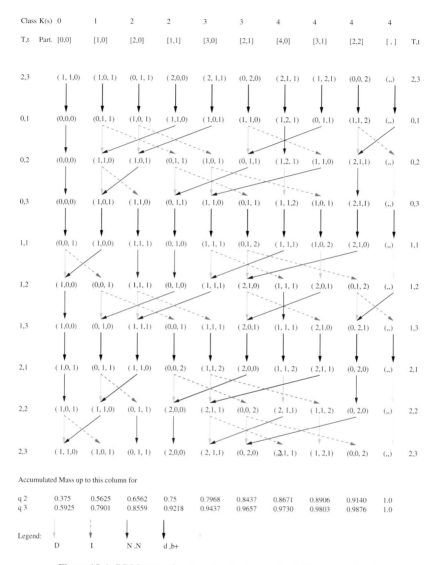

Figure 18.4  BDM states for $M = 2$ with class up to 4. (See color plate.)

Whenever $b_t > d$, both D and I may occur, whose probabilities sum up to 1. From the allowed transitions, we can read off the state transition matrix of the BDM, an infinite stochastic matrix $\mathcal{T}$ indexed by $s, s' \in S$.

Every row of $\mathcal{T}$ either includes an I and a D, or else one of $N_=$, $N_<$, $d_-$, or $b_+$. Each column either has one of I, $N_=$, $d_-$, $b_+$ or both D and $N_<$.

### 18.3.3  Class $K(s)$ of the state $s$

Each state $s$ has an associated *class $K(s) \in \mathbb{N}_0$*.

(1) Static definition: $K(s) := -\pi_s + M \cdot T + 2 \cdot \sum_{m=1}^{M+1} \tilde{b}_m \cdot (M + 1 - m)$, where $\pi_s$ is the minimum number of transpositions between neighbors necessary to sort $(b_1, \dots, b_{t-1}, d, b_t, \dots, b_M)$ into nonincreasing order as $(\tilde{b}_1, \dots, \tilde{b}_{M+1})$, with $\tilde{b}_i \geq \tilde{b}_{i+1}$ for $1 \leq i \leq M$. Observe that the place of $d$ in the initial sequence depends on $t$.

(2) Dynamic definition: let $s_0 \xrightarrow{\alpha_1 \dots \alpha_k} s$ be a path from the initial state $s_0$ to $s$. Let $\#I = \#\{1 \leq i \leq k \,|\, \alpha_i = I\}$ and $\#N_< = \#\{1 \leq i \leq k \,|\, \alpha_i = N_<\}$. Then $K(s) = \#I - \#N_<$ (see [29, Theorem 8(ii)], [33, Theorem 3]).

(3) Meaning: the limit mass distribution $\mu_\infty$ on $S$ satisfies $\frac{\mu_\infty(s)}{\mu_\infty(s')} = q^{K(s')-K(s)}$ for any two states $s, s' \in S$ (see [29, Theorem 11], [33, Theorem 6]). Here, $\mu_\infty$ refers to the mass distribution in the steady state for $n \to \infty$, normalized to $\sum_{s \in S, T(s)=T_0, t(s)=t_0} \mu_\infty(s) = 1, \forall T_0, t_0$.

### 18.3.4  The main result: asymptotic distribution of $d$

Let the asymptotic $(n \to \infty)$ mass on states with drain (l.c.dev.) $d$ be $\gamma(d, T, t) = \sum_{s \in S(T,t,d)} \mu_\infty(s)$, distinguished according to the timesteps $(T, t)$. Then we have the following result.

**Theorem 18.15** [29, Theorem 22] *For $1 \leq M \leq 8$ or $M = 16$, $0 \leq T \leq M$, $1 \leq t \leq M + 1$, and any finite field $\mathbb{F}_q$, let $\Delta := (M \cdot T + t) \mod (M + 1), 0 \leq \Delta \leq M$. Then for every linear complexity deviation $d \in \mathbb{Z}$, with $\gamma(d, \Delta) := \gamma(d, t, T)$ we have*

$$\gamma(d, \Delta) \doteq \sum_{h=1}^{M} C_{M,h} \cdot \frac{q^{\varepsilon_{\text{sgn}(d)}(\Delta,h)}}{q^{h \cdot (M+1) \cdot |d|}}$$

*where*

$$C_{M,h} = (-1)^{h+1} \frac{(q^{(M+1)h} - 1)}{q^{(M+1)h-h^2}} \frac{\prod_{k=M-h+1}^{M}(q^k - 1)}{\prod_{k=M+1}^{M+h}(q^k - 1)}$$

*is independent from $d$ and $\Delta$ and has degree zero ($C_{M,h} = (-1)^{h+1} + O(1/q), \forall M, h$), and where $\varepsilon_{\text{sgn}(d)}(\Delta, h)$ depends only on the sign of $d$ and is given by*

$$\varepsilon_-(\Delta, h) = h \cdot (M - \Delta) - \binom{h}{2},$$
$$\varepsilon_+(\Delta, h) = h \cdot \Delta - \binom{h}{2},$$
$$\varepsilon_0(\Delta, h) = \min\{\varepsilon_+(\Delta, h), \varepsilon_-(\Delta, h)\}.$$

*Furthermore, $\doteq$ means equality with precision at least $q^{-(1200-100 \cdot M)}$, or $q^{-200}$ in the case of $M = 16$.*

More on how we obtained this formula can be found in [29], including Sloane's OEIS repository, Victor Shoup's NTL library, and "educated and not-so-rewarding guessing" until successful.

### 18.3.5 Parsimony

**Conjecture 18.16** *For all prime powers q, for all $M \in \mathbb{N}$, for all $d \in \mathbb{Z}$, for all $0 \leq \Delta \leq M$ (with $\varepsilon_{-,0,+}(\Delta, h)$ as before) we have the unified closed formula:*

$$\gamma(d, \Delta) = \sum_{h=1}^{M} (-1)^{h+1} \cdot \frac{q^{\varepsilon_{\mathrm{sgn}(d)}(\Delta,h)}}{q^{h \cdot (M+1) \cdot |d|}} \cdot \left( \frac{(q^{(M+1)h} - 1)}{q^{(M+1)h-h^2}} \cdot \frac{\prod_{k=M-h+1}^{M}(q^k - 1)}{\prod_{k=M+1}^{M+h}(q^k - 1)} \right).$$

**2006, Vielhaber and Canales [31]** Applying the BDM, we have the *law of the logarithm* for the linear complexity deviation of multisequences [31, Theorem 27], [29, Theorem 26]:

$$\lim_{n \to \infty} \left\{ \begin{matrix} \sup \\ \inf \end{matrix} \right\} \frac{d(n, a)}{\log n} = \frac{1}{(M+1)\log q} \cdot \left\{ \begin{matrix} +1 \\ -1 \end{matrix} \right.$$

(already in [10] for $M = 1$) and we also obtain from the BDM [31, Theorem 29], [29, Theorem 28],

$$\lim_{n \to \infty} \frac{L(n, a)}{n} = \frac{M}{(M+1)},$$

already shown by Niederreiter and Wang in [18, 35].

**2007, Moshchevitin and Vielhaber [8]** We have the following results.

**Theorem 18.17** [8, Theorem 1] *For all $M \geq 1$, we find*

$$E_M(L(n)) = \left\lceil \frac{M}{M+1} \cdot n \right\rceil + O(1)$$

*(improving [35] for $M \geq 4$, while the cases $M = 1, 2, 3$ are already known).*

Based on [29], we have the conjecture [8, Conjecture 5] that for all $M \geq 1$

$$\left\lceil \frac{M}{M+1} \cdot n \right\rceil - 1 + O\left(\frac{1}{n}\right) \leq E_M(L(n)) \leq \left\lceil \frac{M}{M+1} \cdot n \right\rceil + 1 + O\left(\frac{1}{n}\right).$$

**2007, Niederreiter and Wang [19]** This paper is a follow-up to [35] and treats the case $M = 3$ in detail.

**Theorem 18.18** [19, Theorem 5]

$$-\frac{1}{M+1} \le \liminf_{n\to\infty} \frac{L_M(n,a) - n\cdot\frac{M}{M+1}}{\log_q(n)}$$

$$\le \limsup_{n\to\infty} \frac{L_M(n,a) - n\cdot\frac{M}{M+1}}{\log_q(n)} \le 1$$

and $N_3(n,L)$ is given in closed form.

We can extract repeated terms in [19, Theorem 11] to obtain a shorter representation as follows:

$$A = (q^3-1)q^{4L-3}, \ B = \frac{(q^2-1)(q^3-1)(q^4+1)}{q^5-1}q^{8L-3n-5},$$

$$C = \frac{(q^3-1)^2}{(q^2+1)(q^5-1)}q^{2n-2L},$$

$$D = \frac{q^3-1}{q^2+1}q^{6n-6L+1}, \ E = \frac{(q-1)(q^3-1)(q^4-q^2+1)}{q^5-1}.$$

**Theorem 18.19** [19, Theorem 11]

$$N_3(n,L) = \begin{cases} 1 & L=0 \\ A & 1 \le L < \frac{1}{2}n \\ A - B + C + \frac{(q^2-q+1)(q^3-1)}{q^2+1}q^{6L-2n-4} & \frac{1}{2}n \le L < \frac{2}{3}n \\ A - B + C - D + E\cdot q^{12L-6n-6} & \frac{2}{3}n \le L < \frac{3}{4}n \\ A\cdot q^{6n-8L+3} - B\cdot q^{12n-16L+6} + C - D+ \\ \quad +E\cdot q^{12n-12L+3} + \frac{(q+1)(q^2-1)(q^3-1)}{q^5-1}q^{4n-3L} & \frac{3}{4}n \le L \le n. \end{cases}$$

A remarkable result!

**Theorem 18.20** [19, Theorem 12] *For any prime power $q$ and $M = 3$,*

$$E_3(L(n)) = \frac{3}{4}n + O(1).$$

**2011, Niederreiter *et al.* [20]**

$$\limsup_{n\to\infty} \frac{L_M(n,a) - \frac{M}{M+1}\cdot n}{\log_q(n)} \le \frac{1}{M+1}.$$

The corresponding lower bound $\limsup \cdots \ge -1/(M+1)$ was already shown in [19].

As a consequence, the expected value for the linear complexity is

$$E_M(L(n)) = \frac{M}{M+1} \cdot n + O(1), \quad \forall M \in \mathbb{N}.$$

## 18.4 Partitions, bijections, conjectures

In this section, we collect recent results, numerically verified, interesting in themselves, asking for mathematical confirmation.

### 18.4.1 Finite strings

Here, we consider finite prefixes $(a_1, \ldots, a_n) \in (\mathbb{F}_q^M)^n$ of $a$ and we restrict ourselves to timesteps $\tau \equiv 0 \mod (M+1), m = M$, and $t = M+1$, at the end of a full column of symbols from $a$.

*Experimentally*, it turns out that the following holds.

(1) A state $s$ with corresponding partition $(\tilde{I}_1, \ldots, \tilde{I}_M)$ acquires positive mass at the $\tilde{I}_1$th column, i.e., after reading $a_{\tilde{I}_1, M}$.

(2) This initial mass is higher than the asymptotic value by $\prod_{k=M_1(s)}^{M} \frac{q^k}{q^k-1}$ (a factor), where $M_1(s) = M+1-\#\{1 \le m \le M \mid b_m = \max\{b_1, \ldots, b_M, d\}\}$. In particular, if $d$ is larger than all batteries, $M_1(s) = M+1$ and the factor is 1. Also, $M_1(s) = 1$, if all batteries and $d$ have the same value. In this case (e.g. for $s_0$) the factor is $\prod_{k=1}^{M} \frac{q^k}{q^k-1} = \mathcal{P}(q, M) = \sum_{K \in \mathbb{N}_0} P_M(K)q^{-K}$, the generating function in $q$ of the number of partitions of $K \in \mathbb{N}_0$ into at most $M$ parts.

(3) The mass drops immediately to its asymptotic value $q^{-K(s)}/\prod_{k=1}^{M} \frac{q^k}{q^k-1}$ at the next recurrence, at $\tau = \tilde{I}_1 + M + 1$, and stays there forever (no gradual convergence).

(4) From (1), it follows that all partitions with highest part $n$ are active from column $n$ on. There are $\binom{n+M}{M}$ of these, and the effort is thus $O(n^M)$.

(5) We obtain the state corresponding to a partition by the algorithm we give next.

These findings have been verified numerically for $M = 1, \ldots, 6$ and $n \le 400 - 50M$. We conjecture that they are indeed valid for all $M, n \in \mathbb{N}$.

**Algorithm 18.21** *From partition to state:*

```
0.  Given (p₁,…,p_M) ∈ ℕ₀^M,  and any T,t,  set p_{M+1} = 0.
```

1. Set $\tilde{p}_1 = \left\lfloor \frac{p_1+M-T}{M+1} \right\rfloor \cdot (M+1) + T, \tilde{b}_1 = \left\lfloor \frac{\tilde{p}_1}{M+1} \right\rfloor, b_1 := \tilde{b}_1.$

2. For $m = 2, 3, \ldots, M+1$:
$$\tilde{p}_m = \tilde{p}_{m-1} - \left\lfloor \frac{\tilde{p}_{m-1} - p_m}{M+2-m} \right\rfloor \cdot (M+2-m), d_{m-1} = \left\lfloor \frac{\tilde{p}_{m-1} - \tilde{p}_m}{M+2-m} \right\rfloor,$$
$\tilde{b}_m = \tilde{b}_{m-1} - d_{m-1}, b_m := \tilde{b}_m.$

3. For $m = M, M-1, \ldots, 2, 1$, for $i = 0, \ldots, \tilde{p}_m - p_m - 1$ do:
swap$(b_{m+i}, b_{m+i+1})$.

4. For $m = t, t+1, \ldots, M$ do:
swap$(b_m, b_{m+1})$.

5. Finally, set $d := b_{M+1}$. Now $(b_1, \ldots, b_M)\} \in \mathbb{Z}^M$, and $(b_1, \ldots, b_M, d; T, t)$ is the state corresponding to the input partition.

### 18.4.2 Correspondence between the NW and VC approaches

We will first obtain a connection between the approach of Niederreiter and Wang [18], whose partition of $L$ into $(\tilde{I}_m)$ we will call the "NW-partition," and the BDM approach taken by Vielhaber and Canales [29], with its corresponding state $(b_1, \ldots, b_M, d; T, t)$, called the "VC-partition."

Each (nonordered!) NW-partition $(I_1, \ldots, I_M)$ is equivalent to the BDM state with $b_m = \lfloor \frac{n}{M+1} \rfloor - I_m$, i.e., $(b_1, \ldots, b_M) = \lfloor \frac{n}{M+1} \rfloor \cdot (1, \ldots, 1) - (I_1, \ldots, I_M)$.

Wang and Niederreiter [35] use only the ordered (nonincreasing) NW-partition $(\tilde{I}_1, \ldots, \tilde{I}_M)$. There are up to $M!$ BDM states subsumed into this NW-partition, all those with $\{b_1, \ldots, b_M\} = \{I_1, \ldots, I_M\}$ (as a multiset).

The corresponding partitions apparently are (no proof yet) those with the highest part not exceeding $k$ (VC-partitions of the BDM states), and those with the sum of the $M$ parts not exceeding $k$ (NW-partitions).

The growth rates are accordingly as follows. There are exactly $\binom{n+M}{M} = \frac{n^M}{M!} + O(n^{M-1})$ active states or VC-partitions in the BDM up to column $n$, but only $P_M(n) = \frac{n^M}{M!M!} + O(n^{M-1})$ partitions in the NW approach.

We obtain the NW-partition from the BDM state and vice versa (see above) and the BDM state from the VC-partition by the previous algorithm.

We do not yet know how to get from a BDM state/NW-partition to the VC-partition, except by walking through the transitions, starting from $s_0$, and counting inhibitions.

Some examples for the correspondence of VC-partitions to NW-partitions are as follows. From VC-partition (e.g. [00]) to NW-partition (e.g. 33), for $M = 2, n = 9$, see Table 18.2a. Dropping every $(M+1)$st row, and then alternatingly distributing rows and columns onto two parts, we have four sub-matrices, which now exhibit more structure, see Table 18.2b. For $M = 3, n = 22$, see Table 18.2c.

Table 18.2 *VC-partition to NW-partition,* **a, b** *for* $M = 2, n = 9$, **c** *for* $M = 3, n = 22$

| a | 0] | 1] | 2] | 3] | 4] | 5] | 6] | 7] | 8] | 9] |
|---|----|----|----|----|----|----|----|----|----|----|
| [0 | 33 |    |    |    |    |    |    |    |    |    |
| [1 | 43 | 43 |    |    |    |    |    |    |    |    |
| [2 | 42 | 32 | 52 |    |    |    |    |    |    |    |
| [3 | 42 | 32 | 52 | 22 |    |    |    |    |    |    |
| [4 | 44 | 53 | 53 | 62 | 62 |    |    |    |    |    |
| [5 | 41 | 51 | 31 | 61 | 21 | 71 |    |    |    |    |
| [6 | 41 | 51 | 31 | 61 | 21 | 71 | 11 |    |    |    |
| [7 | 54 | 54 | 63 | 63 | 72 | 72 | 81 | 81 |    |    |
| [8 | 50 | 40 | 60 | 30 | 70 | 20 | 80 | 10 | 90 |    |
| [9 | 50 | 40 | 60 | 30 | 70 | 20 | 80 | 10 | 90 | 00 |

| b | 1] | 3] | 5] | 7] | 9] |
|------|----|----|----|----|----|
| [0 |    |    |    |    |    |
| [2/3 | 32 | 22 |    |    |    |
| [5/6 | 51 | 61 | 71 |    |    |
| [8/9 | 40 | 30 | 20 | 10 | 00 |

| | 0] | 2] | 4] | 6] | 8] |
|------|----|----|----|----|----|
| [0 | 33 |    |    |    |    |
| [2/3 | 42 | 52 |    |    |    |
| [5/6 | 41 | 31 | 21 | 11 |    |
| [8/9 | 50 | 60 | 70 | 80 | 90 |

| | 1] | 3] | 5] | 7] |
|------|----|----|----|----|
| [1 | 43 |    |    |    |
| [4 | 53 | 62 |    |    |
| [7 | 54 | 63 | 72 | 81 |

| | 0] | 2] | 4] | 6] |
|------|----|----|----|----|
| [1 | 43 |    |    |    |
| [4 | 44 | 53 | 62 |    |
| [7 | 54 | 63 | 72 | 81 |

| c | 0] | 1] | 2] | 3] | 4] | 5] |
|-----|-----|-----|-----|-----|-----|-----|
| [00 | 111 |     |     |     |     |     |
| [10 | 211 |     |     |     |     |     |
| [11 | 211 | 211 |     |     |     |     |
| [20 | 221 |     |     |     |     |     |
| [21 | 221 | 311 |     |     |     |     |
| [22 | 221 | 311 | 311 |     |     |     |
| [30 | 220 |     |     |     |     |     |
| [31 | 210 | 310 |     |     |     |     |
| [32 | 210 | 310 | 110 |     |     |     |
| [33 | 320 | 320 | 410 | 410 |     |     |
| [40 | 220 |     |     |     |     |     |
| [41 | 210 | 310 |     |     |     |     |
| [42 | 210 | 310 | 110 |     |     |     |
| [43 | 320 | 320 | 410 | 410 |     |     |
| [44 | 300 | 200 | 400 | 100 | 500 |     |
| [50 | 220 |     |     |     |     |     |
| [51 | 210 | 310 |     |     |     |     |
| [52 | 210 | 310 | 110 |     |     |     |
| [53 | 320 | 320 | 410 | 410 |     |     |
| [54 | 300 | 200 | 400 | 100 | 500 |     |
| [55 | 300 | 200 | 400 | 100 | 500 | 000 |

As can be observed, for $M = 3$ and $n = 22$, the outer VC-partitions are those $[abc]$ with $5 \geq a \geq b \geq c \geq 0$, while the inner NW-partitions are those $[abc]$ with $5 \geq a \geq b \geq c \geq 0 \wedge 5 \geq a + b + c \geq 0$.

### 18.4.3  Summing up

(1) The refined LBRMS and mSCFA algorithms are equivalent and are also equivalent to the Feng–Tzeng [6] algorithm (see [36]) concerning the development of degrees/valuations of the approximation.
(2) The Niederreiter–Wang and Vielhaber–Canales (BDM) approaches are equivalent, the former having less (by a factor $M!$) states, the latter having a smaller (almost trivial) recurrence equation (state transition).
(3) The results of Niederreiter and Wang are mathematically proven and give closed formulas for $M = 1, 2, 3$ (at the moment), while the result of Vielhaber and Canales gives a single closed formula valid for every $M \in \mathbb{N}$, but this has only been verified numerically for $M \leq 8$ and $M = 16$.

### 18.4.4  From $\mathbb{N}$umber to $\mathbb{Z}$ahlen tuples: bijections between $\mathbb{N}_0^M$ and $\mathbb{Z}^M$

This section describes experimentally obtained findings, verified only up to the precision stated in the previous paragraphs.

The remarkable fact is that the BDM states/NW-partitions and the corresponding VC-partitions apparently induce a 1:1 correspondence between the sets $\mathbb{Z}^M$ and $\mathbb{N}_0^M$, for all dimensions $M$.

While such bijections abound, it is nevertheless surprising to get such an algorithm out of the seemingly unrelated field of multisequence complexity or Diophantine approximation.

There is a conjectured **bijection** from $\mathbb{Z}^M$ to $\mathbb{N}_0^M$: for all $0 \leq T \leq M, 1 \leq t \leq M$, there are $M(M+1)$ different functions from $\mathbb{Z}^M$ to $\mathbb{N}_0^M$, using the $\mathbb{Z}^M$-tuple as battery values, setting the drain according to the invariant $d + T + \sum_m b_m = 0$, and obtaining the (nonincreasing) partition $(p_1, \ldots, p_M)$. The $\mathbb{N}_0^M$-tuple then is $(p_1 - p_2, p_2 - p_3, \ldots, p_{M-1} - p_M, p_M)$.

**Theorem 18.22** *For $1 \leq M \leq 6$, for $n \leq 400 - 50M$, we have that the function is injective for all BDM states with class up to $n$.*

*If the all-zero vectors of $\mathbb{N}_0^M$ and $\mathbb{Z}^M$ map onto each other, only the parameters $(T, t) = (0, t)$ and $(1, 1)$ are permitted.*

*Proof.* By simulation over the mentioned ranges.  □

**Conjecture 18.23** *The preceeding theorem is valid for all $M \in \mathbb{N}, n \in \mathbb{N}$.*

Furthermore, the bijections behave well in the following sense.

Let $(b_1, \ldots, b_M, d; 0, M+1)$ be a state with corresponding set $\{I_m | 1 \leq m \leq M\}$ of inhibitions counts. Let $(\tilde{I}_m)$ be the ordered nonincreasing tuple, and $p_m = \tilde{I}_m - \tilde{I}_{m-1}, 1 \leq m < M, p_M = \tilde{I}_M$.

Then $\mathbf{b} = (b_1, \ldots, b_M) \in \mathbb{Z}^M$ and $\mathbf{p} = (p_1, \ldots, p_M) \in \mathbb{N}_0^M$, and with the three standard norms

$$\|\mathbf{b}\|_1 = \sum_{m=1}^{M} |b_m|, \quad \|\mathbf{b}\|_2 = \sqrt{\sum_{m=1}^{M} b_m^2}, \quad \|\mathbf{b}\|_\infty = \max_{m=1}^{M} |b_m|,$$

and

$$\|\mathbf{p}\|_1 = \sum_{m=1}^{M} |p_m|, \quad \|\mathbf{p}\|_2 = \sqrt{\sum_{m=1}^{M} p_m^2}, \quad \|\mathbf{p}\|_\infty = \max_{m=1}^{M} |p_m|,$$

we have the following conjecture.

**Conjecture 18.24** *For all $M \geq 2 \in \mathbb{N}$, for all $\mathbf{b} \in \mathbb{Z}^M$ (BDM state) and the corresponding $\mathbf{p} = \mathbf{p}(\mathbf{b}) \in \mathbb{N}_0^M$ (VC-partition according to the algorithm):*

$$\frac{2M}{5}\|\mathbf{b}\|_1 + O(1) \leq \|\mathbf{p}\|_1 \leq 2M\|\mathbf{b}\|_1 + O(1)$$

$$\left(\frac{M}{5} + \frac{3}{2}\right)\|\mathbf{b}\|_2 + O(1) \leq \|\mathbf{p}\|_2 \leq (3M - 1.8)\|\mathbf{b}\|_2 + O(1)$$

$$\frac{M+1}{M}\|\mathbf{b}\|_\infty + O(1) \leq \|\mathbf{p}\|_\infty \leq 2M\|\mathbf{b}\|_\infty + O(1)$$

*and all 6 bounds are sharp with $O(1)$ dependent only on $M$.*

## 18.5 Open questions and further research

(1) Derive a closed form for $N_4^{(q)}(L, n)$, starting from Niederreiter and Wang's recursions.
(2) Repeat this for all $M \geq 5$.
(3) Give a proof for the parsimonious formula $\gamma(\Delta, d)$ from [29, Theorem 22] for all $M, d, q, \Delta$.
(4) Prove that the mapping from states to partitions is a bijection onto the set of all partitions into at most $M$ parts.

(5) Give an algorithm for this mapping.
(6) Prove that a state $s$ with partition $(\tilde{I}_1, \ldots, \tilde{I}_M)$ acquires positive mass at the $\tilde{I}_1$th column, for all $s \in S$, all $M$, and give an algorithm for $\tilde{I}_1$ using $(b_1, \ldots, b_M, d)$, not $\tilde{I}_1$. (Question (5) would include this.)

## 18.6 Conclusion

The field of modeling linear complexity, for sequences and multisequences, is both mature and active, with precise results for the expectation and variance in cases $M$ up to 3, and two approaches for higher $M$: a conjectured closed form and a general recurrence relation, both employing partitions.

Our birthday celebrant *Harald Niederreiter* has been shaping this field from the beginning and continues to do so, on starting his eighth decade, by recently giving us closed formulas for the numbers

$$N_2(n, L) \qquad \text{and} \qquad N_3(n, L)$$

of sequences with prescribed length and linear complexity for multisequences of width 2 and 3.

## Acknowledgements

Harald Niederreiter has been my "academic father" ever since accepting me as an external doctoral student in 1994. Hence, for all my publications on stream ciphers and complexity, I would consider him as a "spiritual co-author" (of course without any responsibility for potential typos, something unheard of and unseen in his own papers; in this regard many thanks to the anonymous referee for pointing out not only typos, but numerous stylistic improvements of this paper).

Mónica del Pilar Canales Chacón, my wife, co-investigator, and proofreader, has again managed to point out the weak spots in earlier versions.

Rainer Göttfert, whom I first met in 1993 at $\mathbb{F}$q2 in Las Vegas, has since then been a good friend in many ways and aspects.

I am really lucky to know all three of them.

## References

[1] M. del P. Canales Chacón and M. Vielhaber, Structural and computational complexity of isometries and their shift commutators. *Electronic Colloq. on Computational Complexity, ECCC*, TR04–057, 2004.

[2] Z. Dai, X. Feng and J. Yang, Multi-continued fraction algorithm and generalized B–M algorithm over $\mathbb{F}_2$. In: T. Helleseth, D. Sarwate and H.-Y. Song (eds.), *Proc. SETA 2004, Int. Conf. on Sequences and their Applications, October 24–28, 2004, Seoul, Korea.* Lecture Notes in Computer Science, volume 3486. Springer, Berlin, 2005.

[3] J. L. Dornstetter, On the equivalence of Berlekamp's and Euclid's algorithms. *IEEE Trans. Inf. Theory* **33**(3), 428–431, 1987.

[4] P. Erdős and P. Révész, On the length of the longest head-run. Colloq. Math. Soc. J. Bolyai, Keszthely, 1975. *Top. Inf. Theory* **16**, 219–228, 1975.

[5] X. Feng and Z. Dai, Expected value of the linear complexity of two-dimensional binary sequences. In: T. Helleseth, D. Sarwate and H.-Y. Song (eds.), *SETA 2004, Int. Conf. on Sequences and their Applications, October 24–28, 2004, Seoul, South Korea.* Lecture Notes in Computer Science, volume 3486, pp. 113–128. Springer, Berlin, 2005.

[6] G. L. Feng and K. K. Tzeng, A generalized Euclidean algorithm for multisequence shift-register synthesis. *IEEE Trans. Inf. Theory* **35**, 584–594, 1989.

[7] F. G. Gustavson, Analysis of the Berlekamp–Massey linear feedback shift-register synthesis algorithm. *IBM J. Res. Develop.* **20**, 204–212, 1976.

[8] N. Moshchevitin and M. Vielhaber, On an improvement of a result by Niederreiter and Wang concerning the expected linear complexity of multisequences, arXiv:math/0703655v2, 2007.

[9] H. Niederreiter, Sequences with almost perfect linear complexity profile. *Proc. EUROCRYPT 1987.* Lecture Notes in Computer Science, volume 304, pp. 37–51. Springer, Berlin, 1988.

[10] H. Niederreiter, The probabilistic theory of linear complexity. *Proc. EUROCRYPT 1988.* Lecture Notes in Computer Science, volume 330, pp. 191–209. Springer, Berlin, 1988.

[11] H. Niederreiter, Keystream sequences with a good linear complexity profile for every starting point. *Proc. EUROCRYPT 1989.* Lecture Notes in Computer Science, volume 434, pp. 523–532. Springer, Berlin, 1990.

[12] H. Niederreiter, A combinatorial approach to probabilistic results on the linear-complexity profile of random sequences. *J. Cryptol.* **2**, 105–112, 1990.

[13] H. Niederreiter, The linear complexity profile and the jump complexity of keystream sequences. *Proc. EUROCRYPT 1990,* Lecture Notes in Computer Science, volume 473, pp. 174–188. Springer, Berlin, 1991.

[14] H. Niederreiter, Linear complexity and related complexity measures for sequences. *INDOCRYPT 2003.* Lecture Notes in Computer Science, volume 2904, pp. 1–17. Springer, Berlin, 2003.

[15] H. Niederreiter and M. Vielhaber, Linear complexity profiles: hausdorff dimensions for almost perfect profiles and measures for general profiles. *J. Complexity* **13**(3), 353–383, 1997.

[16] H. Niederreiter and M. Vielhaber, Simultaneous shifted linear complexity profiles in quadratic time. *Appl. Algebra Eng. Commun. Comput.* **9**(2), 125–138, 1998.

[17] H. Niederreiter and M. Vielhaber, An algorithm for shifted continued fraction expansions in parallel linear time. *Theor. Comput. Sci.* **226**, 93–114, 1999.

[18] H. Niederreiter and L.-P. Wang, Proof of a conjecture on the joint linear complexity profile of multisequences. *INDOCRYPT 2005*. Lecture Notes in Computer Science, volume 3797, pp. 13–22. Springer, Berlin, 2005.

[19] H. Niederreiter and L.-P. Wang, The asymptotic behavior of the joint linear complexity profile of multisequences. *Monatsh. Math.* **150**, 141–155, 2007.

[20] H. Niederreiter, M. Vielhaber and L.-P. Wang, Improved results on the probabilistic theory of joint linear complexity of multisequences. *Sci. China Inf. Sci.* **55**(1), 165–170, 2012.

[21] F. Piper, Stream ciphers. *Elektrotech. Machinen.* **104**, 564–568, 1987.

[22] P. Révész, Strong theorems on coin tossing. *Proceedings of the International Congress Mathematicians, Helsinki*, pp. 749–754, 1978.

[23] P. Révész, *Random Walk in Random and Non-Random Environment*. World Scientific, Singapore, 1990.

[24] R. A. Rueppel, *Analysis and Design of Stream Ciphers*. Springer, Berlin, 1986.

[25] M. Vielhaber, Kettenbrüche, Komplexitätsmaße und die Zufälligkeit von Symbolfolgen. Dissertation, University of Vienna, 1997 (PhD Advisor Harald Niederreiter).

[26] M. Vielhaber, A unified view on sequence complexity measures as isometries. In: T. Helleseth, D. Sarwate and H.-Y. Song (eds.), *SETA 2004, Proc. Int. Conf. on Sequences and their Applications, October 24–28, 2004, Seoul, Korea*. Lecture Notes in Computer Science, volume 3488. Springer, Berlin, 2005.

[27] M. Vielhaber, Continued fraction expansion as isometry – the law of the iterated logarithm for linear, jump, and 2-adic complexity. *IEEE Trans. Inf. Theory* **53**(11), 4383–4391, 2007.

[28] M. Vielhaber and M. del P. Canales, The Hausdorff dimension of the set of $d$-perfect $M$-multisequences. *SETA 2006*, pp. 259–270. Springer, Berlin, 2006.

[29] M. Vielhaber and M. del P. Canales, Towards a general theory of simultaneous Diophantine approximation of formal power series: linear complexity of multisequences, arXiv.org/abs/cs.IT/0607030, 2006.

[30] M. Vielhaber and M. del P. Canales, On a class of bijective binary transducers with finitary description despite infinite state set. *CIAA 2005*. Lecture Notes Computer Science, volume 3854, pp. 356–357. Springer, Berlin, 2006.

[31] M. Vielhaber and M. del P. Canales, Simultaneous Diophantine approximation of formal power series: the asymptotic distribution of the linear complexity of multisequences. *Project FONDECYT*, 1040975 manuscript 8, 2006.

[32] M. Vielhaber and M. del P. Canales, The asymptotic normalized linear complexity of multisequences. *J. Complexity* **24**(3), 410–422, 2008.

[33] M. Vielhaber and M. del P. Canales, The linear complexity deviation of multisequences: formulae for finite lengths and asymptotic distributions. *SETA 2012, Int. Conf. on Sequences and their Applications*. Lecture Notes in Computer Science, volume 7280, pp. 168–180. Springer, Berlin, 2012.

[34] M. Z. Wang and J. L. Massey, The characterization of all binary sequences with perfect linear complexity profile. Presented at *EUROCRYPT'86, Linköping, 1986*. www.isiweb.ee.ethz.ch/archive/massey_pub/pdf/BI959.pdf.

[35] L.-P. Wang and H. Niederreiter, Enumeration results on the joint linear complexity of multisequences. *Finite Fields Appl.* **12**, 613–637, 2006.

[36] L.-P. Wang, Y.-F. Zhu and D.-F. Pei, On the lattice basis reduction multisequence synthesis algorithm. *IEEE Trans. Inf. Theory* **50**, 2905–2910, 2004.

Printed in the United States
by Baker & Taylor Publisher Services